ROBOTIC INTELLIGENCE

World Scientific Encyclopedia with Semantic Computing and Robotic Intelligence

ISSN: 2529-7686

Published

World Scientific Encyclopedia with Semantic Computing and Robotic Intelligence – Vol. 2

ROBOTIC INTELLIGENCE

Editor

Phillip C-Y Sheu

University of California, Irvine

World Scientific

NEW JERSEY · LONDON · SINGAPORE · BEIJING · SHANGHAI · HONG KONG · TAIPEI · CHENNAI · TOKYO

Published by

World Scientific Publishing Co. Pte. Ltd.

5 Toh Tuck Link, Singapore 596224

USA office: 27 Warren Street, Suite 401-402, Hackensack, NJ 07601

UK office: 57 Shelton Street, Covent Garden, London WC2H 9HE

British Library Cataloguing-in-Publication Data

A catalogue record for this book is available from the British Library.

World Scientific Encyclopedia with Semantic Computing and Robotic Intelligence — Vol. 2
ROBOTIC INTELLIGENCE

Copyright © 2019 by World Scientific Publishing Co. Pte. Ltd.

ISBN 978-981-120-347-3

For any available supplementary material, please visit
https://www.worldscientific.com/worldscibooks/10.1142/11361#t=suppl

Desk Editor: Catherine Domingo Ong

Typeset by Stallion Press
Email: enquiries@stallionpress.com

Foreword

In the first volume of this World Scientific Encyclopedia with Semantic Computing and Robotic Intelligence, we covered mainly Semantic Analysis (Layer 1), Semantic Data a Integration (Layer 2), and Semantic Services (Layer 3). In this volume, we will focus on Semantic Service Integration (Layer 4), and Semantic Interface (Layer 5), to cover integrates services with different sources such as robotic intelligence, and realize the user intentions to be described in natural language, or express in some other natural ways.

This volume aims to provide a reference to the development of robotic intelligence, build upon Semantic Computing, in the terms of "action", as the realization of "context", and "intention" formuated by Semantics Computing during the "thinking" or reasoning process to address the three following core areas:

- Ability to interface and interaction with human in the form of natural languaes, and map them onto Semantic Services.

- Understanding the (possibly naturally-expressed) intentions (semantics) of users and develop into plans to devlier the outcomes by menas of robotics.
- Improve on the sophistication in big data environment, to develop capability to process and analyzie massive data on-the-fly.

Finally, I would like to thank members of the editorial board for their kind support, and thank Dr. K. K. Phua, Dr. Yan Ng, Ms. Catherine Ong and Mr. Mun Kit Chew of World Scientific Publishing for help and support to publish Volume Two of this Encyclopedia.

Phillip Sheu
Editor-in-Chief
University of California, Irvine

CONTENTS

(Continued)

CONTENTS — (*Continued*)

Part 1

Understanding Semantics

Goodness of machine learning models

Joseph R. Barr

Trust Science, Irvine, CA 92618, USA

joseph.barr@trustscience.com

In this context, a model is an algorithm or a procedure that applies to data resulting in a functional relation τ between "input space" $\mathcal{X}$ and "output space" $\mathcal{Y}$. In this short paper, we will delineate objective criteria which help to disambiguate and rate models' credibility. We will define pertinent concepts and will voice an opinion on the matter of good versus bad versus so–so models.

Keywords: Machine learning; cross-validation; complexity; parsimony; variable selection; AIC; cost function; deviance.

1. Introduction

A *model* is a mathematical abstraction of a physical phenomenon. A *statistical*, or *machine learning*, model is one which *learns* "patterns" in the data and is able to *predict* new instances. The term *data science* is sometime used to describe that domain.

A model is inherently inaccurate because any procedure used to produce a model is based on heuristics and short of an exhaustive search through an "infinite set," it (an optimal model) could never be found. Thus a somewhat redundant observation in the spirit of Turing's "Halting Problem" is that *there is no effective computational method to determine whether a model is optimal.* This however does not abrogate one from being able to assess a model.

It is generally accepted that a model is *good* if it has some application or serves some purpose. There is no consensus, however, as to what it means for a model to be *correct*.

To paraphrase George Box's pronouncement that a model is only as good as its utility, practicality often helps distinguish goodness or badness of a model, therefore, a good model is one that is able to predict *accurately enough*. For example, Kepler's laws of planetary motion are good not because they give the right answer like the shape of the orbit of earth around the sun. Kepler's laws are good because they are rather accurate as laws of classical mechanics which are able to predict celestial events "accurately enough." There are application, however that require better accuracy. At times Kepler's equations are not "accurate enough" because the equations of Newtonian mechanics do not tell the whole story. A case in point, the *Global Positioning System* (GPS) uses relativistic corrections to "fix" for gravitational "distortions" of time and space.

2. Data and Predictive Algorithms

At the risk of oversimplifying, statisticians, [*sic*] "machine learners" do one of the two things: They classify items as one of two types, say 0 or 1, and they associate a numerical (or vector) value y to an input X, a vector. Often there are additional layers associated with a model which further result in additional insights.

Lying at the basis of the pyramid is the data, a set of vectors $(X,y) = (x_1, x_2, \ldots, x_p, y)$, $x_j \in \mathbb{R}$, $y \in \mathbb{R}$. In machine learning the data is also referred to as the *examples*. Clearly not every dataset has "neat" form. In fact, production data like log files, surveys, financial transactions, etc. contain disparate forms of data including alphanumeric, textual, dates, IP addresses, images, voice and so on. Practitioners are deft at transforming production data into sets of vectors whose entries are numerical. Numerical values may be real (continuous), Boolean (0/1), categorical (many-value ordinal). Each vector (X,y) is called a *record*, or in the context of statistical modeling/machine learning, an *example*. The two archetypical cases are $\mathcal{Y} = \mathbb{R}$ and $\mathcal{Y} = \{0,1\}$ where $\mathcal{Y}$ represents the set of all possible output values. The former is related to *regression* and the latter to *classification*. In the presence of three classes $\mathcal{Y} = \{0,1,2\}$, it is common to classify every $(x_1, x_2, \ldots, x_p)$ as $y = 0$ or as $y = 1$ and the further classify the "1" class (as class 1 or 2). Whether a model is statistical (like generalized linear models) or strictly "machine learning" (like "deep" neural networks), or a hybrid, the person who builds a model shall be henceforth known as a *statistical modeler*, or sometimes just *modeler*.

3. Models

Over the past century and a half, since Sir Francis Galton introduced regression as a predictive procedure in the 1880s, a plethora of modeling techniques were developed to address diverse prediction problems. A statistical modeler has to straddle a fine line of model's ability to predict and its various (sometime ancillary) attributes. For example, in certain domains it is necessary that models are interpretable, where the influence of inputs on output could be assessed. In the consumer credit space, for instance, where the Fair Credit Reporting Act (FCRA) prohibits discrimination against "protected classes", a credit score must be appended by *reason codes*, a ranking of the inputs and their influence explaining a person's credit rating. At times, models must be complex in order to be effective which invariably diminishes model's interpretability. In image and voice recognition, natural language processing and various other high-dimensional, highly correlated phenomena, complexity is an imperative. Deep neural networks are paragons of complexity. For example, deep learning with all its variants has proven effective in learning complex interactions commonly found in semantic computing and image analysis. However, in any modeling technique, the problem of assessing model performance remains the same: how to discern a good from bad from a "so–so" model?

4. Loss and Risk

Consider some model τ where τ may be viewed as mapping $\mathcal{X}$ into $\mathcal{Y}$ where $\mathcal{X}$ is the set of all 'input' values and $\mathcal{Y}$, 'output' values. A *loss function* is a mapping $\mathcal{L} : \mathcal{X} \times \mathcal{Y} \to [0, \infty)$ designed to measure the departure of a predicted $\tau(x) = \widehat{y(x)}$ from an observed $y(x)$. Notice that we use $\tau(x)$ and $\widehat{y(x)}$ interchangeably. The two most common loss functions are the *quadratic*, or $\mathcal{L}^2$-*loss*, $\mathcal{L}(x, y) = (y(x) - \widehat{y(x)})^2$, and the *mismatch loss*,

$$\mathbb{I}\left(\tau(x) \neq y(x)\right) = \begin{cases} 1 & \tau(x) \neq y(x), \\ 0 & \text{otherwise.} \end{cases}$$

The *empirical risk function* $\mathcal{R}_\tau(D)$ with respect to a model τ and the set of examples $D = \{(X, y), \ X \in \mathcal{X} \subset \mathbb{R}^p, \ y \in \mathcal{Y}\}$ is

$$\mathcal{R}_\tau(D) = \frac{1}{\|D\|} \sum_{(x,y) \in D} \mathcal{L}(x, y),$$

where $\|D\|$ is the number of elements in D. The empirical *mean squared error* (MSE),

$$\text{MSE}_\tau(D) = \frac{1}{\|D\|} \sum_{(x,y) \in D} (y(x) - \tau(x))^2,$$

is common to measuring goodness of a *regression model*.

For the classification of τ, the *empirical error rate* is

$$\text{ER}_\tau(D) = \frac{1}{\|D\|} \sum_{(x,y) \in D} \mathbb{I}\left(\tau(x) \neq y(x)\right).$$

There are other loss functions like *exponential loss* used in boosting, but we will not worry about it here. In the context of mathematical optimization the terms *cost function* is more or less synonymous to risk function and will be used interchangeably as a more generic term.

5. Model Fitting

With a dataset of labeled examples $D = \{(X, y), \ X \in \mathcal{X} \subset \mathbb{R}^p, \ y \in \mathcal{Y}\}$ and a specific modeling framework $\mathcal{M}$, like generalized linear models, neural networks, boosting or whatever, the modeler trains a model by identifying the "best" τ with respect to a subset T of examples in D,

$$\tau^*(T) = \arg \min \left\{ \mathcal{R}_\tau(T) : \tau \in \mathcal{M} \right\}.$$

On the face it, it seems that the minimization procedure searches through the entire space of models $\tau \in \mathcal{M}$ but in practice a procedure only produces a suboptimal model because the algorithm is not designed to search through the entire space $\mathcal{M}$. Additionally, the "optimal" τ^* depends on the *training set* T, causing τ to vacillate in response to varying training sets.

The procedure producing τ^* is often a mix of heuristics and numerical optimization, with a stopping criteria forcing the algorithm to stop after certain conditions are met. It is not unusual that stopping criteria will force the algorithm to stop prematurely, before a global minimum is identified.

Whether optimal or not, the algorithm halts with a model, i.e., a procedure τ^* to calculate $y \in \mathcal{Y}$ from any given $X \in \mathcal{X}$. A procedure that produces y from X can be as simple as a linear function of the form $(x_1, x_2, \ldots, x_p) \xrightarrow{\tau} b_0 + b_1 x_1 + \cdots + b_p x_p = \widehat{y(x)}$ or as complicated as thousands of decision trees, and tens of thousands of leaves and a procedure that produces an estimate.

Often, the model space $\mathcal{M}$ is constrained by additional parameters. The imposition of *shrinkage* reduces complexity, hence reduces the risk of overfitting. In practice, shrinkage is "baked into" $\mathcal{M}$.

Early stopping is yet another proviso to hedge against risk of overfitting (next section) where training stops when a certain condition is met. A commonly encountered example involves constraints on decision trees like *minimal leaf size* or *maximal tree depth* resulting in a parametrized model space

$$\mathcal{M}_{k,d} = \{T \in \mathcal{T} : \max \text{ depth of } T \leq d, \text{ min leaf size } \geq k\}.$$

6. Overfitting

Noise is one mean temptress: it tempts the algorithm to fit noisy data, something the modeler must steadfastly resist.

The celebrated Ockham's Razor provides a guiding principle in the search for a good model where parsimony is balanced off against model's explanatory qualities.

Overfitting is measured by the departure of the risk function on new, yet unseen data. It is possible to fit a model with a risk function equal to say $\mathcal{R} = 0.1$, yet when calculating the cost against new, yet unseen data, one discovers shockingly large risk like $\mathcal{R} = 2$, i.e., twenty-fold greater. This is known as overfitting.

6.1. Cross-validations

The *cross validation* (CV) procedure has the following steps.

- Form a random partition D into training set T and validation set V; more or less equally sized sets.
- Train the model on T and use V to test it.

  The modeler will attempt to minimize training error of the empirical risk function $\mathcal{R}$; i.e., $\tau^* = \arg\min \{\mathcal{R}_\tau(T) : \tau \in \mathcal{M}\}$ where $\mathcal{M}$ is the space of models in some model framework.

- Calculate $\mathcal{R}_\tau(V)$, the validation error.
- Compare the two risk functions with $\delta = \delta_{\tau^*}(T, V)$ the *empirical deviance* [with respect to the pair (T, V)]:

$$\delta = \frac{|\mathcal{R}_{\tau^*}(T) - \mathcal{R}_{\tau^*}(V)|}{\mathcal{R}_{\tau^*}(T)}.$$

6.2. The k-fold cross-validation

The *k-fold cross-validation* is a robust cross-validation technique. The procedure consists of the following steps:

- Partition the dataset D into k sets of more or less equal size, namely $D_1, D_2, \ldots, D_k$.
- For each $1 \le j \le k$, we train a model on the union $U_j = \cup\{D_i : i \ne j\}$; each training results in a model τ_j, which we then test on the set D_j with $\mathcal{R}_{\tau_j}(D_j)$.
- Calculate the deviance for each $1 \le j \le k$:

$$\delta_j = \frac{|\mathcal{R}_{\tau_j}(U_j) - \mathcal{R}_{\tau_j}(D_j)|}{\mathcal{R}_{\tau_j}(U_j)}.$$

- Return *average* deviance:

$$\delta = \frac{1}{k}\sum_{j=1}^{k}\delta_j.$$

6.3. "Leave-one-out" cross-validation

The *leave-one-out cross-validation* (LOOCV) is common when $|D|$ is rather small. It goes like this: Fix some positive integer m (say $m = 500$), and do the following:

- Sample m examples $\{(X_j, y_j)\}$ with replacement.
- For each $1 \le j \le m$ train a model $\tau_{(X_j, y_j)}$ on the complement.
- For each $1 \le j \le m$ calculate the empirical risk function $\mathcal{R}_{\tau_{(X,y)}}(D - (X, y))$ of the model developed in previous step as well as the corresponding deviance δ_{X_j}.
- Repeat m times for the predetermined m, and average out the deviance to produce the average deviance δ.

6.4. On the deviance

In the classical setting of testing hypothesis, the deviance Λ is defined as the likelihood ratio statistic

$$\lambda(X) = \frac{\sup_{\theta \in \Theta_1}\log L(\theta; X)}{\sup_{\theta \in \Theta_0}\log L(\theta; X)},$$

where L is the likelihood of the i.i.d. sample X taken from some distribution $p(x)$ and Θ_1, Θ_0 are subspaces of the parameter space (a subset of $\mathbb{R}^d$). Since machine learning focuses on the algorithm rather than on the distribution, We have taken the liberty to reuse the term [*sic*] deviance in a different setting. In fact, in our context it is possible to define the deviance as a bona fide statistic. There are several contenders for the definition. The setting consists of a set of examples $D = \{(X, y) : X \in \mathcal{X}, y \in \mathcal{Y}\}$, a model τ and an empirical risk function $\mathcal{R}$ (We will deprecate the subscript τ).

Deviance: The deviance $\delta_{\max}$ is the maximum

$$\delta_{\max} = \max_{(T,V)} \frac{|\mathcal{R}(T) - \mathcal{R}(V)|}{\mathcal{R}(T)},$$

where $T, V \subset D$ and $|V| = |T| = \frac{|D|}{2}$ or $|V| = \frac{|D|+1}{2}$ and $T = D - V$, depending on the parity of $|D|$. Clearly the definition is well defined, the deviance always exists with the unlikely exception that $\mathcal{R}(T) = 0$ for all T. This in practice never happens. The deviance $\delta_{\max}$ is impractical because to calculate $\delta_{\max}$ one has to train a model for each of the $\binom{|D|}{|D|/2}$ $= O(|D|^2)$ training sets. Quadratic complexity is far too prohibitive and the k-fold CV is generally preferred.

6.5. Bias–variance

Consider a model τ trained on some examples, a training set T. The *bias* of a model τ is the empirical risk function of τ measured against the training set T: Bias$= \mathcal{R}_\tau(T)$. A complex model τ corresponds to small bias value. In fact, an exceedingly complex model may conceivably reduce bias to 0. The *variance* is the empirical risk function of τ measured against a validation set V: var$(\tau) = \mathcal{R}_\tau(V)$. A complex model results in high variance. We have the *variance-bias axiom*: *The bias is a decreasing function of complexity while the variance is an increasing function of complexity.* Figure 1 depicts the bias–variance phenomenon.

The x-axis represents complexity, measured in model's parameters or degrees of freedom, i.e., complexity: in regression complexity equals the number of parameters, in decision trees complexity is measured in terms of depth of treem or number of splits and in neural networks complexity is measured in terms of number of hidden layers and associated parameters. The bias–variance graph is rather theoretical; to produce such a graph one must fix a model framework, training and validation sets, each containing roughly half of the data; gradually increase complexity and calculate the empirical risk function on training set to get

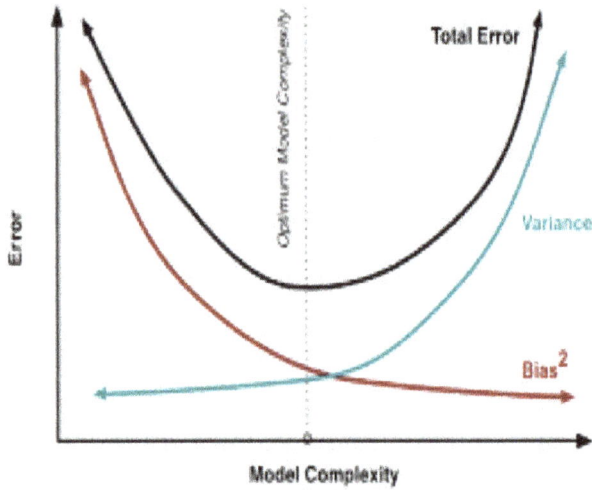

Fig. 1. Bias decreases with complexity while variance increases.

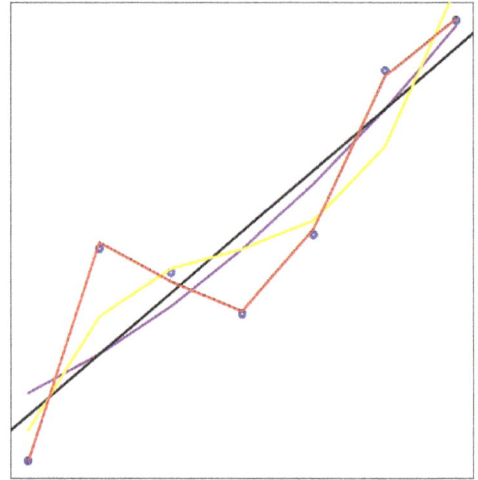

Fig. 2. Comparing splines of various levels of complexity.

bias, and on validation set to get variance. A departure from this graphical depiction is a fluke.

6.6. *Classical bias and variance*

In the domain of statistics bias and variance can be defined directly. In classical settings, the data $Y_1, Y_2, \ldots, Y_n$ is sampled from some probability distribution $p(y, \theta)$, depending on some parameter θ where θ lies in some unknown parameter space $\Theta \subset \mathbb{R}^p$. The sample is considered i.i.d. provided the *likelihood* function "multiplies," i.e., the likelihood $L(y_1, y_2, \ldots, y_n; \theta) = \Pr(Y_1 = y_1, Y_2 = y_2, \ldots, Y_n = y_n)$ is

$$L(y_1, y_2, \ldots, y_n; \theta) = p(y_1; \theta) \times p(y_2; \theta) \times \cdots \times p(y_n; \theta).$$

An *estimator* T is a function of the data used to estimate θ. T itself is a random variable whose value depends on the data $X_1, X_2, \ldots$, and on θ. T is an *unbiased estimator for θ* if $\mathbb{E}(T) = \theta$. The *variance* of an estimator T of a parameter θ is $\mathrm{var}(T) = \mathbb{E}(T - \theta)^2$. In case T and θ are vectors of the same dimension, the variance is the matrix $\mathrm{var}(T) = \mathbb{E}(T - \theta)(T - \theta)^T$ where U^T is U transposed.

6.6.1. *Classic versus modern definitions*

On the face of it the classical and newer "machine learning" definitions of bias–variance seem completely different; one relies on the empirical risk function on training and validation sets while the other is "formulaic." However, on closer look the two definitions are not at all that different. After all, an estimate $\widehat{\beta}$ is a random variable, and having high variance will result in high vacillations when applied on validation sets.

A situation is easily visualized in Fig. 2 where we compare the fit of linear (black), quadratic (purple), cubic (yellow) and high-order piece-wise linear (red). Clearly as one increases complexity, the fit gets better, in other words, the bias decreases. However, a higher-order spline will perform worse against a new data.

7.   Controlling for Complexity

It is axiomatic that good model balances bias and variance, or alternatively, that it balances *fitness* and complexity. Fitness is a classical measure of model's $\tau \in \mathcal{M}$ ability to fit training data, where $\mathcal{M}$ is a model framework, and its complexity is usually measured in terms of the number of unknown parameters used in the model. While variance is correlated with complexity, the correlation between variance and complexity is not altogether straightforward. It is not difficult to simulate data where in response to reduction in complexity, gains in variance are minuscule, if at all perceptible.

For a model $f \in \mathcal{F}(\theta), \theta \in \Theta \subset \mathbb{R}^p$, where $\mathcal{F}(\theta)$ is a family of distributions, the *Akaike Information Criterion* (AIC) has two parts:

$$\mathrm{AIC}(f) = 2p - 2\log(\hat{L}),$$

the first part p is the number of parameters while $\hat{L}$ is the likelihood evaluated at the *maximum likelihood estimator*.

Log($\hat{L}$) represents the fit, and p represents the complexity. Although the AIC is stated in the context of the likelihood settings, it may be applied in a vast range of situations (see Ref. 5), and it encapsulates the principle of balancing bias and variance with a measure of fit and complexity.

Schwarz's Bayes Information Criterion (BIC) achieves a similar objective, albeit with a different emphasis and different weight used to measure complexity.

In classical regression (OLS, logit, proportional hazard, etc.), *variable selection methods* strive to select a subset of the features in order to find the elusive point where bias and variance meet. Variable selection methods employ a variety of cost functions, each emphasizing different facets. In the case of OLS, the oldest is the classical *Fisher F-test* (named after the founder of twentieth century statistics, Ronald A. Fisher). Using mean squared errors as measure of goodness, the algorithm compares the gain of "excised"

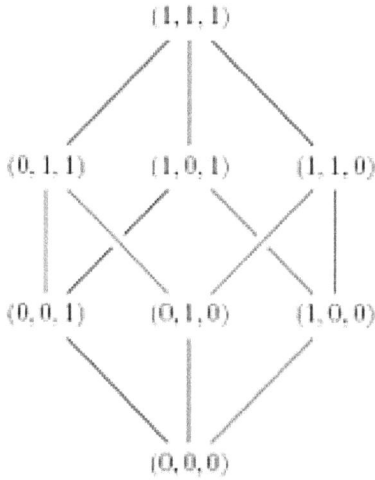

Fig. 3. Hasse diagram of subset with three elements.

model, having one fewer variable, against a "baseline" (saturated) model.

Variable selection is an iterative "Greedy" process. To demonstrate the process, consider the "Hasse diagram" of the partially-ordered set of a set having three features (Fig. 3), where for example, (1,0,1) representing the first and third features.

There are two ways to search for optimal subset of features; "top to bottom" and "bottom to top." Suppose the former. Initially, a baseline model is saturated, containing all three variables [top node $(1, 1, 1)$]; one feature is dropped, trying out the models corresponding to (1,1,0), (1,0,1) and (0,1,1) and the best (with respect to some cost function) is selected, say (1,1,0). Next, starting from (1,1,0), we test a model against the two possibilities (1,0,0) and (0,1,0). If neither provides a lift in terms of lower cost function, then the optimal is (1,1,0), else the process continues until an optimal subset of variables are found.

Note that with n features, an exhaustive search will entail testing 2^n models, and if n is large, the process is far too expensive. Consequently, variable selection algorithms searches for a "path" in the Hasse diagram of subsets of the features. If there are p variables in a saturated model, then the complexity of the variable selection is $O(p^2)$. In most cases quadratic complexity is too high and so selection procedures involve mechanisms to reduce the complexity, e.g., early stopping.

In fact, some implementations employ a "backtracking" which helps "recover" from a local minimum. Clearly, a backtracking would not necessarily result in a global optimum; it only expands the search while keeping it fairly tractable. (In the three feature example, a single backtrack results in a comprehensive search, but not in a 100-feature set). As mentioned earlier, this method does not guarantee a global optimum and may fail to deliver "the best" model when the number of variables is large (say a 1000), however, in most cases, and coupled with heuristics, variable selection works fairly well.

Although details vary, similar variable selection methods are used in the entire universe of regression techniques which

include the generalized linear models, the Cox proportional hazard models as well as nonparametric regression models.

The AIC and the Schwarz's BIC are commonly used for selection criteria.

Virtually all commercial software implement a variety of selection methods. For example, the R system has a function `stepAIC` tracing out the selection process based on AIC, resulting in an "optimal" model with respect to AIC.

7.1. *Regularization or shrinkage methods*

By constraining the model space, shrinkage methods provide a way to controlling complexity. The classic *ridge regression* (sometime *Tikhonov's method*) is one of the earliest attempts at shrinkage. Ridge regression estimates β by minimizing

$$\hat{\beta} = \arg \min_{\beta \in \Theta, \lambda \geq 0} \|X\beta - Y\|^2 + \lambda \|\beta\|^2.$$

The minimization is over both $\beta \in \Theta$ where Θ is the parameter space; and λ is some nonnegative number. If $\lambda = 0$, then the optimal β is the standard OLS solution. If $\Theta \subset \mathbb{R}^p$, then the solution of the convex function $\|X\beta - Y\|^2 + \lambda \|\beta\|^2$ is computed using numeric algorithms.

Another example is the *Lasso* which constrains the OLS by imposing $\mathcal{L}^1$ constraints, $\sum_{j=1}^{p} \|\beta_j\| \leq v$. In other words, "Lasso-constrained OLS" has the form $\min_{\beta \in \Theta} \|X\beta - Y\|^2$ subject to a constraint: $\sum_{j=1}^{p} |\beta_j| \leq v$.

8. Conclusions

Learning from data serves multiple purpose, not least is modeling. One favors a model if it fits training data as well as it predicts validation data. The metric used to measure departure of training-validation model's performance is the deviance δ, as is defined above. However, smallness of deviance is not an altogether conclusive measure of model's goodness or its utility. A good model is one that serves a practical purpose and adheres to "common sense," therefore an unexplained departure from what is considered normal should warn of looming problems.

Experience shows that at times the efforts required to fine-tune a model and effort required to fix problems may exceed the efforts of "initial" development. One should never take things for granted and exercise healthy skepticism before committing to deploying a model into a production system.

Acknowledgments

The author expresses his gratitude to his many teachers, mentors, colleagues and friends over the years who shaped his philosophy and taste with special thanks to Roger Engringer, a graph theorist, and his doctoral advisor, and to the late Peter Enis, a Statistician who got him interested in statistical methodology.

Appendix A

```
install.packages(``car'')
library(car)
library(MASS)
set.seed(123)
dat = read.csv(`/Users/josephbarr/data/concrete.csv', sep=`\t', header=TRUE )
dat[``noise1''] = water + rnorm(1030, 0, 2) # add noise
dat[``noise2''] = super + rnorm(1030, 0, 1) # add noise
dat[``noise3''] = water + cement + rnorm(1030,0,3) # add noise
attach(dat)
sampleRow = sample(1:nrow(dat), 0.6*nrow(dat), replace=FALSE, prob = NULL)
train = dat[sampleRow,]
valid = dat[sampleRow,]
nrow(valid) # number of records validation 412
nrow(train) # number of records training 618
lm1 = lm(strength ~., data=train)
summary(lm1)
vif(lm1) # cement blast fly water super coarse fine age
        # 7.277088 7.303704 6.044700 6.593355 2.795887 4.891411 6.436055 1.128597
        # VIF a little weak on 5 of the features — weak multicollinearity
predT1 = predict(lm1, train)
residT1 = predT1-train$strength
sum(residT1^2)/nrow(train) # Bias = 104.5029
predV1 = predict(lm1, valid)
residV1 = predV1-valid$strength
sum(residV1^2)/nrow(valid) # Variance = 114.0638
        # deviance1 = (114.0638-104.5029)/104.5029
        # Deviance Model1 = 0.09148933
lmAIC = stepAIC(lm1) # super, fly, water, blast, age, cement-AIC=2890.3
lm2 = lm(strength ~ super+ fly + water +blast + age + cement, train)
summary(lm2) # Get a list of optimal variables
vif(lm2) # super fly water blast age cement
        # 2.287370 2.344316 1.849142 1.774580 1.121432 1.911564
        # VIF solid — no collinearity
predT2 = predict(lm2, train)
residT2 = predT2-train$strength
sum(residT2^2)/nrow(train) # Bias2 = 105.0258
predV2 = predict(lm2, valid)
residV2 = predV2-valid$strength
sum(residV2^2)/nrow(valid) # Variance2 = 114.0521
        # Deviance2 = (114.0521-105.0258)/105.0258
        # Deviance2 = 0.08594364
        # Deviance2 < Deviance1
```

Fig. A.1. R code: Demonstrates a parsimonious model results in lower deviance.

Consider the Concrete dataset that you can find at the UCI data repository. Consider the annotated R code in Fig. A.1. You may use `fread("URL")` in the library "`data.table`" instead of loading from a local drive.

The Concrete data contains eight features and one response "`strength`". You will find three additional features by adding random noise (random normal) to `water`, `super` and `water + cement`. There are 1030 records. Generate a random training set with 618 rows, and remaining 412 records to test. Build two linear models, one saturate, having all eight predictors and one optimal with respect to AIC, having six predictors. The deviance of the saturated model is 0.09148933 while for the more parsimonious model it is 0.08594364. This does not "prove" that the parsimonious model is optimal, it merely demonstrates that it is likely better than the saturated model (well, we actually know that).

References

[1] J. R. Barr and J. Cavanaugh, Forensics: Assessing model goodness: A machine learning view, *Encycl. Semant. Comput.* (2018).

[2] J. R. Barr and S. Zacks, Goodness of fit of statistical distributions, *Encycl. Semant. Comput.* (2018).

[3] P. J. Bickel and K. Doksum, *Mathematical Statistics: Basic Ideas and Selected Topics*, 2nd edn., Vols. I and II (Chapman, 2015).

[4] J. Friedman, T. Hastie and R. Tiribishani, *The Elements of Statistical Learning*, 2nd edn. (Springer, 2009).

[5] S. Konishi and G. Kitagawa, *Information Criteria and Statistical Modeling* (Springer, 2007).

Part 2

Data Science

Goodness-of-fit of statistical distributions

Joseph R. Barr*,‡ and Shelemyahu Zacks†,§

*Barr Analytics, Irvine, CA, USA

and

†Department of Mathematical Sciences
Binghamton University, Binghamton, NY 13902, USA
‡barr.jr@gmail.com
§shzacks@outlook.com

Goodness-of-fit is used for the evaluation a model. They are commonly used to compare among competing models. The material is mostly classic. For more on the subject the reader is referred to the References including the two revised volumes Bickel and Docksum (2016).

Keywords: R-squared; deviance and chi-squared; Kolmogorov–Smirnov; loss function; empirical risk function.

1. Introduction

Data analysts try to test which one of the possible statistical models fit the data under investigation. In order to outline what kind of data could be investigated, we start with a general discussion about types of data, and what is a statistical or probabilistic model. The reader is referred to Refs. 1 and 2.

Empirical sciences try to infer from observation laws which govern the phenomenon under investigation. Sometimes, the results of a given experiment can be predicted accurately. In this case we say that the subject matter is *deterministic*. We are not concerned here with deterministic models. Often, the results of trials cannot be determined prior to observations, and in such cases we say that the observations are composed of two components: deterministic and *random*, or "signal" and "noise". A *probability model* is a function that can assess the expected proportion of times that the result of a trial will fall in a given category. In order to construct a probability model, the experimenter has to define the *sample space*, i.e., the set of all possible results of a specific trial. The type of observations, whether categorical like "yes"/"no," or "red"/"green"/"blue," or numerical, must also be defined. A real-valued function of the points in a sample space is called a *random variable*. The actual value of a random variable, before an observation is made, is unknown. Its value under observation is called *sample realization* of a random variable.

Every random variable is associated with a *distribution function*. We generally distinguish between *discrete* random variables and *continuous* ones, where a discrete random variables attains a finite or a *countable* number of values while a continuous may assume any real value. A *distribution function F* of a random variable X is a function whose domain is a set of possible values in the real line, and its range in the unit interval $[0, 1]$, defined as $F(x) = \mathbf{Pr}(X \le x)$. All random variables having the same distributions are *equivalent*. Thus a distribution function assigned to the result of a particular experiment is a *model*, applied for predicting a future outcome. A result of an experiment might yield a mixture of discrete and continuous random variables, and correspondingly, the distribution function is a *mixture* of several distributions. The data analyst should first determine whether the data under consideration is indeed random, and from the values observed he/she must estimate a distribution function for the purpose of predicting future values. The objective of the present paper is to present several tests which can be used to confirm the validity of a given model, or reject it. We call these *tests of goodness-of-fit*. We say that a model fits well the data if the differences between the actual values of the data and the values predicted by the model are not significantly large at any given risk level (p-value.) The classicist's primary concern is with models specified by distribution functions.

2. Categorical Data

A variable is called *categorical* if there is no "natural" order among its values. For example, a trial resulting in two outcomes *success* or *failure*. These values are categorical, and there is no inevitable ordering of the two categories (although you might think that a success is preferred to a failure, hence a success is bigger than a failure). A categorical model with two possible disjoint categories will be called *binomial*. If the observation may fall into one of k categories, where $k > 2$, we call the model *multinomial*. Independent binomial trials

‡Corresponding author.

with the same probability of success in each trial are called *Bernoulli trials*.

2.1. *Chi-squared tests of categorical data*

Suppose we have n Bernoulli trials. Let K be the observed number of successes among these trials. The data can be put in a *contingency table*:

Contingency table for binomial data.

n	Observed	Expected
Success	K	np
Failure	$n - K$	$n(1 - p)$

where p is the probability of success. The *chi-squared* test statistic for contingency tables is of the structure

$$X^2 = \sum (\text{observed} - \text{expected})^2 / \text{expected}.$$

If we wish to test the hypothesis $H : p = p_0$ we can use the statistic

$$X^2(n, p_0) = \frac{(K - np_0)^2}{np_0} + \frac{(n - K - n(1 - p_0))^2}{n(1 - p_0)}$$
$$= \frac{(K - np_0)^2}{np_0(1 - p_0)}.$$

By the *Central Limit Theorem* we can show that asymptotically (for large values of n) $X^2(n, p_0)$ is distributed like a chi-squared random variable with one degree of freedom, χ_1^2. Let α denote the probability of type-I error (rejecting H when it is true), and let $\chi_{1,1-\alpha}^2$ denote the $(1 - \alpha)$ quantile of χ_1^2. Then H is rejected if $X^2(n, p_0) > \chi_{1,1-\alpha}^2$. The exact distribution of $X^2(n, p_0)$ for small sample size can be determined by the binomial distribution of K with parameters (n, p_0).

We obtain a discrete distribution at the set of $n + 1$ points:

$$\xi_j = \frac{(j - np_0)^2}{np_0(1 - p_0)}, \quad j = 0, 1, \ldots, n,$$

having probabilities

$$\mathbf{Pr}(\xi_j) = \binom{n}{j} p_0^j (1 - p_0)^{n-j}, \quad j = 0, 1, \ldots, n.$$

Let

$$\xi_{(0)} \leq \xi_{(1)} \leq \cdots \leq \xi_{(n)}$$

be the ordered values of ξ.

The distribution of $X^2(n, p_0)$ is

$$F_{X^2}(x; n, p_0) = \sum_{j=0}^{n} I\{\xi_{(j)} \leq x\} \mathbf{Pr}\{\xi_{(j)}\},$$

where $I\{\}$ is an indicator random variable. The $(1 - \alpha)$ quantile of this distribution is

$$Q_{1-\alpha} = \inf\{x : F_{X^2}(x; n, p_0) \geq 1 - \alpha\}.$$

The hypothesis H is rejected at a level of significance α if $X^2(n, p_0) \geq Q_{1-\alpha}$.

The multinomial case prevails when the observations are independent and each observation yields a value in one of m disjoint categories. The probability that an observation falls into category j, $j = 1, \ldots, m$, is p_j. Thus, an m-nomial distribution is specified by the parameters (n, π) where $\pi = (p_1, \ldots, p_m)$ is a probability vector. The observation is an m-dimensional vector $J_m = (j_1, \ldots, j_m)$ where $0 \leq j_i \leq n$ and $\sum_{i=1}^{m} j_i = n$. Furthermore, let K_i denote the number of observations in the ith category, then

$$\mathbf{Pr}\{K_1 = j_1, \ldots, K_m = j_m\} = \frac{n!}{\prod_{i=1}^{m} j_i!} \prod_{i=1}^{m} p^{j_i}.$$

The contingency table is as follows:

Category	1	$\ldots$	m
K	j_1	$\ldots$	j_m
p	p_1	$\ldots$	p_m

The corresponding chi-squared statistic for testing the hypothesis H: $p_i = p_i^0$, $i = 1, \ldots, m$, is

$$X^2(n, \pi^0) = \sum_{i=0}^{m} \frac{(j_i - np_i^0)^2}{np_i^0}.$$

A large sample criterion for rejecting H is the chi-squared quantile with $m - 1$ degrees of freedom, i.e.,

$$X^2(n, \pi^0) \geq \chi_{m-1,1-\alpha}^2.$$

A small sample criterion can be approximated by simulations. This multinomial chi-squared test is sometimes used to test whether a dataset originates from some specific distribution F^0. This is done by partitioning the real line into m intervals or *bins* and counting the number of observations, the *frequency* in each interval. The interval $(\xi_{i-1}, \xi_i]$ has the probability $p_i^0 = F^0(\xi_i) - F^0(\xi_{i-1})$.

3. Graphical Tests of Fitness for Continuous Distributions

Let

$$F_n(x) = \frac{1}{n} \sum_{i=1}^{n} I\{X_i \leq x\} \qquad (1)$$

be the *empirical distribution function* of the random sample $X_1, \ldots, X_n$. Notice that if $X_{(1)} \leq X_{(2)} \leq \cdots \leq X_{(n)}$ is the order statistic then

$$F_n(x) = \sum_{i=1}^{n-1} \frac{i}{n} I\{X_i \leq x < X_{i+1}\} + I\{X_{(n)} \leq x\}. \qquad (2)$$

Empirical vs. Null

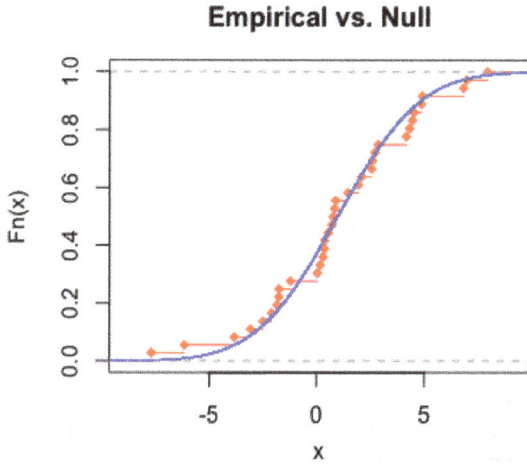

Fig. 1. Empiric distribution versus theoretical.

This is a random step function distributed around the true distribution. The empirical distribution contains all the information in the sample on the true distribution function of the observed random variable. It is known that

$$\lim_{n \to \infty} F_n(x) = F(x), \qquad (3)$$

where $F(x)$ is the true distribution. Thus if the hypothesis is $H_0 : F(x) = F_0(x)$, one can plot F_0 "on top of" $F_n(x)$ and judge whether F_0 fits the data well. In Fig. 1 we see a plot of a normal distribution with mean 1 and variance 3, against the empirical distribution of a random sample of size 36 where the fit seems good.

Another graphical test is the q–q plot, described as follows. Let $X_{(1)} \leq \cdots \leq X_{(n)}$ be the order statistic of a sample $X_1, \ldots, X_n$. The $\frac{i}{n}$th quantile of the empirical distribution is $X_{(i)}$. Let F^0 be the distribution under examination. The $\frac{i}{n}$th quantile of F^0 is $\xi_i^0 = \inf\{x : F^0(x) \geq \frac{i}{n}\}$. A plot of $\{(X_{(i)}, \xi_i^0), i = 1, \ldots, n\}$ is the q–q plot (see Fig. 2). If the hypothesis

Normal Q-Q Plot

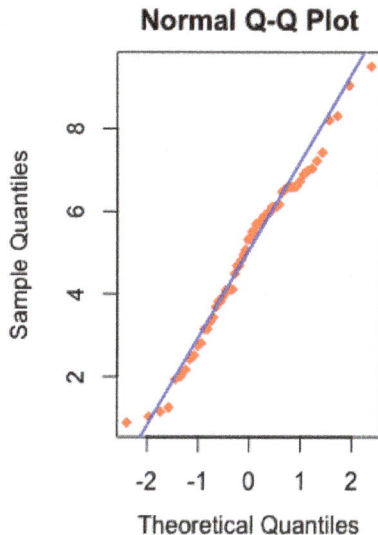

Fig. 2. The q-q plot of normal sample.

under consideration is valid, that F^0 is the true distribution, the points of the q–q plot should scatter at random around a straight line.

4. Some Analytical Tests

Several test statistics are available in the literature. We list a few of these:

(i) The Cramér–von Mises:

$$C_n = \int_{-\infty}^{\infty} (F_n(x) - F^0(x))^2 dF^0(x). \qquad (4)$$

(ii) The Anderson–Darling:

$$A_n = \int_{-\infty}^{\infty} (F_n(x) - F^0(x))^2 (F^0(1 - F^0(x)))^{-1} dF^0(x). \qquad (5)$$

(iii) Kolmogorov–Smirnov:

$$\begin{aligned} D_n &= \sup \left\{ x : |F_n(x) - F^0(x)| \right\} \\ &= \max_{1 \leq i \leq n} \left\{ \max \left(\left| \frac{i-1}{n} - F^0(X_{(i)}) \right|, \right. \right. \\ &\qquad \left. \left. \left| \frac{i}{n} - F^0(X_{(i)}) \right| \right) \right\}. \end{aligned} \qquad (6)$$

If $F^0(x)$ is the true distribution then $F_n(X)$ has the uniform distribution on $[0, 1]$.

Let $U_i = F^0(X_{(i)})$, leaving details to the reader the computation formulae are

$$nC_n = \frac{1}{12n} + \sum_{i=1}^{n} \left(U_{(i)} - \frac{2i-1}{2n} \right)^2, \qquad (7)$$

$$nA_n = -n - \frac{1}{n} \sum_{i=1}^{n} (2i-1)[\log U_{(i)} - \log(1 - U_{(n-i+1)})]. \qquad (8)$$

Similarly,

$$D_n = \max_{1 \leq i \leq n} \left\{ \max \left\{ \left| \frac{i-1}{n} - U_{(i)} \right|, \left| \frac{i}{n} - U_{(i)} \right| \right\} \right\}. \qquad (9)$$

One needs critical values for these tests which are quantiles of the corresponding distributions. For any n, the quantiles of the respective distributions can be approximated by simulations of random samples from the uniform distributions. The following asymptotic results are found in the literature:

$$\lim_{n \to \infty} \mathbf{Pr} \{nC_n \geq x\}$$

$$= \frac{1}{\pi} \sum_{j=1}^{\infty} (-1)^{j+1} \int \left(\frac{-\sqrt{y}}{\sin \sqrt{(y)}} \right)^{1/2} \frac{e^{xy/2}}{y} dy \qquad (10)$$

Table 1. Simulation estimates of critical values for he Kolmogorov–Smirnov test.

n	$\alpha = 0.05$	$\alpha = 0.01$
20	0.2859	0.3438
30	0.2451	0.2926
40	0.2085	0.2484
50	0.1869	0.2211
60	0.1706	0.1945
70	0.1618	0.1782
80	0.1542	0.1738
90	0.1368	0.1633
100	0.1366	0.1667

and

$$\lim_{n\to\infty} \mathbf{Pr}\left\{\sqrt{n}D_n \le x\right\} = 1 - 2\sum_{j=1}^{\infty} (-1)^{j-1} e^{-2j^2 x^2}. \quad (11)$$

The asymptotic distribution of nA_n is like that of $\sum_{j=1}^{\infty} \frac{Y_j}{j(j+1)}$ where $\{Y_j, j \ge 1\}$ are i.i.d. like χ_1^2. The reader is referred to Ref. 3 for large sample statistical theory.

In Table 1 we present simulation estimates of critical values (quantiles) for the Kolmogorov–Smirnov test, i.e., $D_{1-\alpha}$. In all these estimates the number of runs in the simulation is 1000.

In a similar manner estimates of the critical values for the Cramér–von Mises and the Anderson–Darling tests can be determined. Stute *et al.*[4] employed the bootstrapping method to generate critical values for the test statistics. To read on the bootstrapping method, see Chap. 4 of Ref. 1.

5. Nonparametric or Distribution-Free Tests

Suppose that we are given two datasets (random samples) and we wish to test whether they originate from the same unknown distribution. If we denote by $F^{(1)}$ the unknown parent distribution of the first sample, and by $F^{(2)}$ that of the second sample, the hypothesis under consideration is $H_0 : F^{(1)} = F^{(2)}$. Suppose that the two samples are of sizes n_1 and n_2, and let $F_{n_1}^{(1)}$ and $F_{n_2}^{(2)}$ denote the corresponding empirical distributions. Then one could test the hypothesis with a test statistic similar to the Kolmogorov–Smirnov one, i.e.,

$$D^{(2)} = \sup\left\{|F_{n_1}^{(1)}(x) - |F_{n_2}^{(2)}(x)|\right\}. \quad (12)$$

An alternative test is the chi-squared test. Partition the real line into m intervals or "bins." Let n_1 and n_2 be the sizes of the two samples. Denote by N_{ij} the number of cases in the ith sample that belong to the jth bin ($i = 1, 2; j = 1, \ldots, m$). Construct the following contingency table:

	1	$\ldots$	m	$\sum$
1	N_{11}	$\ldots$	N_{1m}	n_1
2	N_{21}	$\ldots$	N_2	n_2
$\sum$	$N_{.1}$	$\ldots$	$N_{.m}$	$N_{..}$

We construct a test of independence. The expected frequency of the (i, j) cell is $E_{ij} = n_i N_j = N_{..}$. The chi-squared statistic is

$$\chi^2(N_{ij}) = \sum_{i=1}^{2} \sum_{j=1}^{m} \frac{(N_{ij} - E_{ij})^2}{E_{ij}}. \quad (13)$$

The hypothesis that the two distributions are equal is rejected provided $\chi^2(N_{ij}) \ge \chi^2_{(m-1),(1-\alpha)}$.

6. Bootstrap Method for Evaluating Goodness-of-Fit

The bootstrap method is a resampling procedure to assess the sampling distribution of statistics, which are difficult to obtain analytically, or when the parent distribution of the observation is unknown (distribution-free). The method was introduced by Bradley Efron (1979), and was later studied in many papers. We illustrate here the bootstrap method applied to testing goodness-of-fit.

Generally, the bootstrapping has the following steps:

(1) Draw a random sample (without replacement) of size n, namely $S = \{X_1, \ldots, X_n\}$.
(2) Compute the value T_S of some statistic T based on the sample S.
(3) Draw a random sample $S^* = \{X_1^*, \ldots, X_n^*\}$ from S *with replacement*, i.e., some X_j^*s may be sampled multiple times.
(4) Compute $T^* = T_{S^*}$ the statistic T based on the resampled data S^*.
(5) Repeat steps (3) and (4) independently M times to obtain $\{T_1^*, \ldots, T_M^*\}$.

The empirical distribution $\{T_1^*, \ldots, T_M^*\}$ is called *the empirical bootstrap distribution (EBD)* of T. From the EBD we get approximations to the sampling standard error of T, its quantiles, and many more. In order to test the hypothesis about the distribution of X, say $H : F = F^0$, we draw in step (1), a random sample of size n from F^0; we then perform the bootstrap procedure on this sample. We compute, for the level of significance α, the $(1 - \alpha)$ quantile of EBD, say $Q_{1-\alpha}^*$. The statistic applied in the procedure is one of the above statistics for testing the goodness-of-fit of a distribution. If the value of this statistic is greater than $Q_{1-\alpha}^*$ we reject the null hypothesis H.

Table 2. Power values of the bootstrap test.

n	$Q_{0.95}^*$	p	Power
20	3.809	0.4	0.216
		0.5	0.578
		0.6	0.861
		0.7	0.983
50	6.095	0.4	0.226
		0.5	0.761
		0.6	0.983
		0.7	1.000

We demonstrate this procedure on a sample $X_1, \ldots, X_n$ of n binary random variables, when we wish to test whether the sample is of n Bernoulli trials with $p = 0.30$. We apply the chi-squared, statistic $X^2(n, p_0) = \frac{(K - np_0)^2}{np_0(1 - p_0)}$, where $K = \sum_{j=1}^{n} X_j$. Below, we display R code, a function for computing $X^2(n, p_0)$, labeled "chsqt," and the bootstrap critical value $Q_{1-\alpha}^*$ labeled "bootsh."

```
chsqt = function(x,p) {
    t = sum(x)
    n = length(x)
    res = (t-n*p)^2/(n*p*(1-p))
    res
}
bootsh = function(n, p∅, alpha, M) {
    x = rbinom(n, 1, p∅)
    t = chsqt(x, p∅)
    ebd = c(1:M)
    for(i in 1:M) {
    y = sample(x, n, replace=TRUE)
    ebd[i] = chsqt(y, p∅)
    Qs = quantile(ebd, prob = (1-alpha))
    print(Qs)
    }
}
```

In Table 2 we present the bootstrap critical values, and the power = $\mathbf{Pr}\{\text{reject} \mid p\}$, for $n = 20, 50$, $p = 0.3$, $\alpha = 0.05$, and $M = 100,000$.

The power values given in the table are simulated estimates based on 1000 replicas.

Much is written on bootstrapping. The reader is referred to B. Efron and R. Tibishirani (1998), Chap. 4 of Ref. 1, and C.G.J. Woo (1986).

7. Discussion

Several goodness-of-fit tests were presented for distribution models. The material discussed is just a glimpse of a rich theory readily available, e.g., Ref. 2. We have not discussed the fitness of regression models. There is much literature on this subject, e.g., see Chap. 5 of Ref. 1. In addition, the validity of models for prediction purposes is tested by *cross-validation* in which a portion of the data is used to estimate model parameters and the remainder of the data is used to validate model accuracy. This and other techniques will be discussed in a different paper.

References

[1] R. S. Kenett and S. Zacks, *Modern Industrial Statistics With Applications in R, MINITAB and JMP*, 2nd edn., Statistics in Practice (Wiley, New York, 2014).

[2] R. B. D'Agostino and M. A. Stephens, *Goodness of Fit Techniques* (Marcel Dekker, New York, 1986).

[3] A. Das Gupta, *Asymptotic Theory of Statistics and Probability* (Springer-Verlag, New York, 2008).

[4] W. Stute, W. Gonzáles-Manteiga and M. P. Quindimil, Bootstrap based goodness-of-fit tests, *Metrika* **40**, 243 (1993).

[5] J. R. Barr and S. Zacks, GLiM: Generalized linear models, *Encycl. Semant. Comput. Robot. Intell.* **1**, 1630016 (2017).

[6] R. Shibata, An optimal selection of regression variables, *Biometrika* **68**, 45 (1981).

[7] R. Shibata, Asymptotically efficient selection of the order of the model for estimating parameters of a linear process, *Ann. Stat.* **80**, 147 (1980).

Forensics: Assessing model goodness: A machine learning view

Joseph R. Barr[*,‡] and Joseph Cavanaugh[†,§]

*Barr Analytics, Irvine, CA, USA

†Department of Biostatistics
University of Iowa, Iowa City, IA 52242, USA

‡barr.jr@gmail.com

§joe-cavanaugh@uiowa.edu

It is not unusual that efforts to validate a statistical model exceed those used to build the model. Multiple techniques are used to validate, compare and contrast among competing statistical models: Some are concerned with a model's ability to predict new data while others are concerned with model descriptiveness of the data. Without claiming to provide a comprehensive view of the landscape, in this paper we will touch on both aspects of model validation. There is much more to the subject and the reader is referred to any of the many classical statistical texts including the revised two volumes of Bickel and Docksum (2016), the one by Hastie, Tibshirani, and Friedman [*The Elements of Statistical Learning: Data Mining, Inference, and Predication*, 2nd edn. (Springer, 2009)], and several others listed in the bibliography.

Keywords: Akaike Information Criterion (AIC); Schwarz Criterion; ROC; loss function; empirical risk; confusion matrix.

1. Introduction

There are a variety of ways in which models can be assessed but they broadly fall into two categories: the first is a model's descriptiveness, or how well the model describes the data, and the second is a model's predictive power, or how well the model predicts new, not yet seen values. Although those two views are not necessarily orthogonal, they are different inasmuch as they may result in diverging consequences. Although it is not likely, it is possible that a reasonably good descriptive model will systematically fail to adequately predict new values. *Goodness-of-fit* employs *risk functions* $\mathcal{R}$ to measure the discrepancy between the data, or observed $Y(x)$ at a level x, and predicts a model's values $\hat{Y}(x)$, aggregated for all input values x via some *loss function* $\mathcal{L}$ and sampling probability distribution $\mathbf{Pr}(x)$. Loss functions may vary according to a model's objectives. The generic *squared error*, $\mathcal{L}(Y) = (Y - \hat{Y})^2$, is an archetypical loss function. The *mean squared error* (MSE), the average square error across all possible observations, i.e., $\mathcal{R} = E(Y - \hat{Y})^2$, is the most common risk function. In case $\hat{Y}$ is binary, designating a $(0, 1)$ class membership, the loss function $(Y - \hat{Y})^2$, that is the difference between the observed and the predicted, is 0 if $\hat{Y} = Y$, and 1 otherwise. The risk function is the expected value $E(Y(x) - \hat{Y}(x))^2$, with respect to some sampling distribution $\mathbf{Pr}(x)$. For data (X, Y) consisting of N points and uniform sampling distribution $\mathbf{Pr}(x) = \frac{1}{N}$ the empirical risk function (ERF) $\mathcal{E}$ is simply the arithmetic mean $\mathcal{E} = \frac{1}{N} \sum_{j=1}^{N} I\{\hat{Y} = Y\}$, where $I\{\hat{Y} = Y\} = 1$ if $\hat{Y} = Y$, and 0 otherwise. There are other loss functions like *exponential loss* and *deviance* but we will not address those here.

The *empirical risk function* is the aggregate error measured with respect to the *uniform sampling distribution* endowing every sample point with equal probability (N points, probability $= 1/N$). For binary data the *empirical error rate* is $\frac{1}{N}|\{j : \hat{Y}(x_j) \neq Y(x_j)\}|$, where $\hat{Y}(x_j)$ is the predicted response at level x_j, it is the fraction of times the model "gets it wrong." It is not uncommon that a model performs well when measured against the dataset on which it is *trained*. In other words, the risk function exhibits good behavior when measured against the very data on which the model's is built while the model's ability to predict new data quickly degrades. The term *overfitting* is associated with the phenomenon. Presumably, overfitting occurs when the model attempts to fit noise (in addition to fitting the "signal"). *Parsimony* is a guiding principle where one measures the model's complexity with, e.g., its degrees of freedom. A model is generally referred as *parsimonious* if it has low complexity. The *bias–variance tradeoff*, generically visualized in Fig. 1, illustrates a general modeling phenomenon of controlling bias which invariably results in uncontrolled variance, and vice versa.

2. R-Squared

Consider the classic ordinary multiple regression model $Y = X\beta + \epsilon$, where $X = (1; X^*)$, X is full rank, the N-column vector consisting of 1s and X^* is an $N \times k$ matrix. The Ordinary Least Squares (OLS) estimate is $\hat{Y}(X) = X\hat{\beta} = X(X'X)^{-1}X'Y$ and the residual vector is

$$e = Y - \hat{Y} = Y - X(X'X)^{-1}X'Y$$
$$= (I - X(X'X)^{-1}X')Y. \qquad (1)$$

---

‡Corresponding author.

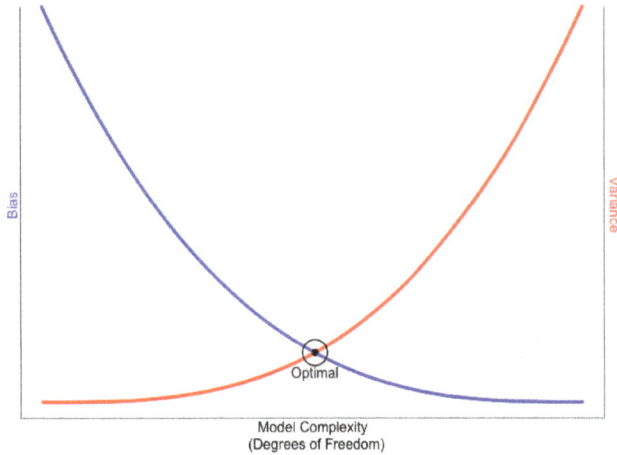

Fig. 1. Bias versus variance.

It is a straightforward exercise to verify that the mean of the residuals $E(e) = 0$ and that the residual vector e and $\hat{Y}$ are uncorrelated. Much of the theory of linear models relies on the fact that $\hat{Y}$ is a *projection* on the column space of the matrix X. The net result is that $\hat{Y}$ and the residual vector $Y - \hat{Y}$ are orthogonal, therefore the sums of squares "add up":

$$\|Y - \bar{Y}\|^2 = \|\hat{Y} - \bar{Y}\|^2 + \|Y - \hat{Y}\|^2, \qquad (2)$$

where $\bar{Y}$ is a projection of Y onto the $(1, 1, \ldots, 1)'$ vector, the "mean vector" $(\bar{Y}, \ldots, \bar{Y})$, with the Euclidean norm $\|(c_1, \ldots, c_m)\|^2 = c_1^2 + \cdots + c_m^2$. Because $\|\hat{Y} - \bar{Y}\|^2 \geq 0$,

$$\|Y - \bar{Y}\|^2 \geq \|Y - \hat{Y}\|^2, \qquad (3)$$

Dividing both sides of (3) by $\|Y - \bar{Y}\|^2$ shows that $\frac{\|Y - \hat{Y}\|^2}{\|Y - \bar{Y}\|^2} \leq 1$. Clearly, *R-squared* ($R^2$) satisfies

$$0 \leq R^2 = 1 - \frac{\|Y - \hat{Y}\|^2}{\|Y - \bar{Y}\|^2} = 1 - \frac{RSS}{TSS} \leq 1. \qquad (4)$$

If the residuals are zero, the model is a perfect fit and so $R^2 = 1$, an unsettling situation, perhaps. A good model is one with as large as possible R^2 against a validation data. It is often the case that R^2 is large against the training set, but falling dramatically against the validation dataset. This can be regarded an indicator that a model overfits the data, i.e., has excessive complexity (has more parameters than are necessary).

3. Kolmogorov-Smirnov

The classical *Kolmogorov–Smirnov* (KS) measures the separation between two distributions and when used properly it provides an objective measure of *lift* that contrasts a binary $(-1, +1)$ classification model's performance against what is known as a "random guess." A classification model based on examples (x, y), $x \in \mathbb{R}^p$ and $y = -1$ or $y = +1$, produces

probabilities $\hat{p}$ that an input vector x belongs to one of the two, $(-1, +1)$ classes. Informally, we write $\hat{p}(x) = \mathbf{Pr}(y = +1|x)$ to represent the probability that x is labeled $+1$. For example, the probability $\hat{p}(x) = 0.97$ overwhelmingly favors x in class $+1$, while $\hat{p}(x) = 0.03$ overwhelmingly favors x in class -1.

Cross validation. In order to control for overfitting, the data is split in two or more subsets, a *training* set, Υ, and *validation* or *test* sets, $\Upsilon_1, \ldots, \Upsilon_k$, where $k \geq 1$. A model $\hat{p}$ (like LOGIT) is trained using Υ to, e.g., minimize an empirical risk function, but the model $\hat{p}$ is tested by calculating the empirical risk function against validation sets $\Upsilon_1, \ldots, \Upsilon_k$. The empirical distribution function of model $\hat{p}$ is defined in the usual manner:

$$\hat{F}(s) = \frac{1}{N} \sum_{\hat{p}(x_t) \leq s} I(y(x_t) = +1), \qquad (5)$$

the proportion of $+1$s among the ys having the corresponding $\hat{p}(x)$ not exceeding s. Clearly, $\hat{F}(1) = +1$ but it is possible that $\hat{F}(0) > 0$, this happens in the unlikely case that $\hat{p}(x) = 0$ for some (x, y) with $y = +1$. $\hat{F}(t)$ is an increasing function with discrete jumps at values t where $y_t = 1$.

The cumulative distribution of the $[0, 1]$ uniform distribution U is the $45°$ line. The uniform distribution represents complete lack of foresight: a random guess, an *uninformed score*. The KS is the maximum difference between $\hat{F}(t)$ and $U(t) = t$ (a uniform distribution):

$$\text{KS}(T) = \max_{0 \leq t \leq 1} |T(t) - U(t)|. \qquad (6)$$

In scoring, a *bad* is an "actor" responsible for some undesirable event usually tagged with $+1$. In credit, for example, a "*bad*" is a "person" who has defaulted on a financial obligation, e.g., to pay off a loan. A *score*, say, credit score, represents the probability that x is "bad," that eventually will default. A tag y is more likely to be $+1$ if $\hat{p}(x)$ is large, and conversely, a tag is likely to be 0 if $\hat{p}(x)$ is small. Informally, *the lift* measures the maximal spread $\max_{0 \leq t \leq 1} |\hat{p}(t) - t|$ between an uninformed score $U(t) = t$ and the model $\hat{p}(t)$. KS, in fact, consists of two metrics, one is $\max_{0 \leq t \leq 1} |\hat{p}(t) - t|$ as well as $t_{\max} = \arg \max_{0 \leq t \leq 1} |\hat{p}(t) - t|$.

KS is *discriminatory* in the sense that for the two models, $\hat{p}_1$ and $\hat{p}_2$, one prefers the model posting a larger of the two, $\max_{0 \leq t \leq 1} |\hat{p}_1(t) - t|$ and $\max_{0 \leq t \leq 1} |\hat{p}_2(t) - t|$.

Consider the *sparse* binary classifications paradigm, with examples (x, y), and vast majority of tags $y \in \{0, 1\}$ consisting of 0s. Sparse data is one with few, sometimes fewer than 1%, labeled with a $+1$. Those paradigms are common in the fraud analytics where each transaction is scored for the likelihood of fraud. In those situations the KS fails to deliver a decisive metric because invariably frauds are rare and are likely to be at the very top end of the score spectrum. Consequently, in sparse data KS is inadequate and an effective measure of lift, common in those situations is the *top-end lift* which measures the percentage of fraudulent cases at the top end of the score scale. Specifically, the top 95% of scores is measured against the test data. An effective model will capture a high percent of the +1s (the frauds) between the 95th and 100th score percentiles. For example, one might consider the score that captures more than 50% of +1 at the top 5% of the scores. The term "top-end lift" is not standard in the literature, but it is as good name as any.

4. The Confusion Matrix and the ROC

With the input space $\mathcal{X} \subset \mathbb{R}^p$, a *soft classifier* is a function $\phi : \mathcal{X} \to [0,1]$. $\phi(x)$ can be interpreted as the probability that x belong to class $+1$. Logistic regression is an example of soft classifier. With a soft classifier it is necessary to select an optimal *threshold* value θ_0 that best "separates" the labels, e.g., -1 and $+1$, in the sense that x is labeled with $+1$ provided $\phi(x) \geq \theta_0$, and -1 otherwise. Below we describe how an optimal θ_0 may be selected. A *false positive* (FP) occurs when an example $x \in \mathcal{X}$ has a high score, say $\phi(x) \geq 0.95$, while in actuality it carries a negative label. One would expect that effective scoring is *discriminating* in the sense that relatively few items with large score values of ϕ are labeled negative, and vice versa. Once a threshold θ_0 is determined, one is able to quickly classify an item: assign the label $+1$ provided $\phi(x) \geq \theta_0$, and -1 otherwise. A *hard classifier* is a function $\eta : \mathcal{X} \to \{+1, -1\}$ predicting a label of $x \in \mathcal{X}$. Many popular machine-learning two-class classification algorithms can be viewed as soft classifiers because the output values are viewed as probabilities of class membership.

There are two kinds of false classification errors, FPs, where an item x is falsely classified as $+1$, and *false negatives* (FNs) where it is falsely classified as -1. Pursuantly, consider classified labeled data, according to a soft classifier ϕ, with a threshold θ, i.e., for each $x \in \mathcal{X}$, $\phi(x)$ is either $+1$ or -1 according to $\phi(x) \geq \theta$ or $\phi(x) < \theta$, respectively. The *confusion matrix* represents all four possibilities, with diagonal values representing ratios of correctly classified examples according to ϕ and a threshold θ_0.

	Predicted	
	+1	−1
Actual +1	TP	FP
Actual −1	FN	TN

True Positive Rate (TPR) or *sensitivity* $= \frac{Predicted \#(+1)}{Actual \#(+1)}$; the True Negative Rate (TNR) or *specificity* $= \frac{Predicted \#(-1)}{Actual \#(-1)}$.

For example, if out of 1000 examples, 600 positives, 400 negatives, a soft classifier ϕ and a threshold θ_0 produce 480 TPs, 340 TNs, 120 FPs and 60 FNs, then the confusion matrix is

	+1	−1
Actual +1	0.80	0.20
Actual −1	0.15	0.85

Since 820 examples out of the 1000 are correctly classified the *match rate* (MR) is therefore $820/1000 = 82\%$. In this case, the false–true positive rate split is $(0.20, 0.80)$: 80% TPR carries a penalty of 20% false positive rate (FPR). Note that the point $Q = (0.2, 0.8)$ lies above the $45°$ line, in the first quadrant of the x-y-coordinate system. With each threshold value θ, consider $t_\theta = \text{TPR}_\theta$ as the true positive rate of the classifier which classifies x as $+1$ provided $\phi(x) \geq \theta$, and $f_\theta = \text{FPR}_\theta$ as the false positive rate accordingly. An optimal threshold value θ is one that maximizes the difference between true and false positive rates, i.e., $\theta_0 = \arg\max\{t_\theta - f_\theta\}$. Since the number of examples is finite, one obtains an optimal value empirically, using the test dataset.

The Receiver Operating Characteristic (ROC). The confusion matrix is a static tabulation of the four possible classification outcomes. We build a model $\phi(x)$ to predict y from x and tabulate the four possible outcomes against a test set using a hard threshold θ. The *Receiver Operating Characteristic* or ROC is an idealized graphical depiction of a classifier's ability to discern positive and negative examples using the full force of the soft classifier ϕ. The graph, which in practice is a step function, scans the examples (x_i, y_i) of the test dataset, and when the threshold value of θ is incremented from 0 to 1, more and more examples are labeled $+1$, and when more examples are labeled $+1$, it is obvious that more negative examples are wrongly classified. We say that the ROC curve is idealized because in reality the ROC curve is step-wise constant while we usually depict it as a smooth curve. This is done using standard smoothing technique. A "good" classifier will maintain a greater proportion of true positive rates than false positive rates for each possible threshold θ. This means that for any θ, the point (f_θ, t_θ) lies above the $45°$ line. There are two special points that lie on the ROC curve. For the threshold $\theta = 0$, every example is positively classified hence 100% of true positives and 100% of

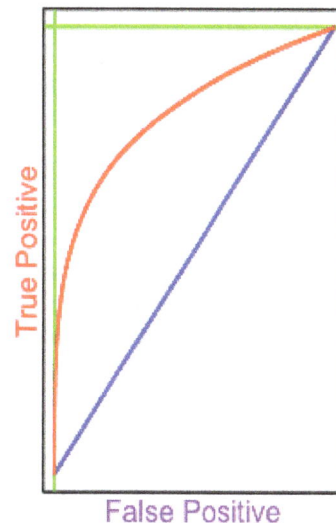

Fig. 2. ROC curves.

false positives are accounted for, hence the point $(1, 1)$ lies on the ROC curve. On the other extreme, for the threshold $\theta = 1.0$, no examples are positively labeled [dismissing the case where for some x we have $\phi(x) = 1.0$] so no example is either falsely or truly classified as positive, and the point $(0, 0)$ lies on the ROC curve. Figure 2 is an idealized depiction of three ROC curves, noninformative in blue, generic in red and perfect score in green.

The *Area Under the Curve* (AUC). The area under the ROC curve measures the overall soft score discriminating power. On the one hand, the area under the noninformative score is 1/2 (the area of the triangle with base $= 1$ and height $= 1$). On the other hand, for the perfect soft score, the area is 1, while for the generic score the area varies between 1/2 and 1. Generally, soft classifiers with greater AUC are preferred to those with smaller AUC.

5. Model Selection Criteria

5.1. *The Akaike information criterion*

In statistical modeling, one the main challenges is to select a suitable model from a candidate collection to characterize the underlying data. Model selection criteria provide a useful tool in this regard. A selection criterion assesses whether a fitted model offers an optimal balance between goodness-of-fit and parsimony. Ideally, a criterion will rule out candidate models that are either too simplistic to accommodate the data or unnecessarily complex.

The Akaike Information Criterion (AIC) was introduced by Hirotogu Akaike in his seminal 1973 paper "Information Theory and an Extension of the Maximum Likelihood Principle."[1] AIC was the first model selection criterion to gain widespread attention in the statistical community. Today, AIC continues to be the most widely known and used model selection tool among practitioners.

The traditional maximum likelihood paradigm, as applied to statistical modeling, provides a mechanism for estimating the unknown parameters of a model once the structure and the dimension of the model have been specified. Akaike proposed a framework wherein both model estimation and selection could be simultaneously accomplished.

For a parametric candidate model of interest, the likelihood function reflects the conformity of the model to the observed data. As the complexity of the model is increased the model becomes more capable of adapting to the characteristics of the data. Thus, selecting the fitted model that maximizes the empirical likelihood will invariably lead one to choosing the most complex model in the candidate collection. Model selection based on the likelihood principle, therefore, requires an extension of the traditional likelihood principle.

To formally introduce AIC, consider the following model selection framework. Suppose we endeavor to find a suitable model to describe a collection of response measurements y. We will assume that y has been generated according to

an unknown density $g(y)$. We refer to $g(y)$ as the *true* or *generating model*. A model formulated by the investigator to describe the data y is called *candidate* or *approximating model*. We will assume that any candidate model structurally corresponds to a parametric class of distributions. Specifically, for a particular candidate model M_k, we assume there exists a k-dimensional parametric class of density functions

$$\mathcal{F}(k) = \{f(y|\theta_k) : \theta_k \in \Theta_k\},$$

a class in which the parameter space Θ_k consists of k-dimensional vectors whose components are functionally independent, in other words $\Theta_k \subset \mathbb{R}^k$.

Let $L(\theta_k|y)$ be the likelihood corresponding to the density $f(y|\theta_k)$, i.e., $L(\theta_k|y) = f(y|\theta_k)$. Let $\hat{\theta}_k$ be the maximum likelihood estimate for θ_k, i.e.,

$$\hat{\theta}_k = \underset{\theta_k \in \Theta_k}{\arg\max}\, L(\theta_k|y).$$

Suppose we formulate a (finite) collection of candidate models $M_{k_1}, M_{k_2}, \ldots, M_{k_L}$ of various dimensions. These models may be based on different subsets of explanatory variables, different mean and variance/covariance structures and even different specifications of types of distribution for the response variable. Our objective is to search among this collection for the fitted model that "best" approximates $g(y)$.

In the development of AIC, optimal approximation is defined in terms of a well-known measure that can be used to gauge the similarity between the true model $g(y)$ and a candidate model $f(y|\theta_k)$: the *Kullback–Leibler information*. The Kullback–Leibler information between two densities $g(y)$ and $f(y|\theta_k)$ with respect to $g(y)$ is defined as

$$I(\theta_k) = \mathbb{E}\left\{\log\frac{g(y)}{f(y|\theta_k)}\right\} = \int \log\left(\frac{g(y)}{f(y|\theta_k)}\right)g(y)dy.$$

Gibbs' inequality states that $I(\theta_k) \geq 0$, with equality if and only if $f(y|\theta_k)$ is "true" density, i.e., identical with $g(y)$. Note that $I(\theta_k)$ is not a metric: it is neither symmetric nor does it satisfy the triangle inequality. Yet, we view the measure in a similar manner to a distance, i.e., $I(\theta_k)$ it can be used to measure the proximity between densities in that growth in disparity between $f(y|\theta_k)$ and $g(y)$ implies growth of magnitude of $I(\theta_k)$. Here, the stated goal is to identify the model $f(y|\theta_k)$, for some k, so that $I(\theta_k)$ is *the smallest among all other values of* $I(\theta_k)$. Since $g(y)$, the truth per se, is not altogether known, one might (and should) ask themselves how could it be possible to ascertain the value of $I(\theta_k)$. This, in fact, is a nontrivial problem which involves the fine art of approximation and asymptotics.

Onward with the plan we define

$$d(\theta_k) = \mathbb{E}\{-2\log(f(y|\theta_k))\} = -2 \int \{\log f(y|\theta_k)\}g(y)dy.$$

We can write

$$2I(\theta_k) = d(\theta_k) - \mathbb{E}\{-2\log(g(y))\}.$$

Since $\mathbb{E}(-2\log g(y))$ is independent of θ_k (the expectation is taken with respect to the true distribution g), any ranking of a set of candidate models corresponding to values $I(\theta_k)$ would be identical to a ranking corresponding to the values of $d(\theta_k)$. Hence, for the purpose of discriminating among various candidate models, $d(\theta_k)$ serves as a valid substitute for $I(\theta_k)$. We will refer to $d(\theta_k)$ as the *Kullback discrepancy*.

To measure the separation between a fitted candidate model $f(y|\theta_k)$ and the generating model $g(y)$, we consider the Kullback discrepancy evaluated at the MLE $\hat{\theta}_k$:

$$d(\hat{\theta}_k) = \mathbb{E}\{-2\log f(y|\theta_k)\}|_{\theta_k=\hat{\theta}_k}.$$

In principle, $d(\hat{\theta}_k)$ would provide an attractive means for comparing various fitted models for the purpose of discerning which model is closest to the truth. As was noted earlier, exact evaluation of $d(\hat{\theta}_k)$ is not possible, since doing so requires knowledge of the true distribution $g(y)$. However, the work of Akaike suggests that $-2\log f(y|\hat{\theta}_k)$ serves as a biased estimator of $d(\hat{\theta}_k)$, and that in certain large-sample settings, the bias adjustment

$$\mathbb{E}\{d(\hat{\theta}_k)\} - \mathbb{E}\{-2\log f(y|\hat{\theta}_k)\}$$

can be estimated as twice the dimension of θ_k. So assuming that $\dim\theta_k = k$ (θ_k is a k-parameter vector), for fitted candidate models that are correctly specified or overfit, it can be shown that the expected value of

$$\text{AIC} = -2\log f(y|\hat{\theta}_k) + 2k$$

will asymptotically approach the expected value of $d(\hat{\theta}_k)$, say

$$\Delta(k) = \mathbb{E}\{d(\hat{\theta}_k)\}.$$

In practical applications AIC is computed for each of the models $M_{k_1}, M_{k_2}, \ldots, M_{k_L}$, and the model corresponding to the minimum value of AIC is generally favored. In AIC the empirical log-likelihood term $-2\log f(y|\hat{\theta}_k)$ is called the *goodness-of-fit term*. The bias correction $2k$ is called the *penalty term*. Intuitively, models that are too simplistic to adequately accommodate the data at hand will be characterized by small goodness-of-fit terms yet large penalty terms. Models that provide a desirable balance between fidelity to the data and parsimony should correspond to small AIC values, with the sum of the two AIC components reflecting this balance.

5.2. *Bayesian information criterion*

The *Bayesian Information Criterion*, BIC, was introduced by Gideon Schwarz as a competitor to the AIC. The criterion is also known as the *Schwarz Information Criterion* (SIC), or simply as the *Schwarz criterion* (SC). Schwarz derived BIC to serve as an asymptotic approximation to a transformation of the Bayesian posterior probability of a candidate model. In large-sample settings, the fitted model favored by BIC ideally corresponds to the candidate model which is a *posteriori* most probable, i.e., the model which is rendered most plausible by the data at hand.

In the setting of the previous subsection, the Bayesian information criterion for candidate model M_k is defined as

$$\text{BIC} = -2\log f(y|\theta_k) + k\log n.$$

In practice, BIC is computed for each model $M_{k_1}, M_{k_2}, \ldots, M_{k_L}$. The model corresponding to the minimum value of BIC is often chosen, based on the premise that this model should have the highest posterior probability.

Since BIC approximates a transformation of a model's posterior probability, one can perform model evaluation by transforming BIC back to a probability. Specifically, if $\text{BIC}(k)$ denotes the value of BIC corresponding to model M_k, then the posterior probability on M_k can be approximated as

$$\mathbb{P}(k|y) \approx \frac{\exp\left\{-\frac{1}{2}\text{BIC}(k)\right\}}{\sum_{l=1}^{L}\exp\left\{\frac{1}{2}\text{BIC}(k_l)\right\}}.$$

AIC and BIC share the same goodness-of-fit term, but the penalty terms differ: BIC employs complexity penalization of $k\log n$ as opposed to $2k$. Since the penalty terms of BIC and AIC can be quite large, BIC will often choose fitted models that are more parsimonious than those favored by AIC. The difference in selected models may be especially pronounced in large-sample settings. This distinction is further discussed in the subsection that follows.

5.3. *Properties of AIC and BIC*

AIC and BIC are applicable in a broad array of modeling frameworks, since their justifications only require conventional large-sample properties of maximum likelihood estimators.

The framework of hypothesis testing is often used for model comparison. Likelihood ratio test (LRT) procedures are ubiquitous for this purpose. In most testing settings, including that for the LRT, the null model is nested within a larger alternative model. The latter model is assumed to be true, and the test is conducted to determine whether the simpler, more parsimonious model can also be deemed suitable. AIC and BIC, however, can be used to compare nonnested models. As emphasized in p. 88 of Ref. 3, "A substantial advantage in using information-theoretic criteria is that they are valid for nonnested models. Of course, traditional likelihood ratio tests are defined only for nested models, and this represents another substantial limitation in the use of hypothesis testing in model selection."

AIC and BIC can also be used to compare models based on different probability distributions, e.g., normal versus gamma. However, if the models in the candidate collection are based on different distributions, all terms in each empirical likelihood must be retained when the values of criterion are evaluated. (If the models in the candidate collection are based on the same distribution, terms in the empirical likelihood that do not depend on the data may be discarded in the criterion computations.) AIC and BIC cannot be used to compare models based on different transformations of the response variable.

As previously mentioned, BIC imposes a more stringent penalization for model complexity than AIC, meaning that BIC tends to favor more parsimonious fitted models than AIC. From a practical perspective, this distinction between AIC and BIC might be characterized as follows: AIC could be advocated when the primary goal is that the modeling application is predictive, i.e., to build a model that will effectively predict new outcomes. BIC could be advocated when the primary goal of the modeling application is descriptive, i.e., to build a model that will feature the most meaningful factors influencing the outcome, based on an assessment of relative importance. As the sample size grows, predictive accuracy improves as subtle effects are admitted to the model. AIC will increasingly favor the inclusion of such effects; BIC will not.

The predictive/descriptive modeling delineation of AIC and BIC may be viewed as a practical manifestation of the large-sample optimality properties of the criteria. Suppose that the generating model is of a finite dimension, and that this model is represented in the candidate collection under consideration. A *consistent* criterion will asymptotically select the fitted candidate model having the correct structure with probability one. On the other hand, suppose that the generating model is of an infinite dimension, and therefore lies outside of the candidate collection under consideration. An *asymptotically efficient* criterion will asymptotically select the fitted candidate model which minimizes the mean squared error of prediction. AIC is asymptotically efficient yet not consistent, whereas BIC is consistent yet not asymptotically efficient. From a machine learning perspective, where structure takes a back seat to predictability, AIC is generally preferred over BIC.

6.   Issues with Multicollinearity

For simplicity consider the zero-intercept linear model $Y = X\beta + \epsilon$. Multicollinearity occurs when the columns of X are not independent, i.e., when $\det(X'X) = 0$. Frequently, the problem is not that the columns are dependent (not independent), rather, the problem is that the columns are "nearly" dependent. This happens when $\det(X'X)$ is quite small (but discernibly not zero). Multicollinearity results in unstable estimates. By way of analogy, one may compare the two systems of equations

$$\frac{1}{6}x + \frac{2}{6}y + \frac{3}{6}z = 1,$$

$$\frac{1}{3}x + \frac{1}{3}y + \frac{2}{3}z = \frac{4}{3}, \tag{7}$$

$$0.4999x + 0.4999y + z = 1.999,$$

with solutions $x = 1, y = 1, z = 1$; and

$$\frac{1}{6}x + \frac{2}{6}y + \frac{3}{6}z = 1,$$

$$\frac{1}{3}x + \frac{1}{3}y + \frac{2}{3}z = \frac{4}{3}, \tag{8}$$

$$0.4999x + 0.4999y + z = 1.998,$$

with solutions $x = 10, y = 10, z = -8$. The terms *ill conditioning* and *multicollinearity* are used interchangeably to describe a situation where small changes in inputs lead to dramatic changes in outputs. Notice that although the coefficient matrix B below is not quite singular, the determinant $\det(B) \approx -0.00001$,

$$B = \begin{bmatrix} 1/6 & 2/6 & 3/6 \\ 1/3 & 1/3 & 2/3 \\ 0.4999 & 0.4999 & 1 \end{bmatrix}.$$

6.1.   *Variance inflation factor*

Consider the standard full rank linear model $Y = X\beta + \epsilon$, the *Variance Inflation Factor* (VIF,) VIF_i of the ith column of X is defined in terms of R_i^2 of the regression of the ith column on the other columns: $\text{VIF}_i = \frac{1}{1-R_i^2}$.

Since $0 \leq R^2 \leq +1$, $\text{VIF}_j \geq 0$. One would expect low column correlation if X is "robustly" nonsingular. To see how VIF works, consider without loss of generality the zero-intercept linear model $Y = X\beta + \epsilon$ with X being an $N \times p$ matrix, $p \geq 2$ and $\epsilon \sim N(0, \sigma^2)$. The OLS solution is $\hat{\beta} = X'(X'X)^{-1}X'Y$; the variance of the unbiased estimator $\hat{\beta}$ is $\text{var}(\hat{\beta}) = \sigma^2(X'X)^{-1}$ and the variances of individual components of $\hat{\beta}$ are the diagonal elements. Put $S_j^2 = \sum_{i=1}^{N} x_{ij}^2$, the variance of the jth column of X. Let X^* be the $N \times (p-1)$ matrix consisting of the columns $2, 3, \ldots, p$ of X and $x^{(1)}$ be the first column of X. The projection of $x^{(1)}$ on the columns of X^* is $X^*[X^{*\prime}X^*]^{-1}X^{*\prime}x^{(1)}$ and the squared norm of its orthogonal vector $x^{(1)} - X^*[X^{*\prime}X^*]^{-1}X^{*\prime}x^{(1)}$ is

$$(x^{(1)} - X^*[X^{*\prime}X^*]^{-1}X^{*\prime})'(x^{(1)} - X^*[X^{*\prime}X^*]^{-1}X^{*\prime})'$$
$$= x^{(1)\prime}x^{(1)}(1 - R_1^2).$$

Therefore, $R_1^2 = \frac{1}{x^{(1)\prime}x^{(1)}}x^{(1)\prime}X^*[X^{*\prime}X^*]^{-1}X^*x^{(1)}$. We state without a proof the following theorem.

Theorem 1. *Consider the zero-intercept linear model with $\sigma^2 = 1$, then $\text{var}(\hat{\beta}_1) = \frac{1}{x^{(1)\prime}x^{(1)}(1-R_1^2)}$. Similar formula holds for $\text{var}(\hat{\beta}_j), \forall j = 1, 2, \ldots, p$.*

This result demonstrates that the variance of the jth weight $\hat{\beta}_j$ "inflates" due to multicollinearity. A common practice is to remove a column that exhibits high correlation with the other columns. Some statisticians apply the rule that a column of X with VIF greater than 5 indicates high correlation and therefore it should be removed.

7. Cross-Validation

Cross-validation (CV) is used extensively in machine learning to assess a model's goodness in terms of resistance to over-fitting. Overfitting in the scourge of statistical modeling, responsible for poor "out of sample" performance. CV helps inoculate against the likelihood of overfitting by examining the discrepancy of goodness-of-fit metric measured against training and validation sets. Metrics vary depending on modeling goal: for example, in regression the MSE is used, and in classification MR is common. However, there are other metrics one could consider including the true positive rate measured against false positive rate via the ROC, etc. There are several essential cross-validation techniques. For example, one can consider the standard *train-and-validate* approach, the *k-fold* cross-validation and the *leave-one-out* cross-validation. The train-and-validate splits the dataset in to two, usually a 50–50 split; train a model on the first and validate on the second. In regression MSE calculated on the training set will likely be smaller than the MSE calculated on the validation set. A model is stable if MSEs calculated on the training set Υ_1 and on the validation set Υ_2 are close in magnitude and this is measured with the R^2-like metric

$$\rho^2 = 1 - \frac{\text{MSE}(\Upsilon_1)}{\text{MSE}(\Upsilon_2)}.$$

Since $\text{MSE}(\Upsilon_1) \leq \text{MSE}(\Upsilon_2)$, it follows that $0 \leq \rho^2 \leq 1$. *Example:* If $\text{MSE}(\Upsilon_1) = 84$, $\text{MSE}(\Upsilon_2) = 88$, then $\rho^2 = 1 - \frac{84}{88} = \frac{4}{88} \approx 4.5\%$. A benchmark for ρ^2 is determined empirically. The *k-fold cross-validation* involves partitioning the data into $k, k \geq 2$ subsets, keeping one for validation and training the remaining subsets. This results in k pairs of MSE values, one for training and one for validation. The average of those values is used to calculate ρ^2. *Leave-one-out cross-validation* is a process of training the model on all but one value and validating on a single value and repeating for each sample point. Both training and validation values are averaged and ρ^2 is reported. To summarize, cross-validation is not intended to assess a model's performance, it is rather used to assess a model's stability, whether it is robust or not.

8. Summary

Building a good predictive model is not easy. One must ensure that inputs provide sufficient signal for a response, that the modeling framework has been carefully chosen to represent the specific paradigm and that the model strikes a healthy balance between parsimony and precision, to ensure the model generalizes (does not overfit). Also important for an overall model's utility is to discover and articulate a model's "band of validity" and its limitations. The techniques outlined in this short paper provide a technical framework but their implementations vary: different paradigms are often measured with different measuring tools. It is easy to make modeling mistakes and so the modeler must take extra care to balance between all the (valid) priorities. In industry, the modeler reports goodness-of-fit metrics. Business tends to pay attention to one goodness metric over another. In credit, for example, where logistic regression is used to predict probability of default, model discriminatory power is measured in terms of Kolmogorov–Smirnov; not surprisingly, business does not care much about things like *p*-values.

References

[1] H. Akaike, Information theory and an extension of the maximum likelihood principle, *Proc. 2nd Int. Symp. Information Theory* (Akadémia Kiadó, Budapest, 1973), pp. 267–281.

[2] H. Akaike, A new look at the statistical model identification, *IEEE Trans. Autom. Control* **19**, 716 (1974).

[3] K. P. Burnham and D. R. Anderson, *Model Selection and Multimodel Inference: A Practical Information-Theoretic Approach*, 2nd edn. (Springer, New York, 2002).

[4] J. E. Cavanaugh, Unifying the derivations of the Akaike and corrected Akaike information criteria, *Stat. Probab. Lett.* **33**, 201 (1997).

[5] R. Christensen, *Plane Answers to Complex Questions: The Theory of Linear Models* (Springer, 2011).

[6] R. B. D'Agostino and M. A. Stephens, *Goodness-of-Fit Techniques* (Marcel Dekker, New York, 1986).

[7] A. Das Gupta, *Asymptotic Theory of Statistics and Probability* (Springer-Verlag, New York, 2008).

[8] T. Hastie, R. Tibshirani and J. Friedman, *The Elements of Statistical Learning: Data Mining, Inference, and Prediction*, 2nd edn. (Springer, 2009).

[9] R. S. Kenett and S. Zacks, *Modern Industrial Statistics: With Applications in R, MINITAB and JMP*, 2nd edn., Statistics in Practice (Wiley, New York, 2014).

[10] S. Kullback, *Information Theory and Statistics* (Dover, New York, 1968).

[11] S. Kullback and R. A. Leibler, On information and sufficiency, *Ann. Math. Stat.* **22**, 76 (1951).

[12] P. H. Kvan and B. Vidokovic, *Nonparametric Statistics with Applications to Science and Engineering* (Wiley, New York, 2007).

[13] G. Schwarz, Estimating the dimension of a model, *Ann. Stat.* **6**, 461 (1978).

[14] D. Sengupta and S. R. Jammalamadaka, *Linear Models: An Integrated Approach* (World Scientific, 2003).

[15] R. Shibata, An optimal selection of regression variables, *Biometrika* **68**, 45 (1981).

[16] R. Shibata, Asymptotically efficient selection of the order of the model for estimating parameters of a linear process, *Ann. Stat.* **80**, 147 (1980).

[17] W. Stute, W. Gonzales-Manteiqa and M. Presedo Quindimil, Bootstrap based goodness of fit tests, *Metrika* **40**, 243 (1993).

[18] W. N. Venables and B. D. Ripley, *Modern Applied Statistics with S-Plus*, 2nd edn. (Springer, New York, 1997).

Diffusion analysis

Mira Kim[*,‡] and Masahiro Hayakawa[†]

*Department of Electrical Engineering and Computer Science
University of California Irvine, Irvine, CA, USA

†NEC Solution Innovators, Ltd., 1-18-7 Shinkiba
Koto-ku, Tokyo 136-8627, Japan
‡mirak1@uci.edu

Graph is a widely used scheme for representing complex datasets in terms of graphical illustration comprised of nodes and edges. Diffusion is a paradigm of propagating or transmitting substances or knowledge pieces from nodes to nodes. Diffusion analysis takes a slightly different approach from the reachability-based graph analysis; it takes the phenomenon as diffusion problems. In this paper, we present a technical survey on literatures of diffusion analysis.

Keywords: Graph; diffusion analysis.

1. Introduction

Nowadays study on influence maximization problem is becoming an emerging research topic in graph database. Social influence research is a fundamental topic that has been studied for many years. Initially the topic was categorized as social sciences, and its main objective was to explore relationships of human groups.[1] Many social scientists and psychologists have begun to show interest in the topic of social influence. In their research, a person can be influenced by other people. When politicians make a speech, their purpose is to spread their influence to win an election or acquire support.[2] Businessmen construct their business network in order to sell their products to customers.[3] Different people have different influences; some people have very strong influence, thus being able to influence a vast number of people. Influence diffusion ranges also vary between people. Some people have strong influence and can influence a lot of people in a larger range and time period. However, some people have weaker influence and cannot spread their influence or have limited influence in a shorter time span.

With the development of the Internet, many social media applications have been developed in recent years. People have constructed contact relationship through these social media applications or communication applications. Through email networks, we are able to contact whoever we want very easily. Through telephone networks, we can contact numerous people without voice or text message. Online retailers, such as Amazon, can construct large purchase networks among products, buyers, and sellers. Social network applications such as Twitter, Facebook, and Instagram can construct friends' networks for people all over the world.[4,5]

LinkedIn has constructed a very large contact network by using users' emails and constructed connections among the users. DBLP has set up a big collaboration network for researchers all over the world. Online gaming websites have constructed big game networks for game players. Analyzing these social networks, especially the dynamic social networks' influence among these networks' attendees, is becoming an emerging topic.[6–8]

This paper presents a comprehensive survey on diffusion analysis and also the challenges associated with diffusion analysis. Section 2 presents the definition of diffusion analysis as well as a comprehensive survey on existing works. Section 3 addresses some challenges associated with diffusion analysis with a conclusion.

2. Diffusion Analysis

The term diffusion in dictionaries is generally known as the state of being spread out or transmitted especially by contact. Hence, the diffusion occurs in the direction from a high concentrated area to a low concentrated area. Examples can be the diffusion of ink dropped into water as shown in Fig. 1.

The patterns of diffusion can vary depending on the domain, purpose of analysis, and the constraints. For example, the way to analyze diffusion in SNS would considerably be different from that in gene–disease relevance network.

As stated earlier in this paper, diffusion analysis refers to a specific method/way of analyzing the diffusions and their impacts in a representation of the target dataset. It can be applied to analyzing diffusions in various network domains such as social network or a gene–disease relevance network.

‡Corresponding author.

Fig. 1. Drop of ink diffusing in water.

Rogers stated diffusion as "the process by which an innovation is communicated through certain channels overtime among the members of a social system."[9] The concept of diffusion process has been around for many decades. It was thought to be a linear process in the earlier research. However, the later studies found that diffusion is not a simple binary process but should take into account unanticipated consequences such as innovation and a social system, and adoption of variables to form a curve, instead of a straight line. The methods and optimization of diffusion have been progressing and growing since.

Dasgupta *et al.* examine the communication patterns of mobile phone users to study the underlying social network and the related role of social ties in formation and groups in a communication network.[10] To research an activation-based technique for figuring out potential churners, they study the tendency of a subscriber to churn out a service provider's network based on the number of friends who have already been churned. Their work can solve queries such as which group of friends should be the next set of churners in a communication network.

Scripps *et al.*[11] study the effects and consequences that preprocessing and different network forces such as selection and influence have on the modeling of dynamic networks. Some of the preprocessing decisions they study are the effect of accumulating links or attributes over time, the utilization of historical data, the variability of importance of attributes over time, and how influence and selection occur in dynamic networks. Additionally, they thoroughly research metrics to measure the forces of selection and influence dynamic network data based on changes in the adjacency and data metrics of such network data. Their work is not industry-specific and can be applied to solve queries in various use cases and domains such as utilizing historical data in social networks to determine how influence occurs in their social dynamic networks.

Gomez-Rodriguez *et al.*[12] study diffusion and virus propagation and address challenges in these applications and underlying networks. They propose a method for sourcing and tracing paths of diffusion and influence through networks in which these contagions spread. Their method is based on an approach for identifying the most optimal network that explains the observed infected times at times when a node adopts information or becomes infected by viruses. They resolve this optimization problem of NP-hardness by proposing an approximation algorithm and prove the effectiveness by tracing information diffusion in a set of 170 million blogs and news articles. Their work can solve queries such as finding effectiveness of various technology articles or finding nodes that have the most tendency to become infected by virus in a security domain.

In social networks, knowing how to find the information's diffusion rules and mastering the information's diffusion models are very important in assisting people to master how the information spreads in social network.[14,15]

Centola[14] investigated the effect of network structure on diffusion by studying the spread of health behavior through artificially structured online communities. Individual adoption is more likely when participants receive social reinforcement from multiple neighbors in the social network. The behavior spreads farther and faster across clustered-lattice networks than across corresponding random networks. Their work can solve queries that compare the effectiveness and speed of option when there are, say, three neighbors versus seven neighbors surrounding the participant in a social network.

Cha *et al.*[15] analyze large-scale traces of information dissemination on social networks. They analyze 2.5 million users on their 11 million photos to evaluate how popular pictures are distributed across a social network using social links. Through their research and analysis, they find that most information does not spread widely and spreads slowly through the network. Their study is very useful for viral marketing campaigns as well as for finding the information that social network websites promote. Their research can answer the following types of queries: how widely and quickly information spreads in the Flickr social network and does information in Flick flow along its links in their network?

Wang *et al.*[16] present a community-based greedy algorithm for mining top-K influential nodes. It detects communities in social networks by considering information diffusion and a dynamic programming algorithm for selecting communities to find influential nodes. Their algorithm first detects communities and initializes the set of influential nodes discovered in their network as well as in each community. Using dynamic programming, they detect and choose which community to mine the influential nodes. They show their algorithm is more than an order of magnitude faster than the conventional greedy algorithm for finding the top-K influential nodes. Their work can solve queries in a social network domain on selecting the right communities to find top influential nodes.

Jalali *et al.*[17] model the mechanisms for examining diffusion process of an online petition using system dynamics modeling. Their calibration approaches include maximum-likelihood estimation, the Akaike Information Criterion, and likelihood ratio test. They compare the relative strengths of Push (i.e., sending announcements) and Pull (i.e., sharing by signatories) and discover that diffusion is heavily more reliant on Pull rather than Push. They conclude that targeting the right cohort is an essential driver in diffusing and spreading information. Their use-case is an online petition system, but the model can be applied to other domains to solve queries involving diffusion processes using system dynamics modeling.

Yang and Leskovec[18] develop a linear influence model which focuses on modeling the influence of a node on the rate of diffusion throughout a network. This approach is different in that other researches in this area have examined the knowledge of a social network first then model the diffusions by prediction nodes' influence on others. Their work is validated and proven accurate for predicting the temporal dynamics of information diffusion by using a set of 500 million tweets and a set of 170 million news articles as well as blog posts. Their work would be ideal for studying not only the result of diffusion itself but also the rate of diffusion in a network.

Dhamal *et al.*[19] focus on selecting and activating seed nodes in multiple phases. They first convey an objective function for their two-phase diffusion by generating algorithms for finding seed nodes in the two phases. Then, their research focuses on solving two problems: budget splitting for splitting total budget between two phases and scheduling an optimal delay to start the second phase. Their work would be ideal for finding *n* top seed nodes that have a budget constraint.

Galuba *et al.*[20] utilize the LT model to predict diffusions. They analyze 15 million URLs among 2.7 million users from a Twitter dataset. Their propagation model takes content popularity, user influence, and the rate of propagation which become the unknown parameters. They aim to find the parameter values that maximize the number of correctly predicted URLs mentioned in their Twitter dataset. Their model can achieve over 50% accurate predictions on the URLs mentioned in the test dataset.

Yang *et al.*[21] propose an approach for information propagation on a Tensor product graph that achieves the same computational complexity and the amount of storage of the propagation on the original graph. They achieve 99.9% retrieval score on the MPEG-7 shape dataset with unsupervised learning.

Wang and Daniels[22] propose a graph method for applying diffusion and graph spectral methods for network forensic analysis. Their method can be applied to address attack scenario extraction and attack profiling. Attack scenario extraction infers the set of entities and events related to an attacker. Attack case profiling assumes attackers have the tendency to show repeated behavioral patterns. The propagation of suspicion in an attack is modeled with heat diffusion in physical terms. They transform evidence graph analysis into approximations of steady-state energy diffusion problems.

In biomedical networks, knowing how to find the most influential genes or diseases in a large biomedical dataset is essential in preventing malignant spread of diseases. Raj *et al.*[23] use network diffusion to study progressions of dementia. Patterns of dementia are known to fall into dispersed brain networks, meaning the disease is transmitted along neuronal pathways. They use the brain's connectivity network provided by tractography of 14 healthy-brain MRIs. Then, they mathematically model this transmission using the connectivity network. Their work can solve queries involving patterns of progressions of dementia but since biomedicine is an area where networks and neuron pathways differ vastly among various diseases, it would be difficult to apply their work to other types of diseases.

Scientists have developed what they call the "human disease network" which is a visual map of all human diseases with known underlying genetic associations and details the genetic connections between these diseases.[24] The map is made up of nodes and edges where each node represents a disease, and the size of the node reflects the number of genes known to be associated with that disorder. The thickness of the edges that connect various nodes is a reflection of the number of genes shared by the connected diseases. The idea behind the map is that most diseases with known genetic associations seem to share most of their genes with other diseases. Despite the uncertainties associated with environmental factors, their work can help physicians better understand the role of genetics in human diseases using their human disease network without having to go through hundreds of journal or conference articles.

3. Conclusions

Diffusion analysis is an analytics approach that applies the paradigm of diffusion. With diffusion, transmission of substances or knowledge pieces from nodes to nodes takes place. Hence, diffusion analysis is different from the conventional reachability-based analysis in the graph model.

In this paper, we examine existing works on diffusion analysis in domains such as social networks, biomedical networks, and general modeling for diffusion. An interesting observation is that many of the existing works do not consider the natural phenomenon of diffusion. In many of the diffusion problems and examples in the real domains, a transmission of a substance or a knowledge piece loses some degree of influence power when a diffusion occurs. Hence, the subsequent diffusion occurrence would have a weakened influence than its immediately preceding diffusion occurrence.

Acknowledgments

The research is supported in part by NEC Solution Innovators, Ltd., Japan.

References

1 S. Chaudhuri and R. Dayal, An overview of data warehousing and OLAP technology, *ACM SIGMOD Rec.* **26**, 65 (1997).

[2]E. Thomsen, *OLAP Solutions: Building Multidimensional Information Systems*, 2nd edn. (John Wiley & Sons, Inc., 2002).

[3]H. Plattner, A common database approach for OLTP and OLAP using an in-memory column database, *Proc. 2009 ACM SIGMOD Int. Conf. Management of Data* (2009).

[4]R. Dunbar, Neocortex size as a constraint on group size in primates, *J. Hum. Evol.* **20**, 469 (1992).

[5]R. M. Bond, C. J. Fariss, J. J. Jones, A. D. I. Kramer, C. Marlow, J. E. Settle and J. H. Fowler, A 61-million-person experiment in social influence and political mobilization, *Nature* **489**, 295 (2012).

[6]P. Domingos and M. Richardson, Mining the network value of customers, *Proc. Seventh ACM SIGKDD Int. Conf. Knowledge Discovery and Data Mining* (2001), pp. 57–66.

[7]Lithium, Lithium company overview (2017), http://klout.com.

[8]P. Moore, Why I deleted my Klout profile (2011), https://www.socialmediatoday.com/content/why-i-deleted-my-klout-profile.

[9]E. Rogers, *Diffusion of Innovations* (Free Press, New York, 1962).

[10]K. Dasgupta, R. Singh and B. Viswanathan, Social ties and their relevance to churn in mobile telecom networks, *Proc. 11th Int. Conf. Extending Database Technology: Advance in Database Technology* (2008), pp. 668–677.

[11]J. Scripps, P.-N. Tan and A.-H. Esfahanian, Measuring the effects of preprocessing decisions and network forces in dynamic network analysis, *Proc. 15th ACM SIGKDD Int. Conf. Knowledge Discovery and Data Mining* (2009), pp. 747–756.

[12]M. Gomez-Rodriguez, J. Leskovec and A. Krause, Inferring networks of diffusion and influence, *Proc. ACM SIGKDD Int. Conf. Knowledge Discovery and Data Mining* (2010), pp. 1019–1028.

[13]E. Bakshy, B. Karrer and L. A. Adamic, Social influence and the diffusion of user-created content, *Proc. 10th ACM Conf. Electronic Commerce* (2009), pp. 325–334.

[14]D. Centola, The spread of behavior in an online social network experiment, *Science* **329**, 1194 (2010).

[15]M. Cha, A. Mislove and K. P. Gummadi, A measurement-driven analysis of information propagation in the Flickr social network, *Proc. 18th Int. Conf. World Wide Web* (ACM, 2009), pp. 721–730.

[16]Y. Wang, G. Cong, G. Song and K. Xie, Community-based greedy algorithm for mining top-*K* influential nodes in mobile social networks, *Proc. 16th ACM SIGKDD Int. Conf. Knowledge Discovery and Data Mining* (2010), pp. 1039–1048.

[17]M. Jalali, A. Ashouri, O. Herrera-Restrepo and H. Zhang, Information diffusion through social networks: The case of an online petition, *Proc. 33rd Int. System Dynamics Conf.* (2015).

[18]J. Yang and J. Leskovec, Modeling information diffusion in implicit networks, *Proc. 2010 IEEE Int. Conf. Data Mining* (2010), pp. 599–608.

[19]S. Dhamal, K. J. Prabuchandran and Y. Narahari, Information diffusion in social networks in two phases, *IEEE Trans. Netw. Sci. Eng.* **3**, 197 (2010).

[20]W. Galuba, D. Chakraborty, K. Aberer and Z. Despotovic, Outtweeting the Twitterers: Predicting information cascades in microblogs, *Proc. 3rd Workshop Online Social Networks* (2010).

[21]X. Yang, L. Prasad and L. Latecki, Affinity learning with diffusion on tensor product graph, *IEEE Trans. Pattern Anal. Mach. Intell.* **35**, 28 (2013).

[22]W. Wang and T. Daniels, Diffusion and graph spectral methods for network forensic analysis, *Proc. 2006 Workshop New Security Paradigms* (2006).

[23]A. Raj, A. Kuceyeski and M. Weiner, A network diffusion model of disease progression in dementia, *Neuron* **73**, 1204 (2012).

[24]L. Pray, Genome-wide association studies and human disease networks, *Nat. Educ.* **1**, 220 (2008).

[25]B. Chou, M. Hayakawa, A. Kitazawa and P. Sheu, GOLAP: Graph based online analytical processing, *Encycl. Semant. Comput.* (2018).

[26]Y. Zhang *et al.*, Information diffusion model based on online social network, *Phys. J. China* **60**, 050501 (2011).

[27]R. M. Anderson and R. M. May, *Infectious Diseases of Humans*, Vol. 1 (Oxford University Press, Oxford, 1991).

[28]W. Chen, Y. Wang and S. Yang, Efficient influence maximization in social networks, *Proc. 15th ACM SIGKDD Int. Conf. Knowledge Discovery and Data Mining* (2009), pp. 199–208.

[29]K. Jung, W. Heo and W. Chen, IRIE: Scalable and robust influence maximization in social networks, *Proc. 12th Int. Conf. Data Mining* (2012), pp. 918–923.

[30]A. Goyal, F. Bonchi and L. V. S. Lakshmanan, A data-based approach to social influence maximization, *Proc. VLDB Endow.* **5**, 73 (2011).

[31]G. Tong, W. Wu, S. Tang and D. Du, Adaptive influence maximization in dynamic social networks. arXiv:1506.06294 [cs.SI].

[32]H. Zhuang, Y. Sun, J. Tang, J. Zhang and X. Sun, Influence maximization in dynamic social networks, *Proc. 13th Int. Conf. Data Mining* (2013), pp. 1313–1318.

[33]B. Liu, G. Cong, Y. Zeng and X. Sun, Influence spreading path and its application to the time constrained social influence maximization problem and beyond, *IEEE Trans. Knowl. Data Eng.* **26**, 1904 (2014).

[34]S. Bharathi, D. Kempe and M. Salek, Competitive influence maximization in social networks, *Proc. Int. Workshop Web and Internet Economics* (2007), pp. 306–311.

Part 3

Data Integration

Network analysis and GOLAP

Jennifer Jin[*,‡] and Masahiro Hayakawa[†]

*Department of Electrical Engineering and Computer Science
University of California at Irvine, Irvine, CA 91761, USA

†NEC Solution Innovators, Ltd.
1-18-7 Shinkiba, Koto-ku, Tokyo 136-8627, Japan
‡jenniyk2@uci.edu

Online Analytical Processing (OLAP) is an effective approach to analyzing various complex business problems, and graph is considered as a common scheme to represent the business datasets. Network analysis is a broad analytics scheme for exploring the connectivity and deriving useful analytics results. However, network analysis for graph-based OLAP presents a set of more specific analytics methods by utilizing graph model, network property, and OLAP principles. In this paper, we present a comprehensive survey on network analysis conducted on graph model for the purpose of OLAP, and we summarize the current research focus, paradigms, and the future needs on the target technology.

Keywords: OLAP; graph model; network analysis.

1. Introduction

Online Analytical Processing (OLAP) is an approach to analyzing the dataset for a business domain with semantic-level analytics as well as conventional data-driven value-level analysis.[1] In a traditional OLAP, a data cube is related to a set of dimensional values where different records are treated as mutually independent. These records can be summarized using aggregate functions like count, sum and average. Multilevel summaries can be achieved if a concept hierarchy is related to each attribute. Users are able to navigate through different dimensions using roll-up, drill-down, and slice/dice procedures.

In recent years, more data sources are represented in graph forms beyond conventional spreadsheets. For example, influence in a social network has been a hot topic. In the area of science, an author/scientist may influence another author or a scientist. At the same time, a publication can influence another publication. The research of social influence has extensive applications such as social science and marketing. For example, a rumor can spread all over a social network and cause a significant social impact. In marketing, some products can sell better through a group of people via word of mouth to each other.

Due to increased structural complexity of graph data, traditional OLAP operations such as "roll-up/drill-down" and "slice/dice" cannot be used on graphs. An extension of traditional OLAP on graphs is called graph-based OLAP (GOLAP). With more and more data being represented in graph forms such as social networks, GOLAP can be very useful in analyzing graph data. GOLAP can be a useful tool for network analysis, whether it is a social network or gene network. It can help view different dimensions of a graph. Some GOLAP operations such as "Rotate" and "Stretch" can be useful in switching to different views (nodes and edges). The users should be able to run analysis in real time with a few clicks.

For example, in a gene/disease network, the nodes represent diseases, the edges represent a relationship between diseases, and the edge weights represent the number of common genes the diseases share. Figure 1 shows an example disease network. Disease 1 and disease 2 share four common genes and disease 2 and disease 5 share three common genes. Through analysis, we could find n diseases that are most closely related to a set of genes. In another graph, a node could represent a gene, an edge represents disease, and its weight represents the relevance strength of two genes which are connected by the edge (disease).

With their expressive advantages, graphs have been used for representing a lot of datasets that hold structural information. Given the myriad amount of graph data amassed in different applications, there is a great need for network analysis from different viewpoints with different granularities.

The paper is organized as follows. We first define the key terms and fundamental concepts related to this survey in Sec. 2. In Sec. 3, we present a survey on network analysis and in Sec. 4, we introduce network analysis in the context of graph-based OLAP. In Sec. 5, we conclude the paper with a summary.

---

‡Corresponding author.

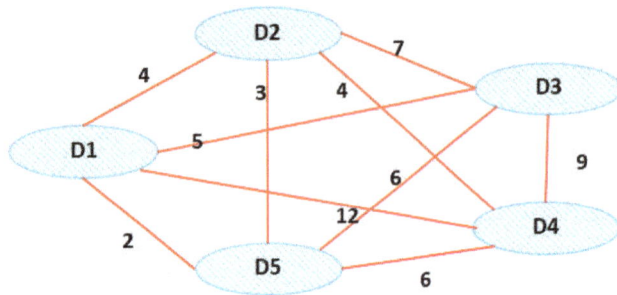

Fig. 1. Graph representation of a disease network.

2.   Key Terms and Fundamental Concepts

In this section, we define the terms and provide the conceptual foundation for network analysis.

2.1.   *Traditional graph problems*

Before the emergence of social networks, graphs were used to model different types of relations and processes in physical and biological systems. Graphs are used to represent networks of communication, data organization, computational devices, the flow of computation, etc. For example, the link structure of a website can be represented as a directed graph, where the nodes represent web pages and the directed edges represent links from one page to another. A similar approach can be taken to problems in travel, biology, computer chip design, mapping the progression of neuro-degenerative diseases,[2] etc. There are numerous practical problems that can be solved utilizing graphs. In this subsection, we discuss some of these traditional graph problems before the social networks appeared.

2.1.1.   *Four-color theorem*

The four-color theorem states that any map in a plane can be colored using four colors in such a way that regions sharing a common boundary (other than a single point) do not share the same color. Three colors are adequate for simpler maps, but an additional fourth color is required for some maps, such as a map in which one region is surrounded by an odd number (greater than one) of other regions that touch each other in a cycle.[3]

2.1.2.   *Hamiltonian path problem*

A Hamiltonian path is a path in an undirected or directed graph that visits each vertex exactly once. The Hamiltonian path problem and the Hamiltonian cycle problem are problems of figuring out if a Hamiltonian path or a Hamiltonian cycle occurs in a given graph. These are both NP-complete.[4]

There is a simple relation between the problems of finding a Hamiltonian path and a Hamiltonian cycle. In one direction, the Hamiltonian path problem for a graph G is equivalent to

the Hamiltonian cycle problem in a graph H obtained from G by adding a new vertex and connecting it to all vertices of G. Thus, finding a Hamiltonian path cannot be significantly slower than finding a Hamiltonian cycle.

2.1.3.   *Minimum spanning tree*

A minimum spanning tree is a subset of the edges of a connected, edge-weighted (un)directed graph that connects all the vertices together, without any cycles and with the minimum possible total edge weight. It is a spanning tree whose sum of edge weights is as small as possible. More generally, any edge-weighted undirected graph has a minimum spanning forest, which is a union of the minimum spanning trees for its connected components.[5]

2.1.4.   *Chinese Postman Problem*

In graph theory, a branch of mathematics and computer science, namely the Chinese Postman Problem (CPP), postman tour or route inspection problem, is to find a shortest closed path or circuit that visits every edge of an undirected graph. When the graph has a Eulerian circuit, that circuit is an optimal solution. Otherwise, the optimization problem is to find the smallest number of graph edges to duplicate (or the subset of edges with the minimum possible total weight) so that the resulting multigraph does have a Eulerian circuit.[6]

2.1.5.   *Seven Bridges of Königsberg*

The Seven Bridges of Königsberg is a historically notable problem in mathematics. Its negative resolution by Leonhard Euler in 1736 laid the foundations of graph theory and started an early idea of topology.[7,8]

The city of Königsberg in Prussia (now Kaliningrad, Russia) was set on both sides of the Pregel River and included two large islands which were connected to each other, or to the two mainland portions of the city, by seven bridges. The problem was to design a walking tour of the city in which you cross all of the seven bridges exactly once.

2.1.6.   *Three-cottage problem*

The classical mathematical puzzle known as the three utilities problem; the three cottages problem or sometimes water, gas, and electricity can be described as follows:

Suppose there are three cottages on a plane (or sphere) and each needs to be connected to the gas, water, and electricity companies. The problem is to decide if there is a way to make all nine connections without any of the lines crossing each other without using a third dimension or sending any of the connections through another company or cottage.[9]

2.1.7. *Traveling Salesman Problem*

The Traveling Salesman Problem (TSP) asks the following question: "Given a list of cities and the distances between each pair of cities, what is the shortest possible route that visits each city and returns to the original city?" It is an NP-hard problem in combinatorial optimization, important in operations research and theoretical computer science.

TSP can be modeled as an undirected weighted graph, such that cities are the graph's vertices, paths are the graph's edges, and a path's distance is the edge's weight. It is a minimization problem starting and finishing at a specified vertex after having visited each other vertex exactly once. Often, the model is a complete graph. If no path exists between two cities, adding an arbitrarily long edge will complete the graph without affecting the optimal tour.[10]

2.1.8. *Max-flow min-cut theorem*

The max-flow min-cut theorem states that in a flow network, the maximum amount of flow passing from the source to the sink is equal to the total weight of the edges in the minimum cut.[11]

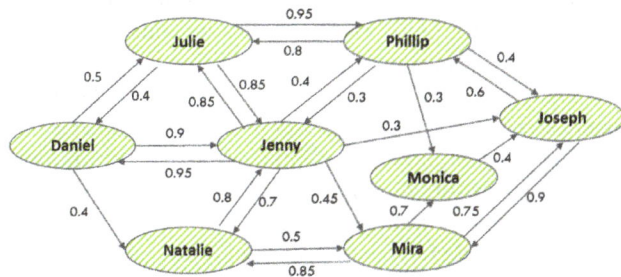

Fig. 2. Graph representation of a social network.

2.2. *Graph model*

A common approach in graph-based problem solving is to represent the data elements and their relationships as a graph model and to apply various analytics on the model using graph properties. Figure 2 shows an example of a social network, where the nodes represent people and edges represent influence. The edge weight represents the strength of influence. For example, Natalie has an influence on Mira with a strength of 0.5 while Mira has an influence on Natalie with a strength of 0.85. Daniel has an influence on Natalie with a strength of 0.4 while Natalie has no direct influence on Daniel. However, since Natalie has an influence on Jenny and Jenny has an influence on Daniel, there is an indirect influence from Natalie to Daniel.

In graph theory, there are some terms that are specific to the subject. The terms and descriptions are shown in Table 1.

3. Network Analysis

Network analysis can mean different things in different contexts; there is no single widely accepted consensus on its meaning. In the context of analyzing datasets for business applications, network analysis can be defined as an approach to performing analytics on a (graph-based) data representation by applying the paradigm of connecting one to another directly or indirectly.

There has been a high demand for network analysis in many different fields. Marketing, advertising, and business network could infer useful knowledge utilizing network analysis tools. Healthcare network, patient network, and gene co-expression network are some other examples where abundant network datasets are being analyzed.

Han *et al.*[12] introduce database-oriented network methods and show how information network can be used to improve

Table 1. Terminology in graph theory.

Term	Description
Node	A point in a graph. It represents an entity in the domain. For example, in a social network, a node represents a person. In a gene–disease network, a node represents a gene.
Edge	An arch which connects a pair of nodes. It represents a relationship between nodes. In a gene–disease network, it represents a relevance between two genes for a particular disease.
Types of node	It specifies the classification of nodes. The type of a node can be different depending on the context. For example, in a social network, different types of nodes could represent different races, genders, age groups or political parties.
Weight of edge	It represents the strength of what an edge represents. For example, in a social network, it represents the strength of the relationship between two persons.
Direction of edge	It specifies the directivity of edges. For analytics applications, it often takes the form of a directed edge. For example, if an edge is going from one node (source node) to another (target node), the source node can influence the target node with the strength of the edge weight.
Influence	The influence from a node to another node is depicted as an edge weight ranging between 0 and 1. The edge weight $\inf(A, B)$ represents the strength of influence from node A to node B.
The strongest influence path (SIP)	SIP from a node to another node is the path that has the strongest influence value (calculated by multiplying the influence strengths along the path). There are multiple paths between a node to another node but there exists at least one strongest influence path among all these paths.

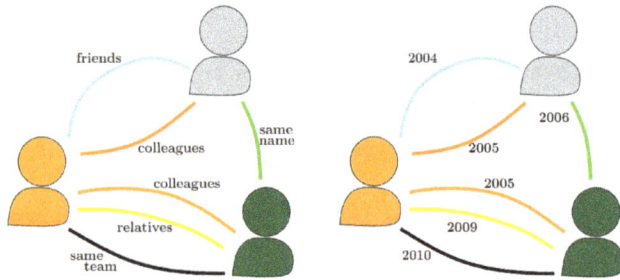

Fig. 3. Examples of multidimensional networks.[14]

data quality and consistency, facilitate data integration, and generate interesting knowledge on DBLP and Flickr networks.

Huang and Jin[13] modify the SIR model to describe rumor propagation on networks and apply random immunization and targeted immunization to the rumor model in a small-world network. They show that these strategies are effective in preventing rumor spreading in a small-world network with a large average degree.

Berlingerio *et al.*[14] present a framework for multidimensional network analysis by an additional degree of complexity that multidimensionality provides in real networks. Figure 3 shows different dimensions that may reflect different types of relationship, or different values of the same relationship. In Fig. 3, on the left there are different types of links, while on the right there are different values (years) for one relationship (for example, co-authorship).[14]

3.1. *Social network analysis*

Social networks are platforms where billions of users interact regularly and share diverse digital contents with one another. People express their emotions and thoughts on every topic. These opinions convey significant value for personal, academic, and commercial applications, but the speed and the volume at which these are produced make it a difficult task for researchers and the fundamental technologies to offer useful insights to such data.[15]

Among social network research, influence analysis is among one of the most popular topics. Research on influence has attracted researchers' interest from the last century since 1950. With the development of the Internet, various business networks and social networks have been constructed in the past decades. More and more scientists and researchers begin to realize the importance of influence in the social networks. In the early ages, several theories have been proposed, such as the theory of six degrees of separation, the theory of four degrees of separation, and small-world phenomenon.[14]

Researchers have tried to mine new information from social networks. Through research on social influence, it can be used to develop numerous new applications, such as advertising, sentiment diffusion, link prediction, recommended systems, and social communities finding.

In social networks, it has a great significance to find the most influential node, e.g., the most important blogger and opinion leader. In social networks, some nodes have bigger influence than other nodes. How to find the most influential node is a very popular area of research in social networks.

Degree centrality[16] is generally used to measure a node's importance. In general, the higher the node's degree is, the higher is the node's importance. One algorithm is the PageRank algorithm,[16] which is used to find the importance of nodes in social networks. Betweenness centrality[16] is another very important factor to measure a node's importance. It measures the extent to which a vertex lies on the paths between other vertices. Vertices with a high betweenness may have considerable influence within a network by their control over information passing between others. Closeness centrality[16] has also been typically used to measure a node's importance. Other than these three methods to measure the importance of a node, many other methods are proposed as well. Some examples include HITS, ARC, K-Shell, randomly walk algorithm, and their variants.[16]

Liu *et al.*[17] propose a generative graphical model and focus on quantitative learning influence between users in heterogeneous networks. Goyal *et al.*[18] propose models and algorithms for learning the model parameters and for testing the learned models to make predictions. Backstrom *et al.*[19] explore a large corpus of thriving online communities. Using metadata from groups, members, and individual messages, they identify users who post and are replied-to frequently by multiple group members. They classify these high-engagement users based on the longevity of their engagements. Sun and Tang[20] show how social influence can help real applications by focusing on opinion leader finding and influence maximization for viral marketing.

Hangal *et al.*[21] propose that search in social networks can be made more effective by incorporating weighted and directed influence edges in the social graph. If A retweets B, B has influence over A. They define influence as an edge weight metric that is calculated based on relative fractions of interaction between two nodes. They define the best path between two people A and B as the most influential and find that the most influential paths are often not the shortest path. Figure 4 illustrates the results of allocating influence to the edges in a network with undirected interactions, such as DBLP. An intuitive interpretation of this graph runs as follows. Imagine that node A is an adviser, and nodes B, C, and D are her students. The edge weights in Fig. 4(a) depict the number of co-authorships between node pairs.

In order to infer influence from one node to another, with $\text{Papers}(v_j, v_i)$ being the number of papers co-authored by v_j and v_i, they use the following equation:

$$\text{Influence}(v_j, v_i) = \frac{\text{Papers}(v_j, v_i)}{\sum_{v_k} \text{Papers}(v_j, v_k)}.$$

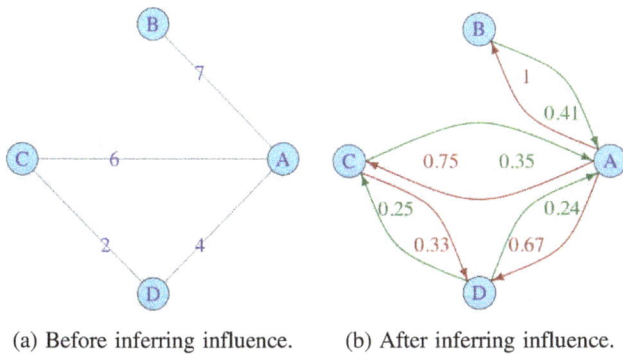

(a) Before inferring influence.   (b) After inferring influence.

Fig. 4. Before and after inferring influence.[21]

In Fig. 4(b) one can see that the adviser holds more influence over her students than her students hold over her. Moreover, student D, who has authored fewer papers than student C, is more influenced by student C because a larger proportion of her total publications involve student C.[21]

Gilbert and Karahalios[22] present a predictive model that maps social media data to tie strength to distinguish between strong and weak ties. Leskovec *et al.*[23] study how the interplay between positive and negative relationships on social media affects the structure of online social networks. They present the positive and negative interactions as positive and negative edges. Xiang *et al.*[24] develop an unsupervised model to estimate relationship strength from interaction activity and user similarity.

Out of all the social network analysis literature, influence maximization is a very popular research area. Influence maximization is the problem of finding a small subset of nodes (seed nodes) in a social network that could maximize the spread of influence. Chen *et al.* propose new degree discount heuristics that improves influence spread which can achieve better performance than the classic degree and centrality-based heuristics.[25] Kempe *et al.* study the influence maximization problem focusing on two propagation models: the linear threshold (LT) model and the independent cascade (IC) model. Since the influence maximization problem is NP-hard,[26] many scholars propose greedy algorithm or heuristics to improve the runtime with acceptable accuracy. Liu *et al.*[27] present the Latency-Aware Independent Cascade (LAIC) model, which encodes the influence propagation latency information into the IC model. The LAIC model considers delayed influence propagation by encoding time into the activation probability of edges in a social network.

Leskovec *et al.*[29] present a general methodology for near-optimal sensor placement in outbreak detection problems. They exploit submodularity to develop an efficient algorithm called CELF that scales to large problems. The idea is that the marginal gain of a node in the current iteration cannot be better than its marginal gain in the previous iterations. It achieves near-optimal placement that is 700 times faster than a simple greedy algorithm.[28] Goyal *et al.*[29] improves the CELF algorithm by Leskovec *et al.* by further optimizing

CELF's submodularity. CELF++ exploits the property of submodularity of the spread function for influence propagation models to avoid unnecessary recomputations of marginal gains incurred by CELF.

Rehman *et al.*[15] extend the established OLAP technology to enable multidimensional analysis of social media data by integrating text and opinion mining approaches into a data warehousing system and by taking advantage of various knowledge discovery techniques to deal with semi-structured and unstructured data from social media. The capabilities of OLAP are extended by semantic enrichment of the underlying dataset to realize innovative measures and dimensions for building data cubes and by supporting up-to-date analysis in addition to historical social media data.

3.2. *Gene network analysis*

Another area of interest to scientists is gene association network analysis. The data comprises nodes and edges where each node represents a disease and the size of the node reflects the number of genes known to be associated with that disorder. The thickness of the edges that connect these nodes reflects the number of genes shared by the connected diseases. The idea behind the map is that most diseases with known genetic associations seem to share most of their genes with other diseases.[30] Many biologists argue that gene–disease networks are a first step not only toward making GWAS data useful for health care, but also toward completely revamping how physicians think about disease — in particular, how they categorize various illnesses.[31]

In gene network analysis, a popular field being widely used is gene co-expression networks. It is an undirected graph where each node corresponds to a gene and a pair of nodes are connected with an edge if there is a significant co-expression relationship between them.

Stuart *et al.*[32] identify pairs of genes that are co-expressed over 3182 DNA microarrays from humans, flies, worms, and yeast. They find 22,163 such co-expression relationships, each of which has been conserved across evolution. This conservation implies that the co-expression of these gene pairs confers a selective advantage and therefore that these genes are functionally related. Many of these relationships provide strong evidences for the involvement of new genes in core biological functions such as the cell cycle, secretion, and protein expression. They experimentally confirm the predictions implied by some of these links and identify cell proliferation functions for several genes. They assemble these links into a gene co-expression network consisting of 12 large components and find several that were animal-specific as well as interrelationships between newly evolved and ancient modules.

Zhang and Horvath[33] introduce several node connectivity measures and provide evidences that they can be important for predicting the biological significance of a gene. They generalize the clustering coefficient to weighted networks.

They provide a model that shows how an inverse relationship between clustering coefficient and connectivity arises from hard thresholding.

Horvath and Dong[34] describe conditions when a close relationship exists between network analysis and microarray data analysis techniques and provide a rough dictionary for translating between the two fields. Using the angular interpretation of correlations, they provide a geometric interpretation of network theoretic concepts and derive unexpected relationships among them. They characterize co-expression networks where hub genes are significant with respect to a microarray sample trait and show that the network concept of intramodular connectivity can be interpreted as a fuzzy measure of module membership.

Over the years, a lot of work has been done for network analysis on different types of networks. However, the existing approaches lack the ability to represent different types of nodes or edges. In a social network where nodes represent people, classifying different types of people based on their gender, political party, age, or race could lead to different interesting analysis. In a gene network where nodes represent genes and edges represent relevance of genes, being able to indicate the type of disease could help visualize and analyze the data. Associating attributes to nodes or edges could allow researchers to post more queries. Many researchers run experiments on real data with some randomly generated attributes. For instance, the available social networks do not have any data on the strength of relationships between people. The experiments would produce more accurate result if these attribute values could be inferred instead of randomly generated.

4. Network Analysis in GOLAP

Network analysis faces another challenge as it lacks easy-to-use tools that are designed to convert graphs into different dimensions. Moreover, very large graphs have emerged in different areas like biological, social, and transportation networks. The size of these networks poses challenges to traditional techniques for storing and analysis of graph data. Not much work has been done for analyzing very large graphs.

In this section, we discuss the research works for graph-based OLAP. Chen *et al.*[35] develop a novel graph OLAP framework, which presents a multidimensional and multilevel view over graphs. They present how a graph cube can be materialized by manipulating a distinct type of measure called aggregated graph and how to implement it effectively. Chou *et al.*[36] describe the existing GOLAP operations including slicing and roll-up/drill-down. They also propose a generalized model of GOLAP in which one can view information at the category level. After drill-down, the details regarding all the nodes inside each category are revealed. According to this framework, given a graph dataset with its nodes and edges associated with respective attributes, a multidimensional model can be built to enable efficient online analytical processing so that any portions of the graphs can be generalized/specialized dynamically, offering multiple, versatile views of the data.

Yin *et al.* design HMGraph OLAP operations named Rotate and Stretch for entity dimensions, which are able to mine relationships between different entities. They propose the HMGraph Cube, which is an efficient data warehousing model for HMGraph OLAP.[37] Shown in Fig. 5 is a co-author network where nodes are authors and the edges between the nodes are papers they co-authored. Using the "Rotate" operation, the nodes and edges have been switched so nodes are representing papers and edges are the authors that wrote the papers connected. This could be useful for researchers for performing operations on graphs in order to switch to different views between nodes and edges. For example, in a gene–disease network, this could be used to switch between a gene-oriented view and a disease-oriented view.

Shown in Fig. 6 is the use of "Stretch" operation. After applying the Stretch operation, the nodes remain the same but the edges, which were papers, have now become nodes but of different shapes/types. This particular operation could give users different options to switch from edge to node to perform different types of network analysis.

Sun and Han[38] view interconnected, multitype data, including the typical relational database data, as heterogeneous information networks. They study how to leverage the rich semantic meaning of structural types of objects and links in a network, and develop a structural analysis approach to mining semi-structured, multitype heterogeneous information networks. They summarize a set of methodologies that can effectively and efficiently mine useful knowledge from such

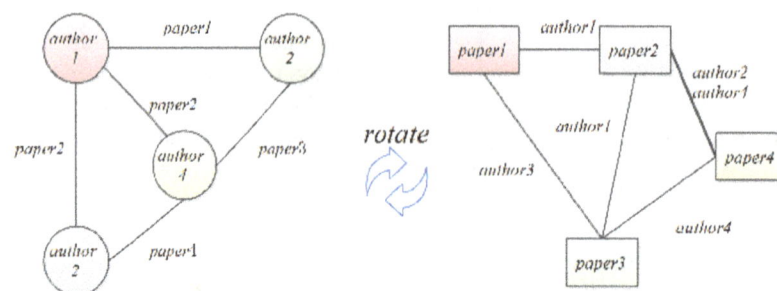

Fig. 5.   Rotate operation for HMGraph OLAP.[38]

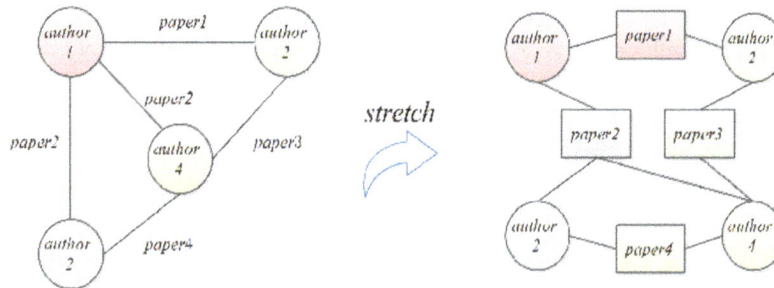

Fig. 6. Stretch operation for HMGraph OLAP.[38]

information networks and point out some promising research directions.

Wang *et al.*[39] propose Pagrol, a parallel graph OLAP system over attributed graphs. In particular, Pagrol introduces a new conceptual Hyper Graph Cube model (which is an attributed-graph analog of the data cube model for relational DBMS) to aggregate attributed graphs at different granularities and levels. The proposed model supports different queries as well as a new set of graph OLAP roll-up/drill-down operations. Furthermore, on the basis of Hyper Graph Cube, Pagrol provides an efficient MapReduce-based parallel graph cubing algorithm, namely MRGraph-Cubing, to compute the graph cube for an attributed graph. Pagrol employs numerous optimization techniques: (a) a self-contained join strategy to minimize I/O cost; (b) a scheme that groups cuboids into batches so as to minimize redundant computations; (c) a cost-based scheme to allocate the batches into; and (d) an efficient scheme to process a bag using a single MapReduce job.

Lu *et al.*[40] propose a set of innovative multidimensional analysis concepts and methods to better understand observations of daily living (ODLs) that people attend in the course of their everyday life. The structure dimension, consisting of three granularities, vertices, edges, and subgraphs, is proposed to integrate into the traditional multidimensional analysis framework. The hierarchy of ODLs Cube is introduced, and the semantics of OLAP operations, roll-up, drill-down, and slice/dice, are redefined to accommodate the structure dimension. The structure dimension and ODLs Cube they propose are useful for multidimensional analysis of ODLs.

Denis *et al.*[41] concentrate on applying OLAP analysis on large, distributed graph data. They describe Distributed Graph Cube, a distributed framework for graph-based OLAP cubes. Jakawat *et al.*[42] focus on bibliographic data which establishes a rich foundation that is the initial point of research on bibliometrics and scientometrics domains. They discuss merging information networks, OLAP, and data mining technologies. They build a framework to materialize this combination, to analyze numerous networks constructed from the bibliographic data representing diverse points of view such as authors networks and citations networks. These frameworks could help researchers manipulate graphs, by merging information and enabling them to access a network from different view.

Zhao *et al.*[43] implement Graph Cube, a new data warehousing model that works on OLAP queries efficiently on big multidimensional networks, by uniting distinct features of multidimensional networks with the current data cube procedures. This tool works effectively for decision support on large multidimensional networks. Beheshti *et al.*[44] present a graph data model for online analytical processing on graphs. This data model allows encompassing decision support on multidimensional networks taking into account data objects and their relationships. Furthermore, they extend SPARQL to support *n*-dimensional computations.

In order to speed up the running time on larger graphs, some scholars use graph data reduction or approximation methods. Graph data reduction removes unimportant nodes or edges to reduce the search space. Approximation methods increases the speed at the expense of accuracy. Sometimes the accuracy of query results may not be strictly required.

Chou *et al.*[36] discuss graph data reduction on GOLAP. They cover syntactical reduction methods that may be applied to all applications such as social networks, map applications, and real-time recommendation engines. They also discuss the specific problem(s) addressed by an application to reduce a graph further. By eliminating information unrelated to an application, one can improve the reduction ratio. Some researchers propose lossy reduction algorithms that can decrease the size of a graph and preserve the outcome of the query of the same semantic queries. This is done at the expense of accuracy with a tolerable error rate to achieve approximation where an error rate is controlled by predefined threshold.

5. Summary

In this paper, we present a survey of traditional graph problems and approaches to network analysis including social network and gene network analysis.

Network analysis, especially on social networks and gene networks, has been worked on by many scholars. The current methods lack the ability to differentiate the types of nodes or edges. Existing data lack some attributes that are essential to analysis, hindering researchers from getting accurate results.

There are no easy-to-use tools that let users convert graphs into different dimensions. Dealing with very large graphs also pose issues for storing and analyzing graphs.

Traditional OLAP is widely used for data analysis. With the abundance of graph data as of late, researchers have been proposing using OLAP on graphs. Several groups of researchers propose their version of OLAP that can work for multidimensional networks. Some apply roll-up, drill-down, and slice/dice to graphs while others create their own graph operations such as "Rotate" and "Stretch".[41] Some create frameworks with dimensions and measures that deal with large, multidimensional data while others extend SPARQL to support OLAP for graphs. For dealing with very large graphs, scholars propose data reduction and approximation algorithms to improve running time.

Acknowledgments

The research is supported in part by NEC Solution Innovators, Ltd., Japan.

References

[1] S. Chaudhuri and R. Dayal, An overview of data warehousing and OLAP technology, *ACM SIGMOD Rec.* **26**, 65 (1997).

[2] F. Vecchio, Brain network connectivity assessed using graph theory in frontotemporal dementia, *Neurology* **81**, 134 (2013).

[3] R. Fritsch and G. Fritsch, *The Four-Color Theorem: History, Topological Foundations, and Idea of Proof* (Springer, 1998).

[4] M. R. Garey and D. S. Johnson, *Computers and Intractability: A Guide to the Theory of NP-Completeness*, Appendix (W. H. Freeman, New York, 1979), pp. 199–200.

[5] R. L. Graham and P. Hell, On the history of the minimum spanning tree problem, *Proc. Ann. Hist. Comput.* **7**, 43 (1985).

[6] F. S. Roberts and B. Tesman, *Applied Combinatorics*, 2nd edn., Chapter (CRC Press, 2009), pp. 640–642.

[7] N. L. Biggs, E. K. Lloyd and R. J. Wilson, *Graph Theory, 1736–1936* (Clarendon Press, Oxford, 1998).

[8] R. Shields, Cultural topology: The Seven Bridges of Königsburg, 1736, *Theory Cult. Soc.* **29**, 43 (2012).

[9] M. Bóna, *A Walk Through Combinatorics: An Introduction to Enumeration and Graph Theory*, Chapter 12 (World Scientific, 2011), pp. 275–277.

[10] K. L. Hoffman, M. Padberg and G. Rinaldi, *Encyclopedia of Operations Research and Management Science*, eds. S. I. Gass and M. C. Fu, Chapter (Springer, Boston, 2013), pp. 697–700.

[11] G. B. Dantzig and D. R. Fulkerson, On the max-flow min-cut theorem of networks, Report No. P-826, RAND Corporation (1964).

[12] J. Han, Y. Sun, X. Yan and P. S. Yu, Mining knowledge from databases: An information network analysis approach, *Proc. 2010 ACM SIGMOD Int. Conf. Management of Data* (2010), pp. 1251–1252.

[13] J. Huang and X. Jin, Preventing rumor spreading on small-world networks, *J. Syst. Sci. Complex.* **24**, 449 (2011).

[14] M. Berlingerio, M. Coscia, F. Giannotti, A. Monreale and D. Pedreschi, Foundations of multidimensional network analysis, *Proc. Int. Conf. Advances in Social Networks Analysis and Mining* (2011).

[15] N. U. Rehman, A. Weiler and M. H. Scholl, OLAPing social media: The case of Twitter, *Proc. 2013 IEEE/ACM Int. Conf. Advances in Social Networks Analysis and Mining* (2013).

[16] E. Yan and Y. Ding, Applying centrality measures to impact analysis: A coauthorship network analysis, *J. Am. Soc. Inf. Sci. Technol.* **60**, 2107 (2009).

[17] L. Liu, J. Tang, J. Han and S. Yang, Learning influence from heterogeneous social networks, *Data Mining Knowl Discov.* **25**, 511 (2012).

[18] A. Goyal, F. Bonchi and L. V. Lakshmanan, Learning influence probabilities in social networks, *Proc. Third Int. ACM Conf. Web Search and Data Mining* (2010), pp. 207–217.

[19] L. Backstrom, R. Kumar, C. Marlow, J. Novak and A. Tomkins. Preferential behavior in online groups, *Proc. Int. Conf. Web Search and Web Data Mining* (2008), pp. 117–128.

[20] J. Sun and J. Tang, Models and algorithms for social influence analysis, *Proc. Sixth ACM Int. Conf. Web Search and Data Mining* (2013), pp. 775–776.

[21] S. Hangal, D. Maclean, M. S. Lam and J. Heer, All friends are not equal: Using weights in social graphs to improve search, *Proc. 4th SNA-KDD Workshop'10 (SNA-KDD'10)* (ACM, 2010).

[22] E. Gilbert and K. Karahalios, Predicting tie strength with social media, *Proc. SIGCHI Conf. Human Factors in Computing Systems* (2009).

[23] J. Leskovec, D. Huttenlocher and J. Kleinberg, Signed networks in social media, *Proc. SIGCHI Conf. Human Factors in Computing Systems* (2010).

[24] R. Xiang, J. Neville and M. Rogati, Modeling relationship strength in online social networks, *Proc. 19th Int. Conf. World Wide Web* (2010).

[25] W. Chen, Y. Wang and S. Yang, Efficient influence maximization in social networks, *Proc. 15th ACM SIGKDD Int. Conf. Knowledge Discovery and Data Mining* (2009), pp. 199–208.

[26] D. Kempe, J. Kleinberg and E. Tardos, Maximizing the spread of influence through a social network, *Proc. 9th ACM SIGKDD Int. Conf. Knowledge Discovery and Data Mining* (2003).

[27] B. Liu, G. Cong, Y. Zeng and X. Sun, Influence spreading path and its application to the time constrained social influence maximization problem and beyond, *IEEE Trans. Knowl. Data Eng.* **26**, 1904 (2014).

[28] J. Leskovec, A. Krause, C. Guestrin, C. Faloutsos, J. VanBriesen and N. Glance, Cost-effective outbreak detection in networks, *Proc. 13th ACM SIGKDD Int. Conf. Knowledge Discovery and Data Mining* (2007).

[29] A. Goyal, W. Lu and L. V. Lakshmanan, CELF++: Optimizing the greedy algorithm for influence maximization in social networks, *Proc. 20th Int. Conf. Companion on World Wide Web* (ACM, 2011), pp. 47–48.

[30] K. Goh, M. E. Cusick, D. Valle, B. Childs, M. Vidal and A. L. Barabasi, The human disease network, *Proc. Natl. Acad. Sci. USA* **104**, 8685 (2008).

[31] L. Pray, Genome-wide association studies and human disease networks, *Nat. Educ.* **1**, 220 (2008).

[32] J. M. Stuart, E. Segal, D. Koller and S. K. Kim, A gene-coexpression network for global discovery of conserved genetic modules, *Science* **302**, 249 (2003).

[33]B. Zhang and S. Horvath, A general framework for weighted gene coexpression network analysis, *Stat. Appl. Genet. Mol. Biol.* **4**, 17 (2005).

[34]S. Horvath and J. Dong, Geometric interpretation of gene coexpression network analysis, *PLoS Comput. Biol.* **4**, e1000117 (2008).

[35]C. Chen, X. Yan, F. Zhu, J. Han and P. S. Yu, Graph OLAP: Towards online analytical processing on graphs, *Proc. Eighth IEEE Int. Conf. Data Mining* (2008).

[36]B. Chou, M. Hayakawa and P. Sheu, Graph online analytical processing, *Encycl. Semant. Comput.* (2018).

[37]M. Yin, B. Wu and Z. Zeng, HMGraph OLAP: A novel framework for multi-dimensional heterogeneous network analysis, *Proc. Fifteenth Int. Workshop Data Warehousing and OLAP* (2012), pp. 137–144.

[38]Y. Sun and J. Han, Mining heterogeneous information networks: A structural analysis approach, *ACM SIGKDD Explor. Newsl.* **14**, 20 (2012).

[39]Z. Wang, Q. Fan, H. Wang, K. Tan, D. Agrawal and A. E. Abbadi, Pagrol: Parallel graph OLAP over large-scale attributed graphs, *Proc. IEEE 30th Int. Conf. Data Engineering* (2014).

[40]J. Lu, B. Zhang, X. Wang and N. Lu, Multidimensional analysis framework on massive data of observations of daily living, in *Proc. Int. Conf. Health Information Science* (Springer, Cham, 2017), pp. 121–127.

[41]B. Denis, A. Ghrab and S. Skhiri, A distributed approach for graph-oriented multidimensional analysis, *Proc. 2013 IEEE Int. Conf. Big Data* (2013).

[42]W. Jakawat, C. Favre and S. Loudcher, *New Trends in Databases and Information Systems*, eds. B. Catania *et al.*, Chapter, Advances in Intelligent Systems and Computing, Vol. 241 (Springer, Cham, 2014), pp. 361–370.

[43]P. Zhao, X. Li, D. Xin and J. Han, Graph cube: On warehousing and OLAP multi-dimensional networks, *Proc. 2011 ACM SIGMOD Int. Conf. Management of Data* (2011), pp. 853–864.

[44]S.-M.-R. Beheshti, B. Benatallah, H. R. Motahari-Nezhad and M. Allahbakhsh, A framework and a language for on-line analytical processing on graphs (GOLAP), *Proc. 13th Int. Conf. Web Information Systems Engineering* (2012).

Part 4

Applications

Automatic analysis of microblogging data to aid in emergency management

Sukanya Manna

Department of Mathematics and Computer Science
Santa Clara University, Santa Clara, CA 95053, USA

smanna@scu.edu

Microblogging platforms like Twitter, in the recent years, have become one of the important sources of information for a wide spectrum of users. As a result, these platforms have become great resources to provide support for emergency management. During any crisis, it is necessary to sieve through a huge amount of social media texts within a short span of time to extract meaningful information from them. Extraction of emergency-specific information, such as topic keywords or landmarks or geo-locations of sites, from these texts plays a significant role in building an application for emergency management. This paper thus highlights different aspects of automatic analysis of tweets to help in developing such an application. Hence, it focuses on: (1) identification of crisis-related tweets using machine learning, (2) exploration of topic model implementations and looking at its effectiveness on short messages (as short as 140 characters); and performing an exploratory data analysis on short texts related to crises collected from Twitter, and looking at different visualizations to understand the commonality and differences between topics and different crisis-related data, and (3) providing a proof of concept for identifying and retrieving different geo-locations from tweets and extracting the GPS coordinates from this data to approximately plot them in a map.

Keywords: Topic models; LDA; machine learning; tweets; short texts; geo-locations.

1. Introduction

Social media platforms like Twitter, Facebook, Instagram, Google+, and so on have become important platforms of communication in recent years. These platforms are largely used for sharing or exchanging various forms of information, not limited to news or messages. Users of these websites have become used to receiving timely updates on important events, both personal and public.

Among these social media platforms, Twitter is a widespread microblogging platform that allows users to broadcast short messages (tweets) through socially-networked channels of listeners. Twitterers subscribe to the tweet broadcasts of other twitterers by following them. Broadcast tweets are sent out to followers in update streams and can be accessed in real-time or stored for later viewing.[1] For example, Twitter has been used to propagate information in real-time during crisis.[2,3]

Tweets can be geo-tagged; which means that enabling location services will allow a user to selectively add location information to their tweets. Once the user enables location services, they will be able to attach a location (such as a city or neighborhood) of their choice to your tweet. With the geo-tag information attached to tweets, Twitter allows the developers to be able to resolve geographic coordinates; there is also a precedent for utilizing Twitter in case of an emergency.[4]

With the said properties, it is often seen that Twitter has become a common platform during a crisis for communication. During such an event, within a short span of time it is necessary to sieve through a huge amount of social media texts to retrieve or extract meaningful information from them. Extraction of emergency-specific information, such as topic keywords or landmarks or geo-locations of sites, from these texts plays a significant role in building an application for emergency management. This paper thus highlights different aspects of automatic analysis of tweets to help in developing such an application.

Researchers from various domains (such as computer science, marketing, geography, and social science) have shown interests in microblogging services, especially Twitter. Previous works mainly focused on quantitative studies on a number of aspects and characteristics of Twitter.[5] For example, Java et al.[6] looked into the topological and geographical properties of Twitter's social network and found that the network has high degree correlation and reciprocity, indicating close mutual acquaintances among users. Krishnamurthy et al.[7] looked into geographical distribution of Twitter users and their behaviors among several independent crawls.

Abdelhaq et al.[8] presented a framework to identify bursty words from Twitter text streams and describe such words in terms of their spatio-temporal characteristics. Others have also used Twitter to look at place semantics[9,10] to investigate the relationship between meetings at places and the structure of the social network.

A number of studies[11–13] have shown that if a retrieval system was able to find good clusters, retrieval performance can be improved over text-based retrieval. Topic models[14] have the ability to discover latent topics in text collections, hence it is one of the goals of this paper to see how distinct topics can be obtained using topic models on tweets so that the result could be better utilized while building such system which could be used during any crisis.

Different researchers are exploiting tweets to extract specific information content from them, but there exist several natural limitations of messages that prevent some standard text mining tools from being employed with their full potentials. Some of these limitations are:

(i) The maximum length of a message on Twitter is only 140 characters (until November 2017). This is substantially different from other textual data commonly used for information retrieval and Web search.
(ii) Within this short length, self-defined notations are used to expand the semantics that are carried out by the messages.
(iii) Self-defined hash tags starting with # are also heavily used to identify certain events or topics. Therefore, from the perspective of length (e.g., in characters), the content in a message is limited while it may convey rich meanings.
(iv) Emojis, abbreviations, and different nonstandard representation of dictionary words are also commonly used, further complicating the filtering process.

In this paper, Twitter is used as the data source to extract and predict different geo-locations from user-generated tweets. A basic prototype is presented here which can be useful for creating application for emergency management in situations where Emergency Response Teams (ERTs) are unreachable; allowing Twitter to become a second viable option to contact someone for help.

Example. Let us analyze a sample tweet: "A 1.4 magnitude earthquake occurred 0.62 mi N of Loma Linda, California." This tweet is related to a *crisis* describing the location of the occurrence of an earthquake. The main goal is to extract the location information from the tweet; in this case, the location information is 0.62 *mi N of Loma Linda* and *California*, and mark it accurately on a map.

This paper also investigates to what extent topics do indeed match substantive crisis issues (e.g., bombing, earthquake, tornadoes, and so on). This is achieved by running a topic model on six different datasets of crisis-related tweets that have been collected over a span of time. Details of the datasets are given in Sec. 6.

Contributions. By way of disclaimer, please note that this paper is rather exploratory in nature. The main goal is to get a more qualitative understanding of the substantive interpretation of Latent Dirichlet Allocation (LDA) topics in terms which are prevalent during any form of emergency or crisis. We have jotted down the main contributions:

(a) Classify tweets whether they are related to a crisis or not using machine learning (ML) algorithms.
(b) Explore different topic model implementations readily available and look at their effectiveness on short messages (as short as 140 characters) in terms of topic words extracted.
(c) Perform exploratory data analysis on crisis data collected from Twitter.
(d) Look at different visualizations to understand the commonality and differences between topics and among any forms of crisis data.
(e) Predict geo-locations and identify different places by extracting information from tweets and correlate them with gazetteer data.
(f) Provide a prototype that can automatically analyze the data from tweets and draw location on maps to spot the locations.

Organization. The paper is organized as follows: Section 2 presents the related works. Section 3 presents tweet classification problem. Section 4 discusses the basic LDA topic model. Section 5 presents the proposed proof of concepts of identifying geo-locations from tweets. Section 6 presents the experiments and analysis done in this paper with crisis data, and Sec. 7 concludes the paper.

2. Related Works

In this section we present some of the related works primarily focusing on the different text classification and topic models and resolution of place names and their respective coordinates from tweets.

Classification of texts: There has been a detailed research on text classification algorithms which can be found in Refs. 15 and 16. Researchers have applied traditional text classification algorithms for classifying social media texts as well. Brynielsson *et al.*[17] described a methodology for collecting a large number of relevant tweets and annotated them with emotional labels. Their method was used for creating a training dataset consisting of manually annotated tweets from the Sandy hurricane. Those tweets have been utilized for building machine learning classifiers able to automatically classify new tweets. Besides there has been other works where the

Topic modeling. Topic modeling has gained attention in different text mining communities, with LDA[14] becoming a standard tool in topic modeling. As a result, LDA has been extended in a variety of ways, and in particular for social networks and social media, a number of extensions to LDA have been proposed. For example, Chang *et al.*[18] proposed a novel probabilistic topic model to analyze text corpora and infer descriptions of the entities and of the relationships between those entities on Wikipedia. McCallum *et al.*[19] proposed a model to simultaneously discover groups among the entities and topics among the corresponding text. Zhang *et al.*[20] introduced a model to incorporate LDA into a community detection process. Similar works can be found in Refs. 21 and 22.

As large collections of short texts collected from social network websites like Twitter are widely available, many people analyze this type of data to find latent topics for various tasks, such as event tracking,[23] content recommendation,[24] and influential users prediction.[25] Initially, due to the lack of specific topic models for short texts, some works directly applied long text topic models such as Refs. 26 and 27.

Java et al.[6] looked into the topological and geographical properties of Twitter's social network and found that the network has high degree correlation and reciprocity, indicating close mutual acquaintances among users. Krishnamurthy et al.[7] looked into geographical distribution of Twitter users and their behaviors among several independent crawls.

Abdelhaq et al.[8] presented a framework to identify bursty words from Twitter text streams and describe such words in terms of their spatio-temporal characteristics. Others have also used Twitter to look at place semantics[9,10] to investigate the relationship between meetings at places and the structure of the social network. Twitter has been used to propagate information in real-time during crisis.[2,3]

Twitter for emergency management. The works of Gelernter and Mushegian[4] and Zheng and Gelernter[28] have mainly influenced the proposed research. Gelernter and Mushegian[4] specifically, discuss the use of Twitter in an emergency situation. They have pointed out how if news sites are down and not reporting during an emergency, Twitter becomes a great source of information and conversation on the emergency. In Ref. 28, Zheng and Gelernter discussed several different methods for resolving locations from tweets, including using a Named Entity Recognizer (NER) in conjunction with a gazetteer, i.e., a list of place names with corresponding latitudes and longitudes. Natural language processing (NLP) techniques: More discussion on NER as well as recognition of objects in language is explained in Refs. 29 and 30. In Ref. 29, Nadeau and Sekine discuss Named Entity Recognition in detail, allowing for the understanding of this approach for text mining and comprehension. Then Ref. 30 discusses geo-parsing and geo-coding, two very important steps in the realm of this project. Additionally, Kinsella et al.[31] discuss the creation of language models for approximate geo-locations.

Geo-parsing and spatial information identification. Geo-parsing is related to the extraction of place names from text, and geo-coding is related to the resolution of those place names (also called toponyms), that is the identification of these names within some sort of database (in our case a gazetteer). Wang[32] discusses many different ways of retrieving spatial information from text as well as the creation of gazetteers. His work is important if gazetteers are to be constructed for specific locations or a specific purpose in mind.

3. Tweet Classification Problem

Document (or text) classification is one of the important and typical task in supervised machine learning. Assigning categories to documents, which can be a web page, library book, media articles, gallery, etc., has many applications like, for example, spam filtering, email routing, sentiment analysis,

etc. In this paper, we can consider *tweets* to be documents and map them with text classification to classify tweets.

Machine learning techniques especially for classification algorithms have been proven useful in text categorization.[33,34] Therefore in this work we investigated two different classification algorithms: *Naive Bayes* (NB) and *Support Vector Machine* (SVM)[35] for classifying crisis-related tweets.

4. Latent Dirichlet Allocation

This section provides a broad description of Latent Dirichlet Allocation,[14] which is the topic model of our consideration. LDA is a popular technique for unsupervised document clustering. Briefly put, LDA fits a generative model to a collection of documents that assumes that each document is composed of a number of topics, which themselves are composed of words. Topic modeling essentially integrates soft clustering with dimension reduction.[36]

Documents are associated with a number of latent topics, which correspond to both document clusters and compact representations identified from a corpus. Each document is assigned to the topics with different weights, which specify both the degree of membership in the clusters as well as the coordinates of the document in the reduced dimension space. The original feature representation plays a key role in defining the topics and in identifying which topics are present in each document. The result is an interpretable representation of documents that is useful for analyzing the themes in document.

Figure 1 illustrates a plate diagram of LDA. The dependencies among the many variables can be captured concisely with the plate notation. The boxes are "plates" representing replicates. The outer plate represents documents, while the inner plate represents the repeated choice of topics and words within a document. M denotes the number of documents, N the number of words in a document. Thus the parameters are explained as follows:

α: Dirichlet prior on the per-document topic distributions.
β: Dirichlet prior on the per-topic word distribution.
θ_m: Topic distribution for document m.
z_{mn}: Topic for the nth word in document m.
w_{mn}: A specific word.

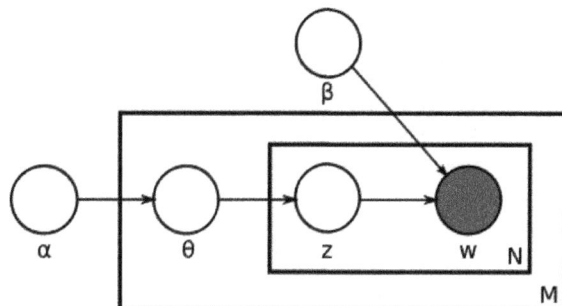

Fig. 1. Plate diagram of LDA.

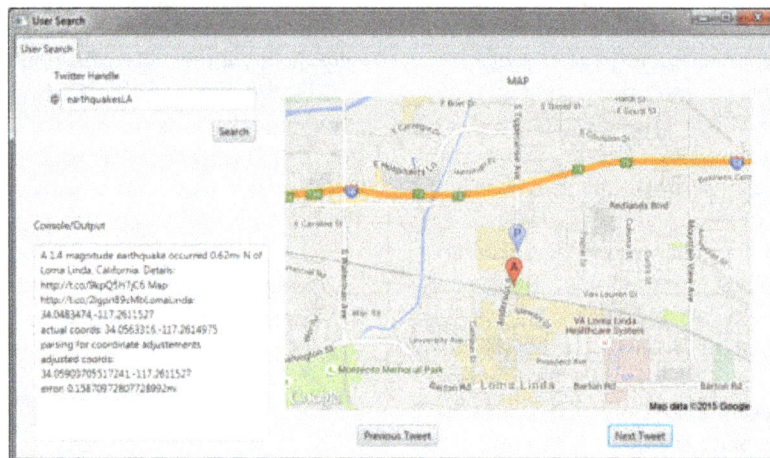

Fig. 2. UI showing the predicted and actual locations on Google Maps.

The words w_{ij} are the only observable variables, and the other variables are latent variables. Further details about the LDA model can be found in Ref. 14.

5. Finding Geo-Locations From Tweets

Tweets contain any geo-tag information (only about 0.87%),[37] making it almost impossible to determine the location of someone based solely on the metadata contained in a tweet. So, in order to approximately identify the geo-locations, NERs[a] were used to extract the proper nouns found within the tweet. They were then matched with gazetteer[b] to see if a location for the place name could be resolved. More details of this portion of the work can be found in Ref. 38. After extractions, the locations were plotted in the Google Maps[c] to visualize the actual locations versus predicted locations using the system. This is illustrated in Fig. 2.

6. Experiments and Analysis

This section presents experiments and analysis done in this research. The analysis is split into three subsections based on the flow of work described in the previous sections.

Datasets. For analysis, CrisisLexT6 dataset[39] is used. It is a collection of English tweet texts related to six large disaster/crisis events in 2012 and 2013: Sandy Hurricane in 2012, Alberta Floods in 2013, Boston Bombing in 2013, Oklahoma Tornadoes in 2013, Queensland Floods in 2013, and West Texas Explosion in 2013. Each dataset contains approximately 10,000 tweets.

Assumptions. Each tweet has been treated as one document, resulting in 10,013 documents per dataset on an average. Characteristics of the datasets are given in Table 1, where

[a]http://nlp.stanford.edu/software/CRF-NER.shtml.
[b]http://geonames.usgs.gov/domestic/download_data.htm.
[c]https://cloud.google.com/maps-platform/.

the abbreviations used are: D: documents, V: vocabulary, and T: tokens.

Data processing. Tweet texts from each dataset are preprocessed before feeding into each topic modeling library. We first used Twokenize (https://github.com/aritter/twitter_nlp/blob/master/python/twokenize.py), a Python-based tokenizer, to tokenize all the texts in datasets. Tokenized texts are then filtered using English stop-words provided by Natural Language Toolkit (NLTK) (http://www.nltk.org/).

For Secs. 6.1 and 6.2 the same datasets, as described in Table 1, are used. But for Sec. 6.3, due to lack of geo-location information, another set of tweets related to "earthquakes" is used.

6.1. *Analysis of twitter classification*

This subsection shows the accuracy with which each of these machine learning algorithms could correctly classify "crisis"-related tweets. It is very clear from Table 2 that SVM persistently outperformed Naive Bayes algorithm.

6.2. *Analysis of topic models*

Though different researchers have proposed a variety of LDA topic models, claiming their applicability, but not all of them are readily available for use. Getting an overview of the existing publicly available topic models and their

Table 1. Characteristics of the datasets.

Dataset	#D	#V	#T
Sandy Hurricane in 2012	10008	18434	82724
Alberta Floods in 2013	10031	20959	98446
Boston Bombing in 2013	10012	18027	89520
Oklahoma Tornadoes in 2013	9992	17619	83193
Queensland Floods in 2013	10033	17804	96925
West Texas Explosion in 2013	10006	15080	87037

Table 2. Classification of crisis-related tweets.

Dataset	Algorithm	Accuracy
Sandy Hurricane in 2012	NB	85.20381168
	SVM	93.19611423
Alberta Floods in 2013	NB	90.60955648
	SVM	95.94249141
Boston Bombing in 2013	NB	89.29421552
	SVM	93.84797928
Oklahoma Tornadoes in 2013	NB	90.25206297
	SVM	93.41435983
Queensland Floods in 2013	NB	94.29961734
	SVM	96.74120141
West Texas Explosion in 2013	NB	95.57285293
	SVM	97.74146893

Table 3. Open-source implementations of LDA.

Name	Language
Stanford Topic Modeling Toolbox	Java
Gensim	Python
Twitter LDA	Java
Scikit-learn	Python

effectiveness on shorter texts is also important. Besides the open-source implementations we have listed here, there are also others available like mallet (http://mallet.cs.umass.edu/topics.php) and jLDADMM (http://jldadmm.sourceforge.net/), which we have not presented in this paper.

Libraries. Here the goal is to explore the effectiveness of LDA-based topic modeling algorithms when applied to short-text datasets on crises context. There have been substantial efforts in developing efficient and effective implementations of LDA. In order to provide a quick hands-on experience, four main implementations are listed (Table 3), that are open-source or publicly accessible which researchers have used in their experiments with larger texts, such as Refs. 40–42, and so on. These are used on CrisisLexT6[39] dataset related to six crises in 2012 and 2013 to see how far they are useful for analyzing short texts like "tweets" which are restricted to 140 characters:

(a) Gensim (https://radimrehurek.com/gensim/index.html): An open-source Python library aimed at finding similar documents in a dataset.
(b) Stanford Topic Modeling Toolbox[43]: A Java-based library and tool written by Stanford NLP group.
(c) Twitter LDA[44]: An algorithm based on the model proposed by Zhao *et al.*
(d) Scikit-learn[45]: Scikit-learn is a simple and efficient tool for data mining and data analysis which is accessible to everybody, and reusable in various contexts. It is built on NumPy, SciPy, and Matplotlib.

Visualization. One of the goals of this paper was to look into the fact of how to get distinct topics through topic models on short texts. Here we perform analysis at both topic level as well as all "crises" dataset level.

For any text mining applications, topic words play a key role. As mentioned earlier, due to shortcoming of tweets, it is important to analyze topic words so that they can be further used to model a system. Figures 3–6, 8, and 9 illustrate sample topic words extracted from tweets and how well these topics generated are distinct.

Topic-level analysis. With repeated experiments, we have decided to choose the number of topics to be 5, with the *assumption* that the topic words could focus on the following abstract ideas:

(i) People's expression or attitude,
(ii) Crisis event description,
(iii) Impact of crisis,
(iv) State of the disaster, and
(v) Other.

With more number of topics, there was substantial overlap between topics. A number of earlier studies[11-13] have shown that if any text-retrieval system was able to find good clusters, retrieval performance can be improved over text-based retrieval. So going with this fact, through experiments on "topic-level analysis" we illustrate how different implementations of topic models can provide us with distinct topics (aka distinct clusters).

Topic distinctiveness computation. The main focus was on how well each algorithm has generated distinct topics. This aspect was evaluated by measuring the distance between topics generated by each open-source implementations, represented by topic–term distribution matrix. Each topic is associated with a probability distribution over all existing words in the dataset of how likely it is for a word to belong to a topic. Discrete Hellinger distance[46] is used as a metric to measure the distinctiveness between two topics,[47] with zero distance meaning the topics are identical and a higher distance meaning the topics are more distinct. Distances between topics of each dataset are visualized using pseudocolor 2D plot functionality of Matplotlib library (https://matplotlib.org/). This is represented using heatmaps, where the small cell represents the distance between two topics.

The Hellinger distance is shown in Eq. (1):

$$H(P,Q) = \frac{1}{\sqrt{2}} \|\sqrt{P} - \sqrt{Q}\|_2, \qquad (1)$$

where $H(P,Q)$ is the Hellinger distance between discrete distributions P and Q, equivalent to the Euclidean norm of the difference of square roots of distributions.

Discussion. As seen in Figs. 3–6 each implementation of LDA has given us different set of results. Though TwitterLDA[44] claimed to perform better for tweets, we found through our experiment that Stanford Topic Modeling Toolbox outperformed it in terms of providing relatively distinct topics in majority. Gensim did not provide satisfactory result

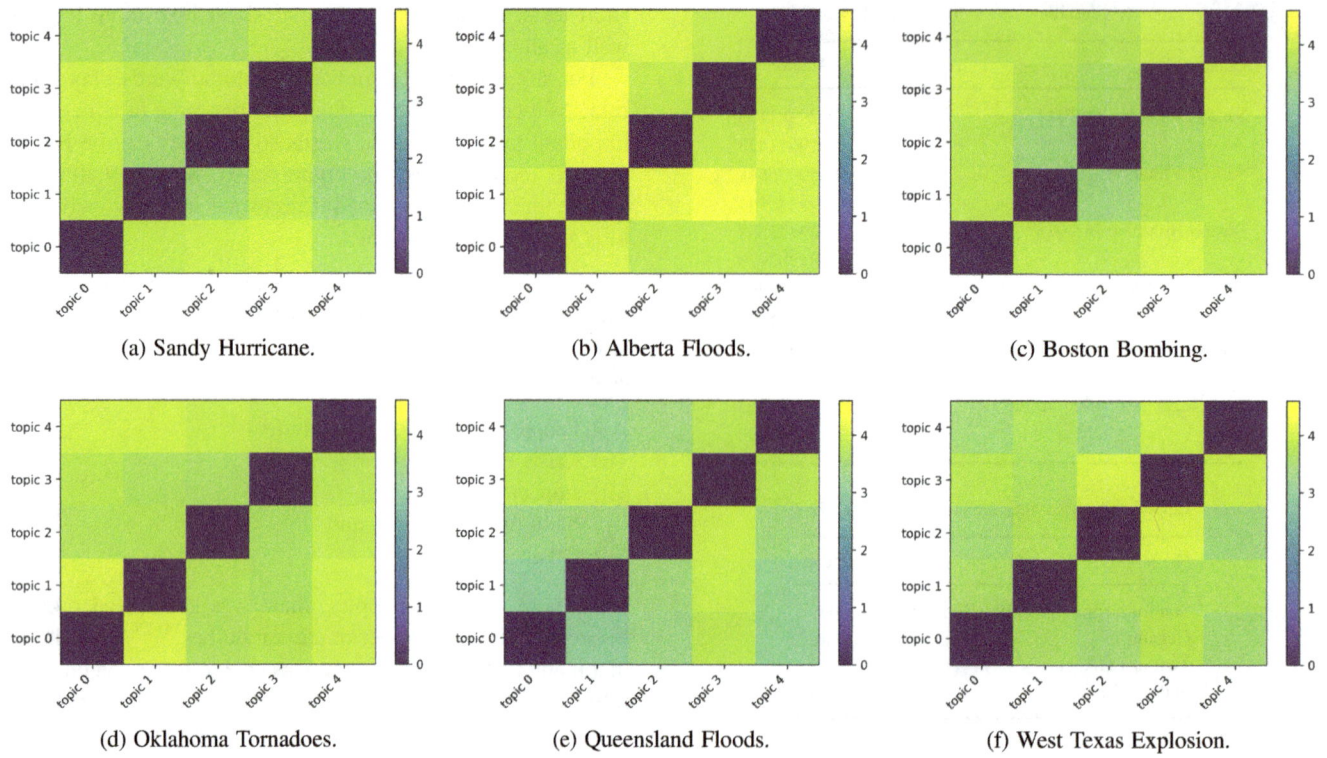

(a) Sandy Hurricane.

(b) Alberta Floods.

(c) Boston Bombing.

(d) Oklahoma Tornadoes.

(e) Queensland Floods.

(f) West Texas Explosion.

Fig. 3. Inter-topic Hellinger distance heatmaps using Stanford Topic Modeling Toolbox.

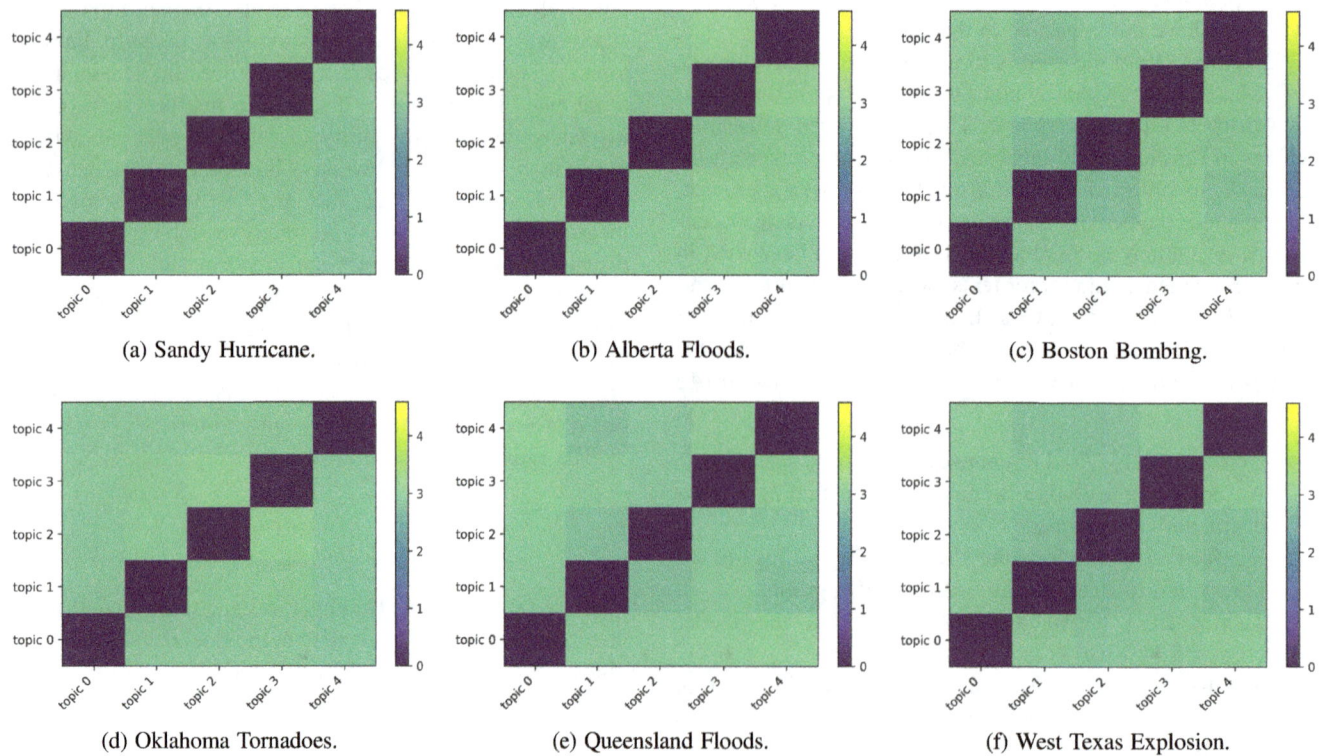

(a) Sandy Hurricane.

(b) Alberta Floods.

(c) Boston Bombing.

(d) Oklahoma Tornadoes.

(e) Queensland Floods.

(f) West Texas Explosion.

Fig. 4. Inter-topic Hellinger distance heatmaps using Gensim toolkit.

(a) Sandy Hurricane.

(b) Alberta Floods.

(c) Boston Bombing.

(d) Oklahoma Tornadoes.

(e) Queensland Floods.

(f) West Texas Explosion.

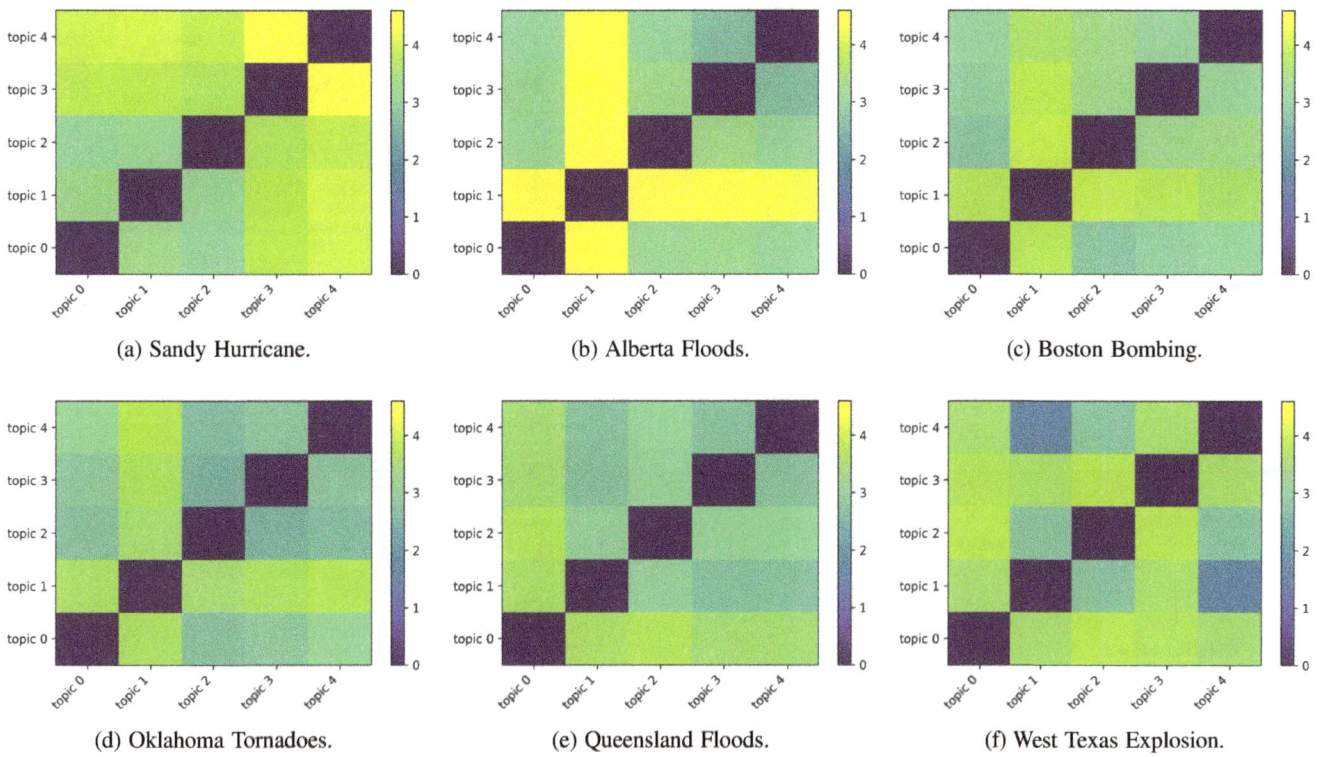

Fig. 5. Inter-topic Hellinger distance heatmaps using Twitter LDA toolkit.

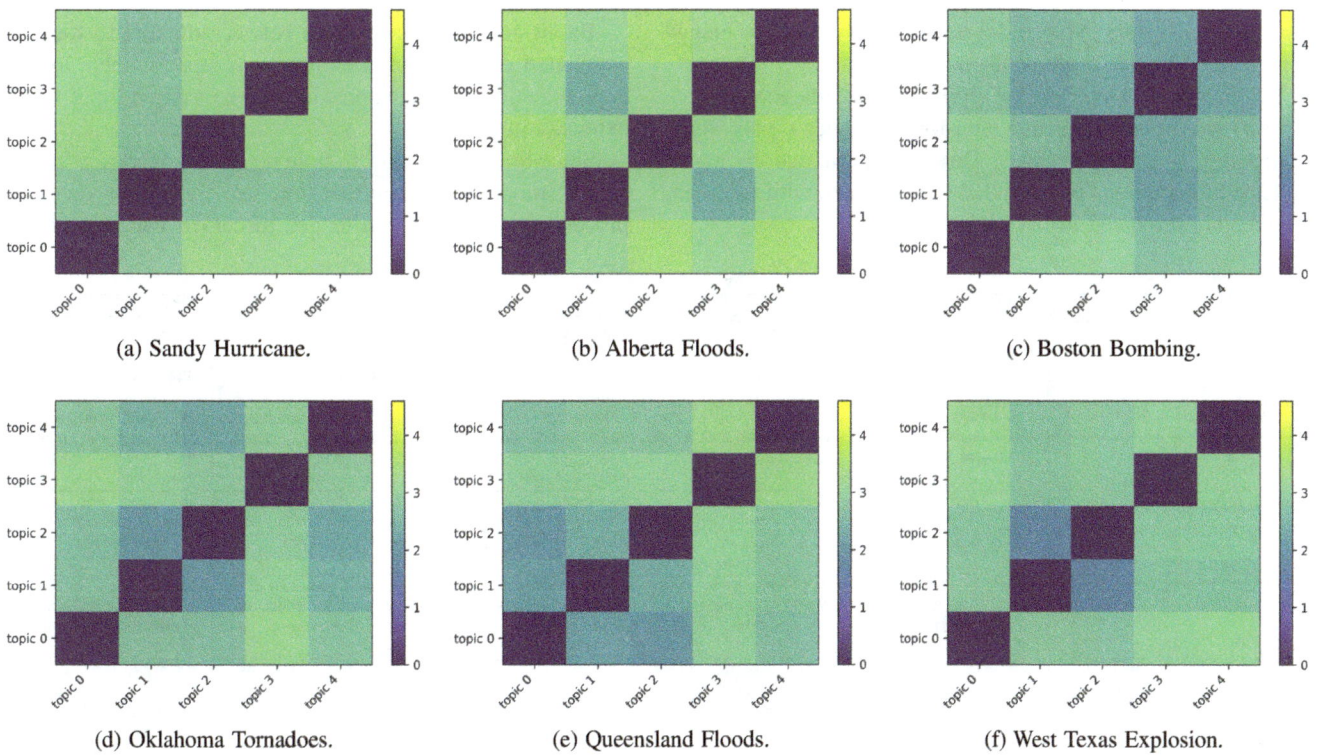

(a) Sandy Hurricane.

(b) Alberta Floods.

(c) Boston Bombing.

(d) Oklahoma Tornadoes.

(e) Queensland Floods.

(f) West Texas Explosion.

Fig. 6. Inter-topic Hellinger distance heatmaps using Scikit-learn toolkit.

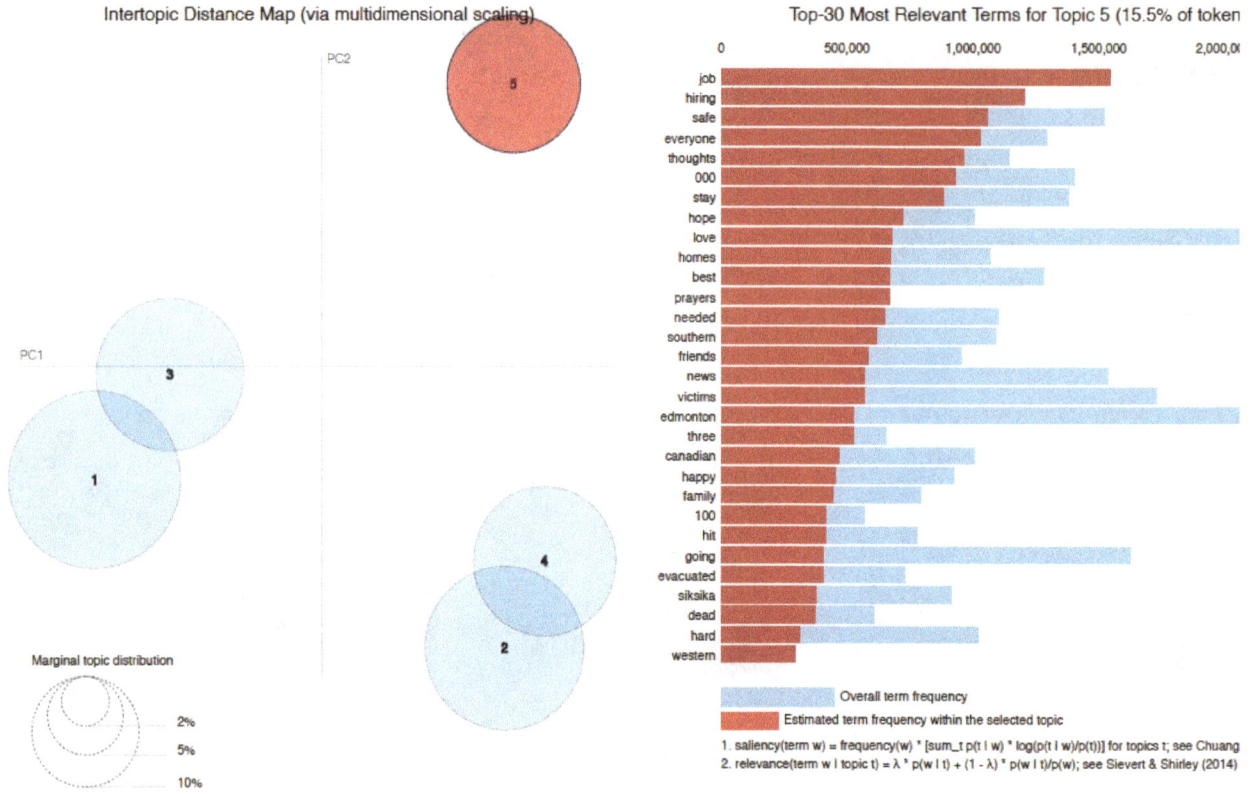

Fig. 7. Sample illustration of pyLDAvis visualization of topic model on Alberta Floods.

with the crisis data. Neither did Scikit-learn in spite of being one of the popular machine learning toolkits.

Using pyLDAvis (https://pyldavis.readthedocs.io/en/latest/) word histograms and inter-topic distance topics were generated, as shown in Fig. 7. Due to page limitations, only selected illustrations are included in this paper, while similar figures are produced for all datasets.

Based on Hellinger distance between the topics, we have concluded that Stanford's toolbox was more suitable for our experiment. So we have provided some sample topic words extracted in Fig. 8.

Since these topic model implementations work on "bag of words" model, it is seen that they heavily rely on data pre-processing. With the challenges discussed earlier with the

Dataset	Topic 1	Topic 2	Topic 3	Topic 4	Topic 5
Sandy Hurricane in 2012	house, others, live, city, last, brooklyn, see, park, water, wind	wish, name, better, god, real, omg, would, bitch, tho, tweets	rain, actually, prayers, news, many, affected, call, twitter, niggas, coast	coast, due, obama, romney, wit, irene, bout, went, take, relief	school, fuck, tomorrow, please, day, want, gonna, think, bad, every
Alberta Floods in 2013	job, hiring, safe, everyone, thoughts, 000, stay, hope, love, homes,	said, time, back, love, night, think, feel, never, ever, see	downtown, power, mission, says, emergency, park, pic, see, update, cityofcalgary	yychelps, volunteers, cityofcalgary, thank, donate, call, clean, use, victims, amazing	red, cross, make, would, redcrosscanada, thanks, time, donate, donations, someone
Boston Bombing in 2013	day, tsarnaev, new, dzhokhar, job, guy, run, family, manhunt, open	died, running, kids, hook, sandy, lol, girl, fuck, shit, going,	everyone, prayers, thoughts, sad, affected, heart, need, city, retweet, safe	obama, finish, texas, line, time, call, bombs, well, think, president	cnn, dead, man, watertown, arrest, video, photos, says, reports, photo
Oklahoma Tornadoes in 2013	million, durant, okc, kevin, dead, state, children, update, weather	got, feel, back, never, would, right, even, make, come, think,	donate, 90999, heart, red, cross, give, support, way, okc, tornados	dog, god, pray, lost, families, news, finds, survivor, kids, video	first, last, someone, night, beautiful, rubble, church, disaster, photo, found
Queensland Floods in 2013	australian, deepens, three, today, floo, sure, coast, areas, abbott, tony	cost, eastern, peaks, counts, deadly, rise, four, worsens, torrential, confirmed,	victims, appeal, qldfloods, help, coverage, live, via, abbott, may, red,	good, love, time, day, one, know, back, would, fire, haha	powerful, stuffed, hoisted, women, photos, warnings, car, hit, 2013, bundaberg
West Texas Explosion in 2013	camera, prayfortexas, via, youtube, aid, cnn, heart, watch, god, another	families, bombing, join, marathon, man, pontifex, obama, help, baylor, week	reported, fertiliser, police, blast, killed, injuries, massive, hundreds, tonight, says	got, good, shit, really, going, think, want, right, fuck, never,	everyone, thoughts, fire, affected, first, responders, safety, youranonnews, killed, pray

Fig. 8. Top 30 topic words for each dataset using Stanford Topic Modeling Toolbox.

Dataset	Top 30 words
Sandy Hurricane in 2012	school, fuck, name, wish, tomorrow, better, god, want, every, house, rain, real, omg, others, coast, please, day, due, gonna, wanna, actually, prayers, many, obama, would, tweets, live, romney, blow, bitch
Alberta Flood in 2013	downtown, job, cross, red, yychelps, hiring, mission, volunteers, safe, cityofcalgary, power, thoughts, said, everyone, says, donate, park, would, emergency, stay, feel love, 000, think ave, best, hope, prayers, bridge, needed
Boston Bombing in 2013	prayers, everyone, thoughts, died, cnn, kids, hook, sandy, girl, obama, arrest, affected, sad, running, heart, lol, dead, watertown, fuck, reports, retweet, photos, finish, shit, tsarnaev, says, day, photo, goes, video
Oklahoma Tornadoes in 2013	donate, dog, 90999, never, got, feel, god, heart, back, families, lost, pray, red, finds, million, fuck, would, durant, even, news, cross, shit, wanna, survivor, sad, kevin, really, idiot, kids, give
Queensland Flood in 2013	good, australian, powerful, stuffed, hoisted, women, cost, eastern, peaks, counts, deepens, love, would, know, three, haha, safe, back, think, victims, appeal, lol, qldfloods, one, deadly, time, day, hope, really, well
West Texas Explosion in 2013	got, camera, reported, fertiliser, bombing, good, shit, families, everyone, join, police, killed, fire, marathon, pontifex, never, really, obama, nigga, affected, thoughts, bitch, injuries, vis, ass, hundreds, blast, baylor, fuck, help

Fig. 9. Overall topic words for each dataset using Stanford Topic Modeling Toolbox.

nature of tweets, it is obvious that lot of irrelevant words can still make the results noisy. Besides, there are still lots of words which are listed as topic keywords which do not make much sense when we manually reviewed the topic words generated. Such words are "tho," "gonna," "bout," "tsarnaev," and so on.

Dataset-level analysis. Here we address two exploratory questions:

(1) What are the topic words at the whole dataset level versus individual topic level?
(2) Is there any commonality among all the crisis data used in this research?

Based on heatmaps illustrated in Figs. 3–6, it is visible that the distinctiveness of topics obtained with Stanford Topic Modeling Toolbox is better than the other two. So in order to answer the above two questions, we have used Stanford Topic Modeling Toolbox.

Jaccard similarity. In order to find out if there is any commonality between each of these crisis datasets, we have computed the Jaccard similarity (aka Jaccard coefficient)[48] as shown in Eq. (2):

$$\text{sim}_{\text{Jaccard}}(\mathbf{v}, \mathbf{w}) = \frac{\sum_{i=1}^{N} \min(v_i, w_i)}{\sum_{i=1}^{N} \max(v_i, w_i)}, \quad (2)$$

where $\mathbf{v}$ and $\mathbf{w}$ are the word vectors from two datasets, respectively, and v_i and w_i are words from each vector.

Figure 10 illustrates the heatmap representation of the individual datasets used in this paper. Higher the score, more similar are the dataset pairs. It is seen that the crisis datasets related to "bombings" and "explosions" have greater similarity. Similar observation is also seen for "hurricane" and "tornado." Since topic model is basically a "bag of words" concept, we still cannot know the underlying semantics, but can have a glimpse of the commonality of the datasets. This will further help us to narrow our research and focus on more domain-specific datasets for our work.

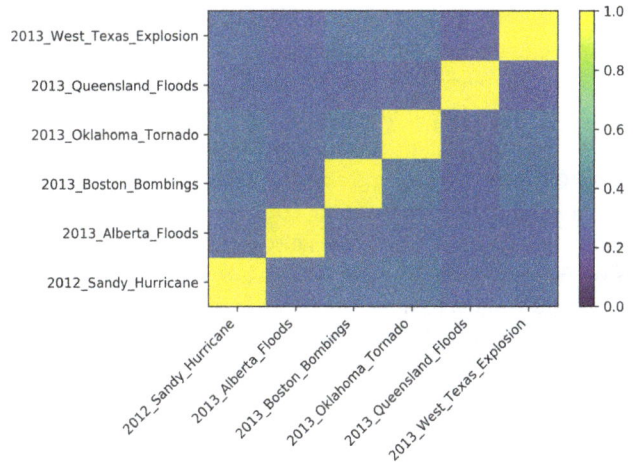

Fig. 10. Similarity across six crisis datasets.

6.3. *Analysis of geo-location*

The analysis was performed on 3008 separate tweets pulled from the user "@earthquakesLA" on Twitter. Each tweet received a pass/fail grade based on the coordinates of the algorithm described in Ref. 38, returned from parsing the text. The distance between these coordinates (known in this paper as predicted coordinates) and the actual coordinates (taken from the tweet) was computed and a pass grade was given to the coordinates for which a location was able to be determined. Out of 3008 tweets, 558 failed to determine a location. The average distance between predicted coordinates and actual coordinates was computed to be 33.7 mi (this is the overall average distance for all 2450 tweets with predicted locations). The median distance (the actual middle distance of our sorted set of determined distances) was 0.66 mi.

The average difference between the actual coordinates and the predicted coordinates is quite high. This however is due to incorrect classifications of place names, which are difficult to track down and fix, since it usually revolves around there being multiple places with the same name. This is why

```
                          Tweet:
A 1.4 magnitude earthquake occurred 0.62mi N of Loma Linda,
California
─────────────────────────────────────────────────────────
Gazetteer lookup (Loma Linda): {34.05,-117.26}
Adjustment (0.62mi N): {0.01, 0}
Predicted coordinates = {34.05,-117.26} +
{0.01, 0} = {34.059,-117.26}
Compared to: (actual) {34.056, -117.26}
```

Fig. 11. An example of a tweet being parsed.

we have included the median in addition to the average difference, to show that there are quite a bit of cases for which the error in the resolution of the coordinates is relatively small.

The difference in predicted versus actual coordinates in reality is very low (over 50% of the coordinates are within 0.66 mi of the actual location), however, the average difference displays a much larger discrepancy. Issues that have led to a high disparity between the correctly identified places are ambiguous names which are able to be classified (?Greater Los Angeles Area? is identified as Los Angeles) and names of places that occur more than once in the state of California. These errors can lead to coordinates that range anywhere from 50 mi to several hundred miles. The overall average will be thrown off by these outliers, even though most values are within 1 mi.

7. Conclusion

The paper presented three different aspects of automatic analysis of microblogging data, *tweets*, related to crisis. Though the paper is exploratory in nature, it has shown how text classification algorithms can be used to identify crisis-related tweets, analyze topic models on short texts, and how can the geo-locations be predicted from the texts.

The paper could highlight the fact that any basic text classification model can pretty well be used for the initial filtration of the data. The accuracy of the classifiers was very promising when tested with crisis-related tweets as illustrated in Sec. 6.

Although this paper does not introduce new topic models, the work sheds some light on how research on topic models can be conducted for short text scenarios especially for analyzing crisis data. It is analyzed how well the topics generated by the already available topic models are distinct and what is the commonality between the nature of data. Extensive exploratory experiments were conducted on three proposed schemes based on standard LDA model. Different visualization techniques were illustrated, focusing on understanding the topics and the word distribution of the topic words.

Finally, the proposed solution of prediction of geo-locations and identification of different places by extracting

information from tweets and co-relating them with gazetteer data seemed to be another useful means to analyze crisis data.

Though there are few shortcomings of this work (as discussed in Sec. 6), here are few plans on extending the work in the future:

(i) Design classifiers more specific towards location extraction.
(ii) Improve the NLP framework to more accurately identify place names within microtext (tweets/Facebook statuses).
(iii) Fine-tune topic models making it more effective for microblogging platforms.
(iv) Perform more quantitative evaluation to perform intrinsic evaluation of the quality of the topics generated.
(v) Processing of real-time messages with cloud streaming services such as Amazon Kinesis.

References

1. K. Starbird and J. Stamberger, Tweak the tweet: Leveraging microblogging proliferation with a prescriptive syntax to support citizen reporting, *Proc. 7th Int. ISCRAM Conf.* (2010).
2. T. Sakaki, M. Okazaki and Y. Matsuo, Earthquake shakes Twitter users: Real-time event detection by social sensors, *Proc. 19th Int. Conf. World Wide Web* (2010), pp. 851–860.
3. M. Okazaki and Y. Matsuo, *BlogTalk 2009: Recent Trends and Developments in Social Software*, Chapter 7 (Springer, 2010), pp. 63–74.
4. J. Gelernter and N. Mushegian, Geo-parsing messages from microtext, *Trans. GIS* **15**, 753 (2011).
5. L. Hong and B. D. Davison, Empirical study of topic modeling in Twitter, *Proc. First Workshop on Social Media Analytics* (ACM, 2010), pp. 80–88.
6. A. Java, X. Song, T. Finin and B. Tseng, Why we twitter: Understanding microblogging usage and communities, *Proc. Joint 9th WebKDD and 1st SNA-KDD 2007 Workshop on Web Mining and Social Network Analysis* (ACM, 2007), pp. 56–65.
7. B. Krishnamurthy, P. Gill and M. Arlitt, A few chirps about Twitter, *Proc. First Workshop on Online Social Networks* (ACM, 2008), pp. 19–24.
8. H. Abdelhaq, M. Gertz and C. Sengstock, Spatio-temporal characteristics of bursty words in Twitter streams, *Proc. 21st ACM SIGSPATIAL Int. Conf. Advances in Geographic Information Systems* (ACM, 2013), pp. 194–203.
9. E. Kim, H. Ihm and S.-H. Myaeng, Topic-based place semantics discovered from microblogging text messages, *Proc. 23rd Int. Conf. World Wide Web* (2014), pp. 561–562.
10. C. Brown, A. Noulas, C. Mascolo and V. Blondel, A place-focused model for social networks in cities, *Proc. 2013 Int. Conf. Social Computing* (2013), pp. 75–80.
11. M. A. Hearst and J. O. Pedersen, Reexamining the cluster hypothesis: Scatter/gather on retrieval results, *Proc. 19th Annu. Int. ACM SIGIR Conf. Research and Development in Information Retrieval* (1996), pp. 76–84.
12. A. Tombros, R. Villa and C. J. van Rijsbergen, The effectiveness of query-specific hierarchic clustering in information retrieval, *Inf. Process. Manage.* **38**, 559 (2002).
13. N. Jardine and C. J. van Rijsbergen, The use of hierarchic clustering in information retrieval, *Inf. Storage Retr.* **7**, 217 (1971).

[14]D. M. Blei, A. Y. Ng and M. I. Jordan, Latent Dirichlet allocation, *J. Mach. Learn. Res.* **3**, 993 (2003).

[15]V. Korde and C. N. Mahender, Text classification and classifiers: A survey, *Int. J. Artif. Intell. Appl.* **3**, 85 (2012).

[16]K. Aas and L. Eikvil, Text categorisation: A survey (1999), https://www.nr.no/~eikvil/tm_survey.pdf.

[17]J. Brynielsson, F. Johansson and A. Westling, Learning to classify emotional content in crisis-related tweets, *Proc. 2013 IEEE Int. Conf. Intelligence and Security Informatics* (2013), pp. 33–38.

[18]J. Chang, J. Boyd-Graber and D. M. Blei, Connections between the lines: Augmenting social networks with text, *Proc. 15th ACM SIGKDD Int. Conf. Knowledge Discovery and Data Mining* (2009), pp. 169–178.

[19]A. McCallum, X. Wang and N. Mohanty, *ICML 2006: Statistical Network Analysis: Models, Issues, and New Directions*, Chapter 3 (Springer, 2007), pp. 28–44.

[20]H. Zhang, C. L. Giles, H. C. Foley and J. Yen, Probabilistic community discovery using hierarchical latent Gaussian mixture model, *Proc. 22nd AAAI Conf. Artificial Intelligence* (AAAI, 2007), pp. 663–668.

[21]Y. Liu, A. Niculescu-Mizil and W. Gryc, Topic-link LDA: Joint models of topic and author community, *Proc. 26th Annu. Int. Conf. Machine Learning* (2009), pp. 665–672.

[22]R. M. Nallapati, A. Ahmed, E. P. Xing and W. W. Cohen, Joint latent topic models for text and citations, *Proc. 14th ACM SIGKDD Int. Conf. Knowledge Discovery and Data Mining* (ACM, 2008), pp. 542–550.

[23]C. X. Lin, B. Zhao, Q. Mei and J. Han, PET: A statistical model for popular events tracking in social communities, *Proc. 16th ACM SIGKDD Int. Conf. Knowledge Discovery and Data Mining* (ACM, 2010), pp. 929–938.

[24]O. Phelan, K. McCarthy and B. Smyth, Using Twitter to recommend real-time topical news, *Proc. Third ACM Conf. Recommender Systems* (ACM, 2009), pp. 385–388.

[25]J. Weng, E.-P. Lim, J. Jiang and Q. He, TwitterRank: Finding topic-sensitive influential twitterers, *Proc. Third ACM Int. Conf. Web Search and Data Mining* (2010), pp. 261–270.

[26]D. Ramage, S. T. Dumais and D. J. Liebling, Characterizing microblogs with topic models, *Proc. Fourth Int. AAAI Conf. Weblogs and Social Media* (2010), pp. 130–137.

[27]Y. Wang, E. Agichtein and M. Benzi, TM-LDA: Efficient online modeling of latent topic transitions in social media, *Proc. 18th ACM SIGKDD Int. Conf. Knowledge Discovery and Data Mining* (ACM, 2012), pp. 123–131.

[28]W. Zhang and J. Gelernter, Geocoding location expressions in Twitter messages: A preference learning method, *J. Spat. Inf. Sci.* **2014**, 37 (2014).

[29]D. Nadeau and S. Sekine, A survey of named entity recognition and classification, *Lingvisticae Investigationes* **30**, 3 (2007).

[30]J. L. Leidner and M. D. Lieberman, Detecting geographical references in the form of place names and associated spatial natural language, *SIGSPATIAL Spec.* **3**, 5 (2011).

[31]S. Kinsella, V. Murdock and N. O'Hare, "I'm eating a sandwich in Glasgow": Modeling locations with tweets, *Proc. 3rd Int. Workshop Search and Mining User-Generated Contents* (ACM, 2011), pp. 61–68.

[32]W. Wang, Automated spatiotemporal and semantic information extraction for hazards, Ph.D. thesis, The University of Iowa (2014).

[33]F. Sebastiani, Machine learning in automated text categorization, *ACM Comput. Surv.* **34**, 1 (2002).

[34]T. Joachims, *ECML 1998: Machine Learning*, Chapter 19 (Springer, 1998), pp. 137–142.

[35]K.-M. Schneider, A comparison of event models for naive Bayes anti-spam e-mail filtering, *Proc. Tenth Conf. European Chapter of the Association for Computational Linguistics*, Vol. 1 (ACM, 2003), pp. 307–314.

[36]S. P. Crain, K. Zhou, S.-H. Yang and H. Zha, *Mining Text Data*, Chapter 5 (Springer, 2012), pp. 129–161.

[37]A. Jaiswal, W. Peng and T. Sun, Predicting time-sensitive user locations from social media, *Proc. 2013 IEEE/ACM Int. Conf. Advances in Social Networks Analysis and Mining* (2013), pp. 870–877.

[38]J. Bassi, S. Manna and Y. Sun, Construction of a geo-location service utilizing microblogging platforms, *Proc. 2016 IEEE Tenth Int. Conf. Semantic Computing* (2016), pp. 162–165.

[39]A. Olteanu, C. Castillo, F. Diaz and S. Vieweg, CrisisLex: A lexicon for collecting and filtering microblogged communications in crises, *Proc. Eighth Int. AAAI Conf. Weblogs and Social Media* (2014).

[40]C. E. Grant, C. P. George, V. Kanjilal, S. Nirkhiwale, J. N. Wilson and D. Z. Wang, A topic-based search, visualization, and exploration system, *Proc. Twenty-Eighth Int. Florida Artificial Intelligence Research Society Conf.* (2015), pp. 43–48.

[41]M. M. Tyler, Topic modeling via scatter/gather clustering, Master's thesis, The University of Texas at Austin (2015).

[42]S. Kaplan, Identifying breakthroughs: Using topic modeling to distinguish the cognitive from the economic, *DRUID 2012: Academy of Management Annu. Meetings Proc.* (2012).

[43]D. Ramage and E. Rosen, Stanford topic modeling toolbox (2011), https://nlp.stanford.edu/software/tmt/tmt-0.4/.

[44]W. X. Zhao, J. Jiang, J. Weng, J. He, E.-P. Lim, H. Yan and X. Li, Comparing Twitter and traditional media using topic models, *Proc. European Conf. Information Retrieval* (2011), pp. 338–349.

[45]F. Pedregosa, G. Varoquaux, A. Gramfort, V. Michel, B. Thirion, O. Grisel, M. Blondel, P. Prettenhofer, R. Weiss, V. Dubourg, J. Vanderplas, A. Passos, D. Cournapeau, M. Brucher, M. Perrot and E. Duchesnay, Scikit-learn: Machine learning in Python, *J. Mach. Learn. Res.* **12**, 2825 (2011).

[46]A. Basu, I. R. Harris and S. Basu, 2 minimum distance estimation: The approach using density-based distances, *Handb. Stat.* **15**, 21 (1997).

[47]A. S. Maiya and R. M. Rolfe, Topic similarity networks: Visual analytics for large document sets, *Proc. 2014 IEEE Int. Conf. Big Data* (2014), pp. 364–372.

[48]P. Jaccard, The distribution of the flora in the alpine zone, *New Phytol.* **11**, 37 (1912).

Applications of natural language processing (NLP) for improving classroom learning experiences using student surveys

Patrick Kuiper* and Karoline Hood

Department of Mathematics, United States Military Academy

646 Swift Road, West Point, NY 10996, USA

*patrick.kuiper@usma.edu

Academic institutions often assess the efficacy of courses by surveying students. These surveys are critical in structuring course content and evaluating instruction. Given the critical function of surveys for academic institutions, it is essential that surveys obtain data which is precise and accurate. Currently most institutions construct surveys employing the Likert scale: questions that require students to map their opinion on precise topics to a discrete, quantitative domain. These surveys uniformly weight response data, irrespective of student interest. We argue greater accuracy may be obtained by building Student-Directed Discussion Surveys (SDDSs) — surveys with several open-ended, student-directed questions, requiring free text responses. SDDSs retain precision by employing several Natural Language Processing (NLP) techniques including word frequency and sentiment analysis. We use SDDSs to improve course content and evaluate survey accuracy by comparing the results of an SDDS to a Likert-scaled survey administered to an overlapping population. We find that the results of these two survey techniques diverge when topics become increasingly significant to respondents. These results, in addition to the documented issues with Likert-scaled surveys, lead to the conclusion that SDDSs may provide more informative and insightful results.

Keywords: Natural language processing; survey design; Likert scale; sentiment analysis.

1. Introduction

1.1. *Problem description and definition*

Educators place great importance in continually improving the courses they deliver. To assess the performance of instruction and the content of courses, surveys are used to obtain feedback. With surveys educators attempt to obtain results which pertain to various topics and accurately reflect the opinion of a student population. These surveys are most commonly built with Likert-scaled questions, reviewing precise topics and requiring respondents to map their opinions on each topic to a discrete, quantitative scale. Given the breadth of material covered in many courses, educators may ask dozens of Likert-scaled questions, which are a simple and direct method to obtain sentiment data on precise topics.

Educators assume that Likert-scaled results are accurate. If students honestly respond to questions in a thoughtful and deliberate manner, than this assumption is valid. A number of factors contradict this assumption, including documented student survey fatigue and the leading nature of these precise topic questions.[1,2] We believe Likert-scaled results are often not accurate. Certainly many of the precise topics covered by Likert-scaled surveys are not important to all students. When students respond to a Likert-scaled question covering a precise topic that does not interest them, the respondent often does not provide thoughtful or deliberate feedback, producing an invalid statistical data point. Given the most common implementation of Likert surveys, all student responses are weighted and analyzed uniformly. This survey technique subsequently introduces a high level of noise into the data, corrupting the precise topics addressed, creating inaccurate results.

1.2. *Previous research*

Leveraging large repositories of data to make informed decisions is becoming standard practice among large organizations. Natural Language Processing (NLP) is a growing subset of the current data revolution, as written language represents an enormous body of useful information.[3] When coupled with statistics, NLP provides a number of tools to make quantitative measurements and inferences from written language. Sentiment analysis is an established subfield of NLP, whose methods have been employed to accurately infer the opinion of a population, even with a limited dataset.[4] We specifically leverage sentiment analysis to measure opinion when analyzing student responses to academic surveys.

Hutto and Gilbert built an effective Python library for NLP sentiment analysis, which we leveraged for the development of Student-Directed Discussion Surveys (SDDSs).[5] Additionally, Hutto and Gilbert provide a succinct analysis of NLP procedures and the algorithmic performance of competing methods in the paper accompanying their software.[5] Further information concerning general NLP methodology and implementations in Python are available in "Natural Language Processing with Python" by Bird *et al.*[6]

*Corresponding author.

In his investigation, "Working with Low Survey Response Rates: The Efficacy of Weighting Adjustments", Dey discusses the historically significant decline in response rates for surveys across general research domains, and evaluates a weighting scheme to account for nonresponsive populations.[7] This analysis highlights the importance of designing surveys that encourage high response rates and the importance of properly weighting survey responses in order to accurately infer the opinion of a large population.

In the marketing research paper "The Likert scale revisited: an alternate version," Albaum explains that Likert-scaled marketing surveys often confound consumer attitude polarity and intensity. Albaum goes on to investigate further methods addressing this shortcoming, but does not progress beyond Likert-based questions.[2] Analyzing survey techniques specifically in academic environments, Porter *et al.* have written a chapter specifically addressing students and surveys which holds particular relevance to our hypothesis.[8] Specifically, Porter's analysis of survey fatigue speaks of the challenges associated with current Likert-scaled surveys, alluding to a demand for an improved survey technique. We believe SDDS provides this technique.

1.3. *Modeling approach and limitations*

To address the issues associated with Likert-scaled surveys, we propose the use of SDDS. With SDDSs, students are asked one to three open-ended, free text discussion questions, which are subsequently analyzed using NLP techniques. NLP provides a universe of tools to leverage when analyzing written language. In our SDDS algorithm, we specifically examine two techniques: word frequency and sentiment analysis. Simple word frequency analysis allows SDDS to determine what topics are most important to students — organically — without surveyor direction. We then employ sentiment analysis to attempt to infer student opinion on precise topics throughout the course.

SDDSs address the issue of properly weighting statistical results by asking students to direct the topics of their responses. The proposed survey technique only employs broad free text questions when gathering responses, which naturally weighs the student responses appropriately — if a student is concerned about a precise topic, they provide a free text response. If a student is not concerned about a topic, they provide no response.

There are several limitations associated with the structure and tools of our investigation. We assume student's response frequency and sentiment establish the importance students place on precise topics. This is not always a correct assumption, as students may not comment on topics which they believe are most important, and discuss other less personally important topics. Additionally, there is always error associated with employing any NLP algorithm to measure sentiment — no sentiment analysis algorithm is perfectly accurate. This ensures some amount of error in our analysis of student

sentiment. Finally, since SDDSs do not request responses from every student on every precise topic of interest, the number of responses for some topics may be low, rendering statistical inference difficult.

1.4. *Summary of contributions*

We began our investigation by administering three identical SDDSs across a single semester to a population of 100 calculus students. We analyzed the results from these three surveys using NLP techniques including word frequency and sentiment analysis, quantifying the opinions of students on specific topics. We find that both of these NLP methods are an effective means to inform a number of pedagogical decisions made in the structure and execution of the course.

Finally, we compare the results of all three SDDSs to a single Likert-scaled survey administered at the completion of the semester. This Likert-scaled survey was administered to over 900 students, including 50% of the students who had taken part in SDDSs. A statistical permutation test is applied to assess the similarity between the distributions of results from each survey technique. Once again, we quantify student opinion with SDDSs by employing NLP sentiment analysis. We find that the distributions of survey results between SDDSs and Likert-scaled surveys are similar when the frequency of SDDS observations for a precise topic is low; however, for more popular topics, the distributions of quantitative sentiment measured between the two survey techniques diverge. This result, along with the significantly higher variance we observed with the Likert-scaled surveys and well documented Likert issues, leads our investigation to conclude that SDDSs are more likely to produce accurate results when sufficient number of observations are produced for a precise topic.[1,2] This accuracy is gained because SDDSs naturally weigh survey data points concerning precise topics more accurately when compared to Likert-scaled surveys.

2. Method

2.1. *Experimentation and data collection*

We believe student surveys should consist of short, open-ended discussion questions and hypothesize that surveys designed with this consideration produce both precise and accurate results when employing NLP. SDDS allows surveyors to leverage the semantic structure of student responses, eliciting more honest and information-rich data when compared to traditional Likert-scaled surveys.

To assess our hypothesis, SDDSs were administered through a web-based application at three consistent periods of time in two freshman calculus courses. The survey consisted of two open-ended questions:

(1) Discuss your thoughts on the structure and execution of this course. Possible points of discussion could

include: Lesson motivation/application, lecture effectiveness, use of technology, in-class exercises, out-of-class assignments, in-class tests, organization of course, etc.

(2) Discuss your thoughts on your preparation and performance in this course. Possible points of discussion could include: homework completion, use of textbook or other resources, problem sets or projects, class attendance/attention, peer collaboration, additional instruction, etc.

2.2. *Analysis*

When completing SDDSs, students had full autonomy and could write as little or as much as they wanted. Respondent data was subsequently downloaded and analyzed employing two widely available NLP Python libraries.

Natural Language Toolkit (NLTK) and the Valence-Aware Dictionary and sEntiment Reasoner (VADER) algorithm were used to analyze word frequency and quantify the sentiment of the SDDS results.[9,5] NLTK is a Python library which we used to quickly tokenize and perform word frequency analysis on SDDS results. We used frequency analysis to identify student focus on precise pedagogical topics.

The VADER sentiment algorithm is an effective application for survey analysis with its focus on microblog or short-response-type entries.[5] Once student sentiment was measured for precise topics, we identified how students' opinions changed over time, varied with Grade Point Average (GPA), and evaluated the similarity of SDDS results with the results of a Likert-scaled survey.

3. Results

SDDSs provided insight into course structure and instruction that would not have been achievable with Likert-scaled surveys. In our results, we analyzed the word frequency, temporal changes in sentiment throughout the semester, and how sentiment relates to GPA of students. Finally, we compare the distribution of the results of SDDS to a Likert-scaled survey administered to an overlapping population.

3.1. *Word frequency*

We began by observing how frequently longer character words appeared in student responses to SDDSs. Evaluating word frequency allows instructors to determine what concepts students consider important. Figure 1 provides the frequencies of longer meaningful words students used in the survey through the three time periods (Blocks) across a semester.

After reviewing the results from the frequency analysis, it is clear that "application" and "understanding" were consistently mentioned. Subsequently, the focus of student responses on these concepts drove how we delivered and structured course content — students clearly communicated

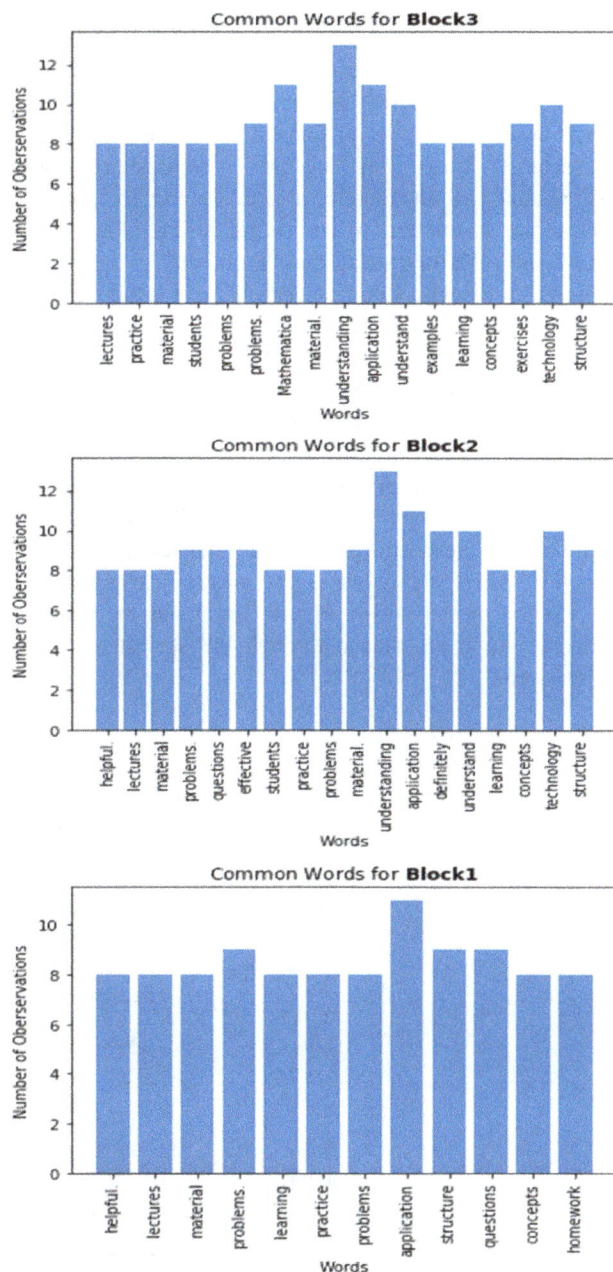

Fig. 1. Blocks 1–3: Survey results.

through SDDSs that application problems and understanding fundamental course concepts were important. Our discussions with students confirmed this conclusion. Typical Likert-scaled surveys would not have identified these areas, unless specifically referenced, while SDDSs allowed students to organically produce this information.

Additionally, it is clear that as time progressed through the semester, student responses produced an increasingly rich dataset. We observed this fact with an increase in the number of high-frequency, meaningful words after the first block. Increased robust word frequency is a positive indicator that students trust SDDSs to provide feedback.

58   *P. Kuiper & K. Hood*

3.2. *Temporal changes in sentiment*

After reviewing what concepts were organically important to students, we considered the sentiment of frequently used words and target words of interest to course stakeholders. These target words were "Lecture," "Webassign," "Quiz," "AI" (Additional Instruction: a term we use for providing one-on-one instruction to a student during office hours), "Textbook," "Mathematica," and "Problem."

To infer student opinion concerning these target words, we used a near operator, which sliced strings centered about each target. Once strings centered upon target words were isolated, we applied the VADER NLP sentiment analysis algorithm.[5] We make the assumption that by quantifying the sentiment of a string surrounding a target word, we will be able to accurately assess student opinion surrounding the precise topic described by the target. The VADER algorithm evaluated the sentiment of these strings within a subjective domain of $[-1, +1]$, where -1 indicates the most negative sentiment and $+1$ the most positive. Figure 2 provides the results of temporal sentiment change through the semester for each target word.

As the semester progressed, we observed that the sentiment towards the target word "Lecture" trended negatively while the sentiment towards "Textbook" and "Mathematica" became more positive. We believe this trend occurred because students became less reliant on in-class lecture and grew more independent, relying on outside classroom resources. These independent resources include the course textbook and technology. Since this survey was administered to freshman, this trend could also demonstrate student's confidence in the use of technology and textbooks increasing throughout the semester. As stakeholders in these calculus courses, we used this information to adjust course content and pedagogical techniques, observing that there are optimal blocks in a semester to place emphasis on external classroom resources.

3.3. *Sentiment relation to grade point average*

In order to compare the priorities of higher and lower performing students, we investigated how student GPA (across all courses) correlated with responses concerning pedagogical topics and course structure. To assess this relationship we

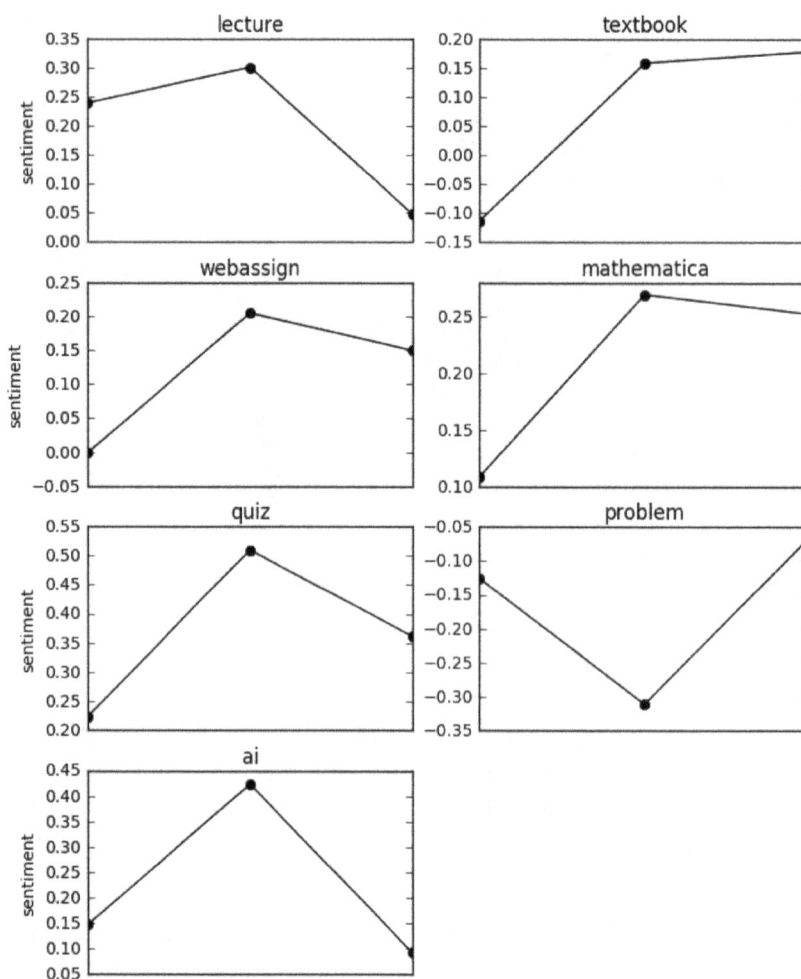

Fig. 2. Sentiment changes for specific words through three different periods in a semester.

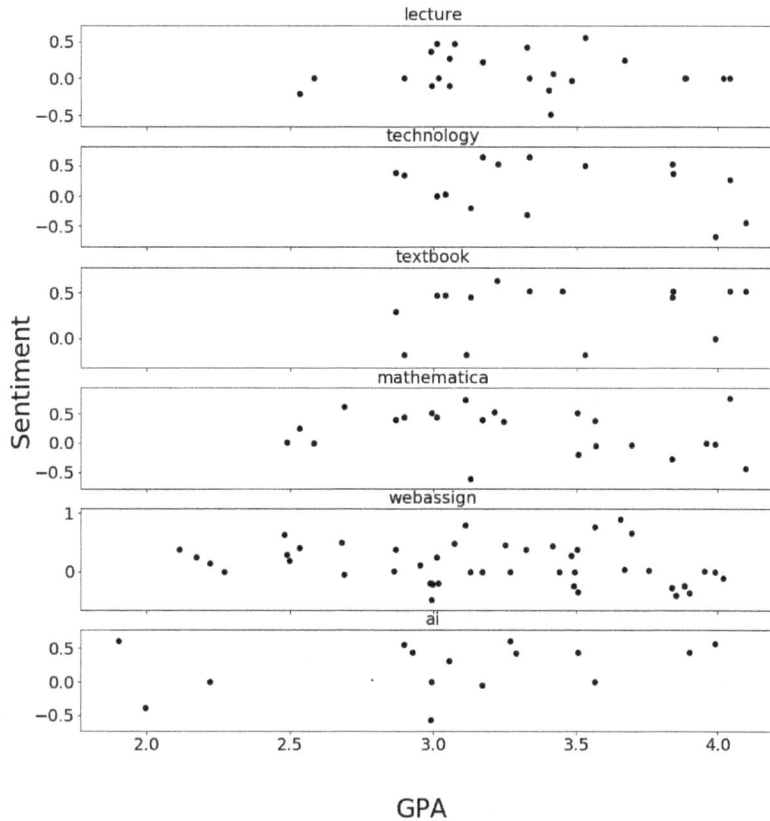

Fig. 3. Sentiment plotted with GPA for specific target words.

evaluated student sentiment towards six specific target words of interest to course stakeholders: "Lecture," "Technology," "Textbook," "Mathematica," "Webassign," and "AI." For each observation of these target words, we identified which student presented the observation, assessed the sentiment of the observation, and compared this sentiment to the student's GPA. Figure 3 provides a plot of this data for all six target words.

We initially performed linear regression and observed no significant positive or negative correlation between GPA and sentiment. We then performed a K-means clustering algorithm on the data and found that no reasonable clusters existed within the dataset. These conclusions are supported by visual inspection of the data in Fig. 3.

Despite the lack of a simple statistical structure, several trends are apparent. We found that clearly students with higher GPAs were more likely to hold opinions on the target words "Lecture," "Technology," and "Textbook" compared to students with lower GPAs. Students with GPAs of roughly 3.0 and higher displayed more sentiment, whether positive or negative, while sentiment towards "Mathematica" and "Webassign" remained evenly distributed across GPA level. Additionally, students with higher GPAs more broadly displayed interest in "AI," representing 80% of the data points, while only several students with low GPAs expressed sentiment towards "AI." These findings informed course stakeholders of what priorities higher and lower performing

students hold. With this knowledge, we tailored pedagogical methods to student performance, specifically when working with individual students.

3.4. *Comparing SDDSs to Likert-scaled surveys*

In Tables 1 and 2 we present the results of a permutation test for each target word using a rejection region of *p-stat* < 0.05. A permutation test is a statistical test employed to measure the similarity of two distributions.[10] Table 1 lists the target words whose distributions of sentiment between SDDSs and Likert surveys are similar, while Table 2 lists words where results from the two survey techniques diverge. For each target word, both tables provide the average sentiment μ and

Table 1. Similar outcomes.

Target	SDDS μ	SDDS σ^2	Likert μ	Likert σ^2	Obs.	p-Stat
Lecture	0.048	0.288	0.035	0.564	22	0.130
Technology	0.163	0.427	0.176	0.594	11	0.691
Online	0.229	0.216	0.178	0.569	6	0.886
Team	−0.361	0.00	−0.201	0.431	1	0.682
Quiz	0.361	0.214	0.367	0.539	8	0.900
AI	0.092	0.372	0.003	0.630	16	0.199
Individual	0.751	0.00	0.386	0.446	1	0.354

Table 2. Not similar outcomes.

Target	SDDS μ	SDDS σ^2	Likert μ	Likert σ^2	Obs.	p-Stat
Textbook	0.178	0.357	−0.532	0.428	38	0.000
Mathematica	0.253	0.408	−0.394	0.552	25	0.000
Problem	−0.064	0.349	0.370	0.518	63	0.000
Webassign	0.150	0.360	0.297	0.569	6	0.007

variance σ^2 (scaled within the subjectivity domain $[-1, +1]$), the number of observations (*Obs.*), and the *p*-statistic associated with the permutation test.

We observe that the distributions of sentiment measured by the Likert survey and SDDSs are statistically similar when the number of observations is low. As the number of observations increase for a target word, the distributions between the sentiment assessed by SDDSs and Likert-scaled surveys tend to diverge. We assert that when there is sufficient evidence to infer the opinion of a population using SDDSs (when enough students have mentioned a target word) the results of the SDDSs and Likert surveys often no longer agree. Given that Likert surveys often require all respondents to provide an opinion on all subjects, there is undoubtedly a significant amount of noise in the measurement of student's responses. Subsequently, we suggest that given a divergence between distributions and sufficient observation size, SDDSs are more likely to infer an accurate opinion of a population when compared to Likert surveys.

4. Conclusion and Future Work

Likert-scaled surveys often provide inaccurate results due to their length and the uniform weights applied to all data points.[1,2] SDDSs, or student surveys with open-ended, free text questions and results analyzed using NLP techniques, have the potential to provide educators with insights not produced by Likert-type surveys. We found that SDDS provides actionable feedback for educators to design effective courses from a structural and pedagogical perspective.

Future analysis includes closely examining the divergence between SDDSs and Likert-scaled surveys with a variety of statistical techniques. Additionally, we will investigate extending SDDSs beyond the academic field as a more general survey technique. For example free text, open-ended questions analyzed with NLP could be used after medical appointments, in consumer products reviews, economic surveys, or to gauge population sentiment in political affairs and military engagements.

References

[1] O. Friborg, M. Martinussen and J. H. Rosenvinge, Likert-based vs. semantic differential-based scorings of positive psychological constructs: A psychometric comparison of two versions of a scale measuring resilience, *Personal. Individ. Differ.* **40**, 873 (2006).

[2] G. Albaum, The Likert scale revisited: An alternate version, *J. Market Res. Soc.* **39**, 331 (1997).

[3] The Economist, Language: Finding a voice (2017), https://www.economist.com/technology-quarterly/2017-05-01/language.

[4] B. Lui, *Sentiment Analysis and Opinion Mining* (Morgan & Claypool, 2012).

[5] C. J. Hutto and E. Gilbert, VADER: A parsimonious rule-based model for sentiment analysis of social media text, *Proc. Eighth Int. AAAI Conf. Weblogs and Social Media* (2014).

[6] S. Bird, E. Klein and E. Loper, *Natural Language Processing with Python: Analyzing Text with the Natural Language Toolkit* (O'Reilly Media, Inc., 2009).

[7] E. Dey, Working with low survey response rates: The efficacy of weighting adjustments, *Res. Higher Educ.* **38**, 215 (1997).

[8] S. R. Porter, M. E. Whitcomb and W. H. Weitzer, Multiple surveys of students and survey fatigue, *New Dir. Inst. Res.* **2004**, 63 (2004).

[9] O'Reilly Media, Inc., Natural Language Toolkit (2017), https://www.nltk.org/.

[10] D. Moore, *The Practice of Business Statistics: Using Data for Decisions*, Chapter (W. H. Freeman, 2003), pp. 14–55.

A semantic recommendation system for cancer-related articles

Charles C. N. Wang[*,‡], Yun-Lung Chung[†], I-Seng Chang[*]
and Jeffrey J. P. Tsai[*]

*Department of Bioinformatics and Medical Engineering
Asia University, 500, Lioufeng Road
Wufeng, Taichung 41354, Taiwan, R. O. China
†Research Assistant Center
Show Chwan Memorial Hospital
Changhua 500, Taiwan, R. O. China
‡chaoneng.wang@gmail.com

There have been an enormous number of publications on cancer research. These unstructured cancer-related articles are of great value for cancer diagnostics, treatment, and prevention. The aim of this study is to introduce a recommendation system. It combines text mining (LDA) and semantic computing (GloVe) to understand the meaning of user needs and to increase the recommendation accuracy.

Keywords: Cancer; semantic computing; text mining; topic model; GloVe, recommendation system; human–computer interaction.

1. Introduction

Cancer is the leading cause of mortality and morbidity worldwide according to the International Agency for Research on Cancer.[1] It is one of the most important study areas for biomedical researchers, and it has been widely studied for more than 100 years. The huge body and rapid growth of text on cancer research provides a valuable resource.[2] As shown in Fig. 1, there are many publications on cancer research and the number of publications keeps increasing every year. We searched PubMed with "cancer" in the abstract and retrieved more than 925,648 publications in this area from 2011 to 2016. It is almost impossible for people to read all of these publications and discover new knowledge. Text mining can help researchers to mine information and knowledge from a mountain of text and it is now widely applied in biomedical research. Many researchers have taken advantage of text mining technology to discover novel knowledge to improve the development of biomedical research especially those pertaining to malignant diseases, such as cancer.[2]

Text mining technology provides a solution for bridging the knowledge gap between free-text and structured representation of related information in cancer research.[2,3] It employs many computational technologies, such as machine learning, natural language processing, and deep learning, to find new exciting outcomes hidden in unstructured, cancer-related articles. There are many applications of text mining on cancer-related articles, such as identifying malignant tumor-related biomedical mentions, finding relationships among biomedical entities such as protein–protein interactions, gene–disease networks, etc., and extracting knowledge from text and generating hypotheses.[2]

There have been a number of text mining applications specifically focusing on extracting cancer-related information; Spasić et al.[3] provide a comprehensive review of these. The review highlights a strong bias towards symbolic techniques, i.e., the use of pattern matching for cancer-related entity extraction. Pletscher–Frankild et al.[4] present a system that uses text mining for extracting disease–gene associations from biomedical abstracts.

The system consists of a highly efficient dictionary-based taggers for named entity recognition of genes and diseases, and it combines a scoring scheme that takes into account co-occurrences both within and between sentences. Gonzalez et al.[5] present an overview of the fundamental methods for text and data mining, as well as recent advances and emerging applications toward precision medicine. Baker et al.[6] provide an extensive Hallmarks of Cancer taxonomy and develop an automatic text mining methodology of cancer-related articles from PubMed into the taxonomy. It offers a great potential to organize and correctly classify cancer-related articles.

In this research, we explore the conceptual content of cancer research based on the abstracts of articles that are extracted from PubMed.

---

‡Corresponding author.

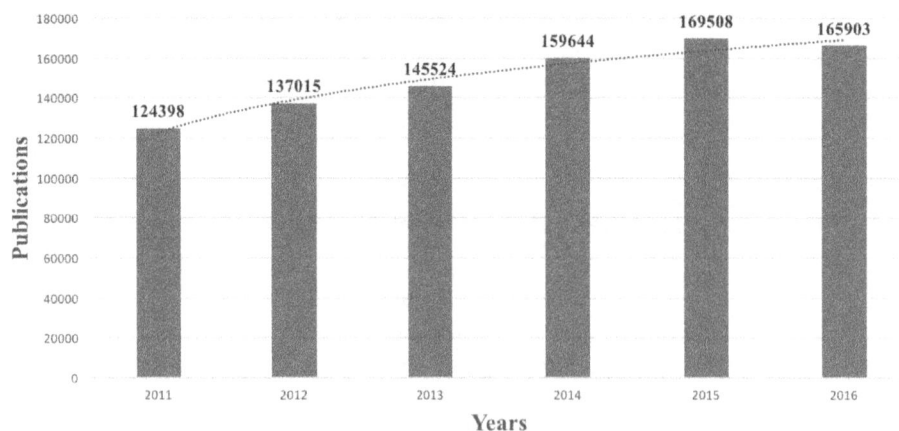

Fig. 1. Number of cancer papers in PubMed search with the keyword "cancer."

2. Methods

2.1. *Semantic computing-based recommendation*

Semantic computing-based recommendation is a computer recommendation system that can recommend articles after processing review data collected from PubMed and extracting knowledge when the user employs exceptional vocabulary expressions other than the predefined input vocabularies. Figure 2 shows the overall systems architecture.

For example, a data crawler collects unstructured data over PubMed by using the keywords "("neoplasms" [MeSH Terms] OR "neoplasms" [All] OR "cancer" [All]) AND ("2011/01: 2016/12/[DP]")" about a cancer-related research topic.

The recommendation knowledge extractor converts the dataset collected by the data crawler into a structured format. The topic model [Latent Dirichlet Allocation (LDA)] and Global Vectors (GloVe) methods are used to extract recommendation knowledge from the converted data, which is stored in the knowledge base. The knowledge base is updated autonomously without additional user events when the data crawler collects data.

Once the knowledge base is constructed and user inputs an exceptional expression that cannot be understood by the computer using predefined vocabularies, the system conducts top–down and bottom–up searches to infer related services. They are recommended to the user first, where the recommendation is provided by the computer to the human user in the form of speech.

2.2. *Data crawler*

The data crawler collects data to extract domain knowledge from PubMed. Data are retrieved and downloaded from PubMed (http://www.ncbi.nlm.nih.gov/pubmed/) that provides free access to biomedical journal citations and abstracts mainly indexed by Medline.[7] The service is administered by the National Center for Biotechnology Information of the United States National Library of Medicine. Search terms such as "("neoplasms" [MeSH Terms] OR "neoplasms" [All] OR "cancer" [All]) AND ("2011/01: 2016/12/[DP]")" are used in the search. If we limit the publication date to be between 2011 and 2016, a total of 925,648 articles are retrieved.

2.3. *Word clouds*

Word clouds display the words that frequently occur in a text. These visualizations are particularly useful when one has no preconceived idea of which concepts should occur in a text.

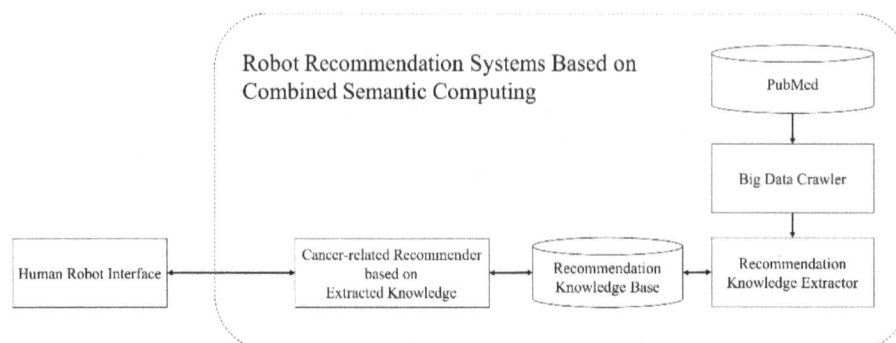

Fig. 2. Architecture of a computer recommendation system.

In word clouds, words that appear more frequently in a text are printed in larger fonts than words that occur less often. The advantage of word clouds is that this visualization is not biased by the use of a predefined set of concepts or an ontology, but is driven by the raw content of the text. As such, they can provide new ideas and insights on a particular concept and can function as a starting point for more specific search.[8]

2.4. Topic model

To incorporate content analysis into cancer-related articles, text mining techniques are applied. Topic-modeling techniques are mostly adapted to identify the topics of a subject area while analyzing that area more abundantly.[9] The Latent Dirichlet Allocation (LDA) is the best received topic-modeling technique. LDA is a populistic generative model that assumes each document is a mixture of latent topics, where the probability distribution of each topic is over all words in the vocabulary. LDA is a three-level hierarchical Bayesian model, in which each item of a collection is modeled as a finite mixture over an underlying set of topics.[10] It treats each document as a mixture of topics, and each topic as a mixture of words. LDA assumes the following generative process for each document W in a corpus D:

(1) Choose a multinomial distribution θ_d for document d, $\text{Dir}(\alpha)$, where $d \in \{1, \ldots, W\}$ and $\text{Dir}(\alpha)$ is a Dirichlet distribution with a symmetric parameter α which typically is spares ($\alpha < 1$).
(2) Choose a multinomial distribution φ_t for topic t, $\text{Dir}(\beta)$, where $k \in \{1, \ldots, T\}$ and β is a Dirichlet distribution parameter.
(3) For each of the work positions i, j, where $j \in \{1, \ldots, N_i\}$ and $i \in \{1, \ldots, W\}$,

    (a) select a topic $z_{i,j}$ from (θ_i);
    (b) select a word $w_{i,j}$ from $(\varphi_{z_{i,j}})$.

In the above generative process, the words in a document are the only observed variables while others are latent variables (θ and φ) and hyper parameters (α and β). In order to infer the latent variables and hyper parameters, the probability of the observed data D is computed and maximized as follows:

$$p(D|\alpha, \beta) = \prod_{i=1}^{W} \int p(\theta_i|\alpha)$$
$$\times \left(\sum_{j=1}^{N_i} p(z_{i,j}|\theta_i) p(w_{i,j}|z_{i,j}, \varphi) p(\varphi|\beta) \right) d\theta_i d_p.$$

In this study, we use Gibbs sampling[11] to estimate the LDA parameters. Gibbs sampling is a Monte Carlo Markov-chain algorithm, which is powerful in statistical inference and a method of generating a sample from a joint distribution when only conditional distributions of each variable can be efficiently computed. Gibbs sampling has been used widely for LSA parameters estimate.[12]

2.5. Learning word embedding

We download the abstracts of 925,648 articles from PubMed. The GloVe for the word representation models are trained based on cancer-related articles. GloVe compute continuous vector representations of words based on neural networks. They take a corpus as the input and produce a vector space. They estimate continuous vector representations of words from a large corpus. Each unique word in the corpus is represented as a vector and the words that share common contexts are positioned closely to each other. The similarity of two words is determined by calculating the cosine similarity of two vectors that represent the two words.[13] The intuitive idea behind the GloVe model is to build a big matrix, X, of the co-occurring words from a corpus. Each cell of the matrix, X_{ij}, represents how many times the row word, i, appears in some context, j. By doing a simple normalization of the values for each row of the matrix, we can obtain the probability distribution of every context given a word. Once the probabilities are calculated, the relationship between words can be calculated by making the ratio between the probabilities of the context given those two words. Also, we give less weight to more distant words, using the below formula:

$$\text{decay} = \frac{1}{\text{offset}}.$$

We define a soft constraint for each word pair:

$$w_i^T w_j + b_i + b_j = \log(X_{ij}).$$

Here w_i is a vector for the main word, w_j is a vector for the context word, and b_i and b_j are scalar biases for the main and context words, respectively. Let us define a cost function

$$J = \sum_{i=1}^{V} \sum_{j=1}^{V} f(X_{ij})(w_i^T w_j + b_i + b_j - \log X_{ij})^2.$$

Here, f is a weighting function that helps us to prevent learning only from extremely common word pairs. The GloVe authors choose the following function:

$$f(X_{ij}) = \begin{cases} \left(\dfrac{X_{ij}}{x_{\max}} \right)^{\alpha} & \text{if } X_{ij} < X_{\text{MAX}}, \\ 1 & \text{otherwise.} \end{cases}$$

3. Results

Of the 925,648 publications for cancer published between 2011 and 2016, Table 1 lists the journals that were published between 2011 and 2016. The top five journals are *PLoS ONE*, *Oncotarget*, *Asia Pac. J. Cancer Preve.*, *Tumor Biol.*, and

Table 1. Top 10 journals that published cancer-related articles between 2011 and 2016.

Title (five-year bioxbio journal impact*)	Frequency	Normalized
PLoS ONE (2.86)	22,543	0.0247
Oncotarget (5.07)	9390	0.2409
Asia Pac. J. Cancer Prev. (1.82)	6616	0.1809
Tumor Biol. (3.33)	6364	0.3056
J. Clin. Oncol. (16.80)	5235	0.1874
Ann. Surg. Oncol. (3.91)	4683	0.2255
Int. J. Cancer (6.77)	4252	0.2851
Sci. Rep. (4.31)	3936	0.0501
Oncogene (6.61)	3012	0.2950
BMC Cancer (3.31)	2969	0. 2086

*Five-year impact factors were based on the 2016 journal citation reports.

J. Clin. Oncol., with counts of 22,543, 9390, 6616, 6364, and 5235, respectively. The five-year impact factors of the journals are ranged from 2.86 to 16.80, with a median of 9.83. The top five most studied cancer types are breast cancer (23.82), lung cancer (10.54), prostate cancer (9.90), rectal cancer (8.44), and ovarian cancer (4.44) (Table 2).

3.1. Word cloud

The term frequencies in the abstracts of the 925,648 articles were visualized as a word cloud where a larger word size represents a higher frequency of appearance among the articles, as shown in Fig. 3. The Top five terms with the highest frequencies over different periods and over the entire period of 2011 to 2016 are "patients," "cancer," "cell," "tumor," and "study." The five terms were then suppressed in the display of the word cloud to allow a better visualization of the remaining terms. The words that ranked the sixth to the tenth in frequency were "expression," "treatment," "survival," "associated," and "clinical."

Table 2. Cancer types in the cancer-related articles between 2011 and 2016 (*N* = 925, 648).

Cancer type	Frequency (%)
Breast cancer	220,523 (23.82)
Lung cancer	97,569 (10.54)
Prostate cancer	91,662 (9.90)
Rectal cancer	78,156 (8.44)
Ovarian cancer	41,062 (4.44)
Cervical cancer	34,451 (3.72)
Pancreatic cancer	31,326 (3.38)
Colon cancer	28,631 (3.09)
Bladder cancer	25,645 (2.77)
Thyroid cancer	15,654 (1.69)
Skin cancer	10,800 (1.17)
Liver cancer	9868 (1.07)

Fig. 3. Plots of terms in the word clouds of highest frequencies from the abstracts of 925,648 cancer-related articles.

3.2. Analysis of high word frequency association

Table 3 presents the results of the analysis of the keyword association for the top 15 most frequently occurred keywords in the cancer-related articles' abstracts. The top three most frequently occurred keywords in the abstracts of the 925,648 articles are "patients," "cancer," and "cell" with 1,445,688, 1,284,140, and 676,924 times, respectively. The remaining 12 keywords have a frequency of occurrence ranged from 590,685 times for "tumor" to 244,349 times for "significantly." Each of the primary keywords is evaluated for its correlations with others in the abstract (secondary keywords). Overall, the correlation coefficients ranged from the highest at 0.52 for survival-overall to the lowest at 0.10 for clinical-benefit and tumor-vitro.

3.3. Topic model

We employ topic modeling to explore the topics as a bunch of words in the abstracts. The topic-modeling analysis on the cancer-related articles is presented in Table 4. As can be seen from the table, we find that the dominant words "cancer" and "patient" are in these texts. In addition, there is a meaningful difference between these collections of words, e.g., "cancer study," "patient treatment," "survival of patients," "risk of patients," "cancer type," and "cancer cells study on genome" in the topics. The topic-modeling process identifies groupings of terms that we can understand as human readers of these description fields.

3.4. Cancer-related articles on GloVe

We adopt GloVe to compute additional vector representations of words in cancer-related articles. Table 5 shows that the top 15 most frequently occurred keywords and the most similar words. The GloVe identify that "cancer" is similar to breast, colorectal, lung, prostate, ovarian, gastric, cervical bladder, pancreatic, and carcinoma; and "patient" is similar to case, treated, enrolled, diagnosed, surgery, undergoing, received, and advance. As "gene expression" is the process by which genetic instructions are used to synthesize gene products, we

Table 3. Analysis of high word frequency association of the corpus for the top 10 most frequently occurring word in the abstract.

Keyword	Frequency	The secondary keyword with highest correlation with the keyword (correlation coefficient)
patients	1,445,688	median (0.36), months (0.31), treated (0.27), rate (0.23), retrospectively (0.23), age (0.22), surgery (0.22), respectively (0.21), chemotherapy (0.21), retrospective (0.21)
cancer	1,284,140	prostate (0.24), lung (0.19), women (0.16), colorectal (0.16), association (0.15), gastric (0.15), colon (0.14), screening (0.13), incidence (0.13), ovarian (0.13), increased (0.12), studies (0.12)
cell	676,924	lines (0.49), proliferation (0.34), line (0.31), cycle (0.29), apoptosis (0.26), growth (0.25), human (0.24), induced (0.21), inhibited (0.21), arrest (0.21), migration (0.21) inhibition (0.21), vitro (0.20), viability (0.20), squamous (0.19), carcinoma (0.18), protein (0.18), stem (0.17), death (0.16), role (0.16)
tumor	590,685	tumors (0.29), growth (0.22), cells (0.17), microenvironment (0.17), mice (0.15), vivo (0.15), metastasis (0.14), suppressor (0.14), antitumor (0.14), progression (0.13), angiogenesis (0.13), xenograft (0.12), vitro (0.10)
study	537,961	aim (0.23), significant (0.16), compared (0.15), age (0.15), purpose (0.15), evaluate (0.15), enrolled (0.15), retrospective (0.15), prospective (0.15), groups (0.14), respectively (0.13), group (0.13), case-control (0.13), assess (0.12), mean (0.12), women (0.11), rate (0.10)
expression	529,424	protein (0.30), mRNA (0.30), gene (0.29), immunohistochemistry (0.27), cell (0.26), correlated (0.25), tissues (0.25), RT-PCR (0.21), upregulated (0.20), overexpression (0.20), downregulated (0.19), role (0.19), human (0.19), proliferation (0.19), factor (0.17), downregulation (0.17)
treatment	501,268	therapy (0.21), treated (0.21), chemotherapy (0.17), efficacy (0.16), radiotherapy (0.15), options (0.15), effective (0.14), combination (0.13), toxicity (0.11), planning (0.11)
survival	356,436	overall (0.52), year (0.34), prognostic (0.33), median (0.32), Kaplan-Meier (0.30), progression free (0.30), disease-free (0.29), months (0.27), multivariate (0.27), stage (0.23), hazard (0.22), independent (0.21), respectively (0.19), chemotherapy (0.19), proportional (0.16), time (0.13)
associated	333,354	multivariate (0.17), odds (0.17), factors (0.16), increased (0.15), regression (0.15), confidence (0.15), polymorphisms (0.15), logistic (0.14), age (0.13), independently (0.13), case-control (0.13), SNP (0.13), ratio (0.12), genotyped (0.12)
clinical	330,148	trial (0.37), practice (0.16), outcome (0.12), preclinical (0.21), review (0.11), therapy (0.11), features (0.11), data (0.10), benefit (0.10)
risk	310,501	association (0.28), case-control (0.27), factors (0.26), confidence (0.23), increased (0.20), associations (0.20), polymorphisms (0.20), women (0.18), controls (0.17), age (0.16), developing (0.16), interval (0.16), population (0.15), regression (0.15), intake (0.15), meta-analysis (0.15), genotype (0.15), history (0.14), ratio (0.13), men (0.12)
breast	309,932	women (0.24), estrogen (0.19), mammography (0.15), mastectomy (0.15), postmenopausal (0.15), Triple-Negative (0.14), ductal (0.13), cancer (0.12), invasive (0.12), ER-positive (0.12), auxillary (0.12), BRCA1 (0.12), mammographic (0.12), tamoxifen (0.12)
analysis	283,821	multivariate (0.35), univariate (0.27), independent (0.20), prognostic (0.20), microarray (0.15), regression (0.15), Kaplan-Meier (0.15), data (0.14), revealed (0.14), identified (0.12), statistical (0.12), quantitative (0.12), gene (0.12), predictor (0.10)
disease	272,925	stage (0.15), median (0.14), progression (0.14), progressive (0.13), advanced (0.13), chronic (0.12), metastatic (0.11)
significantly	244,349	higher (0.28), compared (0.24), groups (0.20), increased (0.19), lower (0.18), incidences (0.17), correlated (0.16), levels (0.16), weights (0.16), decreased (0.15), controls (0.15), females (0.15), significant (0.14), males (0.14), fragrance (0.11)

Table 4. The topic of LDA with six topics.

Topic 1		Topic 2		Topic 3		Topic 4		Topic 5		Topic 6	
Cancer/Study		Patient/Treatment		Cells/Genome		Breast Cancer		Patient/Risk		Patient/Survival	
cancer	0.0205	patient	0.0248	cells	0.0198	cancer	0.0233	patient	0.0346	patient	0.0360
patient	0.0178	cancer	0.0088	cancer	0.0163	patient	0.0097	cancer	0.0170	survival	0.0077
tumor	0.0107	using	0.0056	expression	0.0135	risk	0.0087	study	0.0071	treatment	0.0073
study	0.0085	treatment	0.0054	tumor	0.0070	study	0.0064	risk	0.0068	tumor	0.0063
data	0.0052	study	0.0051	protein	0.0054	breast	0.0054	survival	0.0066	cancer	0.0056
therapy	0.0043	clinical	0.0051	human	0.0054	studies	0.0050	treatment	0.0062	study	0.0052
activity	0.0041	tumor	0.0048	gene	0.0050	women	0.0049	group	0.0053	group	0.0051
expression	0.0041	higher	0.0041	growth	0.0045	data	0.0047	years	0.0044	cases	0.0048
treatment	0.0037	significantly	0.0040	activity	0.0041	associated	0.0046	disease	0.0044	clinical	0.0043
tumors	0.0036	cases	0.0040	study	0.0040	using	0.0039	associated	0.0042	months	0.0042

Table 5. The top 15 most frequently occurred keywords and the most similar words according to GloVe.

Keyword	Top 15 most frequently occurred keywords' most similar words
patients	receiving, cases, treated, enrolled, diagnosed, surgery, undergoing, received, advanced, primary
cancer	breast, colorectal, lung, prostate, ovarian, gastric, cervical, bladder, pancreatic, carcinoma
cell	proliferation, migration, lines, apoptosis, growth, human, viability, stem, vitro, differentiation
tumor	metastasis, metastatic, invasion, growth, size, tissue, malignant, lesion, cell, differentiation
study	conducted, present, retrospective, prospective, carried, examined, evaluated, investigate, aim, review
expression	mRNA, overexpression, downregulation, upregulation, protein, miRNA, levels, furthermore, gene, correlated
treatment	therapy, chemotherapy, adjuvant, regimens, therapies, neoadjuvant, combination, options, radiotherapy, effective
survival	OS (overall survival), overall, disease-free, DFS (disease-free survival), PFS (progression-free survival), progression-free, recurrence-free, disease-specific, outcome, CSS (cause-specific Survival)
associated	related, factors, risk, worse, significantly, correlated, increased, poor, significant, occurrence
clinical	evaluation, outcome, regarding, prognosis, practice, data, review, implications, preclinical, diagnostic
risk	mortality, incidence, associated, occurrence, prevalence, odds, among, predictors, adjusted, disease
breast	endometrial, prostate, ovarian, lung, bladder, colorectal, thyroid, melanoma, metastatic, cervical
analysis	multivariate, Multivariate, Cox, univariate, regression, univariate, logistic, Kaplan-Meier, statistical, determined
disease	failure, relapse, chronic, recurrence, malignancy, progression, metastases, diagnosis, risk, diabetes
significantly	significantly, decreased, reduced, markedly, increased, whereas, higher, correlated, compared, lower

try the words that are most similar to the word "expression." The word "tumor" is similar to metastasis, metastatic, invasion, growth, size, tissue, malignant, lesion, cell, and differentiation. These similar words all describe the morphology of tumor. The word "risk" means risk factor. We identify similar words mortality, incidence, associated, occurrence, prevalence, odds, among, predictors, adjusted, and disease.

We show that text mining findings are reliable, as per the PubMed scale, in that the word–word relationships inferred from the literature-wide findings are similar.

3.5. *Cancer-related recommender based on extracted knowledge*

The recommender uses two search methods to recommend cancer-related articles.

The first search method searches a topic that includes user-mentioned terms within the topic model. Then, the term values in the topic are compared based on a matrix for all cancer-related articles to find the term with the highest mean value. The cancer-related articles with the highest term value of the corresponding term are recommended. The second search method searches for similar words according to GloVe. It takes a text corpus as the input and produces the word vectors as the output. It first constructs a vocabulary from the training text data and then learns the vector representation of the words. The resulting word vector file can be used as the features in many natural language processing systems. A simple way to investigate the learned representations is to find the closest words for a user-specified word. A distance tool serves that purpose.

Consider, for example, given the query:

I would like to find articles related to "breast cancer"

(a) We employ topic modeling to explore the topics as a bunch of words in the abstracts. The topic-modeling

analysis on the cancer-related articles is presented in Table 4. In this example, The topic model will find "breast cancer" in topic 4 which is the priority recommendation.

(b) We adopt GloVe to compute additional vector representations of keyword "breast cancer" in cancer-related articles. Table 5 shows that the "breast" and "cancer" are the most frequently occurred keywords and most similar words. The GloVe identify "cancer" types, such as breast, colorectal, lung, prostate, ovarian, gastric, cervical bladder, pancreatic, and carcinoma, and they capture attribute-related relatedness, such as "breast" similarities for receiving, endometrial, prostate, ovarian, lung, bladder, colorectal, thyroid, melanoma, metastatic, and cervical. We show that the semantic computing findings are reliable, as per the PubMed scale, in that the word–word relationships inferred from the literature-wide findings are similar.

The intersection of the Topic model and GloVe model result in re-construction of cancer-related articles (4095 papers) that are the priority recommendation. Finally, the user can fill out a follow-up survey to feedback the satisfaction and accuracy of the recommendation.

4. Conclusion

In this study, we use various approaches, including topic model, word embedding, and word cloud, to visualize the content of 925,648 cancer-related articles published between 2011 and 2016. We find that breast cancer, lung cancer, prostate cancer, rectal cancer, and ovarian cancer are the top five most studied cancers based on the dataset. In the word cloud visualization, we find that survival of patients, the risk of patients, and cancer cell have the highest frequencies from 2011 to 2016. In the high word frequency association analysis, we find that the top three most frequently occurred keywords in the abstracts of the 925,648 articles are patients, cancer, and cell with 1,445,688, 1,284,140, and 676,924 times, respectively.

This study applies LDA text mining on a vast amount of cancer-related articles. We adopt GloVe to compute additional vector representations of words. We use the Euclidean distance between two word vectors to provide an effective method for measuring the linguistic or semantic similarity of the corresponding words. The result suggests cancer types such as breast, colorectal, lung, prostate, ovarian, gastric, cervical bladder, pancreatic and carcinoma.

In summary, an integrated text mining and semantic computing method is developed to recognize the semantic meaning of a user order and make recommendations based on a corpora created from a large amount of unstructured data over the PubMed literature. In the future research, we will evaluate user satisfaction about the recommendations.

Our study shows that a topic model can accomplish the task of clustering cancer-related articles. Furthermore, each topic is interpreted as a probability distribution over words. We believe that topic model is a promising method with numerous applications for biomedical research. The GloVe result shows that the text mining findings are reliable, as per the PubMed scale, in that the word–word relationships inferred from the literature findings are similar.

Acknowledgments

I-Seng Chang, Charles C. N. Wang, and Jeffrey J. P. Tsai are supported in part under the Grant Nos. MOST 106-2221-E-468-021 and MOST 105-2631-E-468-002 from the Ministry of Science and Technology, Taiwan, and Asia University.

References

[1] B. W. Stewart and C. P. Wild, *World Cancer Report 2014* (IARC Publications, Lyon, 2014).

[2] F. Zhu, P. Patumcharoenpol, C. Zhang, Y. Yang, J. Chan, A. Meechai, W. Vongsangnak and B. Shen, Biomedical text mining and its applications in cancer research, *J. Biomed. Inf.* **46**, 200 (2013).

[3] I. Spasić, J. Livsey, J. A. Keane and G. Nenadić, Text mining of cancer-related information: Review of current status and future directions, *Int. J. Med. Inf.* **83**, 605 (2014).

[4] S. Pletscher-Frankild, A. Pallejà, K. Tsafou, J. X. Binder and L. J. Jensen, DISEASES: Text mining and data integration of disease–gene associations, *Methods* **74**, 83 (2015).

[5] G. H. Gonzalez, T. Tahsin, B. C. Goodale, A. C. Greene and C. S. Greene, Recent advances and emerging applications in text and data mining for biomedical discovery, *Brief Bioinform.* **17**, 33 (2015).

[6] S. Baker, I. Ali, I. Silins, S. Pyysalo and Y. Guo, 2017, "Cancer Hallmarks Analytics Tool (CHAT): A text mining approach to organize and evaluate scientific literature on cancer, *Bioinformatics* **33**, 3973 (2017).

[7] M. Castillo, Is your journal indexed in MEDLINE? *Am. J. Neuroradiol.* **32**, 1 (2011).

[8] W. W. Fleuren and W. Alkema, Application of text mining in the biomedical domain, *Methods* **74**, 97 (2015).

[9] E. Yan, Research dynamics, impact, and dissemination: A topic-level analysis, *J. Assoc. Inf. Sci. Technol.* **66**, 2357 (2015).

[10] D. M. Blei, A. Y. Ng and M. I. Jordan, Latent Dirichlet allocation, *J. Mach. Learn. Res.* **3**, 993 (2003).

[11] T. L. Griffiths and M. Steyvers, Finding scientific topics, *Proc. Natl. Acad. Sci. USA* **101**, 5228 (2004).

[12] H. Jelodar, Y. Wang, C. Yuan and X. Feng, Latent Dirichlet Allocation (LDA) and Topic modelling: models, applications, a survey, arXiv:1711.04305 [cs.IR].

[13] J. Pennington, R. Socher and C. D. Manning, GloVe: Global vectors for word representation, *Proc. 2014 Conf. Empirical Methods in Natural Language Processing* (2014), pp. 1532–1543.

Rapid qualification of mereotopological relationships using signed distance fields

René Schubotz*, Christian Vogelgesang and Dmitri Rubinstein[†]

German Research Center for Artificial Intelligence (DFKI)
Stuhlsatzenhausweg 3, Saarland Informatics Campus 3-2
66123 Saarbrücken, Germany
*rene.schubotz@dfki.de
[†]dmitri.rubinstein.@dfki.de

Although mereotopological relationship theories and their qualification problems have been extensively studied in $\mathbb{R}^2$, the qualification of mereotopological relations in $\mathbb{R}^3$ remains challenging. This is due to the limited availability of topological operators and high costs of boundary intersection tests. In this paper, a novel qualification technique for mereotopological relations in $\mathbb{R}^3$ is presented. Our technique rapidly computes RCC-8 base relations using precomputed signed distance fields, and makes no assumptions with regards to complexity or representation method of the spatial entities under consideration.

Keywords: RCC-8; spatial qualification; signed distance field; mereotopology.

1. Introduction

Qualitative spatial reasoning (QSR) is concerned with knowledge representations of spatial entities and their configurations as well as with reasoning and query processing mechanisms for such spatial configurations. It has applications in areas such as robotic navigation,[1] scene understanding[2] or geographic information systems.[3] During the last two decades a plethora of QSR calculi has been proposed, perhaps the most well known being the region connection calculus (RCC), a theory which uses an axiomatic framework to describe spatial entities by their pairwise mereotopological relations. Howsoever and regardless of the peculiarities of its diverse calculi, QSR research is invariably concerned with the following general problems:

(**P.1**) Representation: How to qualitatively represent spatial entities and their relationships?

(**P.2**) Reasoning: How to infer new relationships from a finite set of qualitative spatial relationships by.

(**P.3**) Qualification: What kind of techniques to apply for detecting qualitative spatial relationships from quantitative spatial descriptions.

In the scope of this paper, we are concerned with the problem of *Qualification* (**P.3**) and wish to automatically turn quantitative descriptions *of three-dimensional spatial configurations* into qualitative descriptions composed of RCC-8 base relations.

For this purpose, a novel qualification technique for mereotopological relations in three-dimensional spatial configurations is presented. Our approach rapidly computes

RCC-8 base relations using signed distance fields as lookup data structures, and it has three main advantages:

(**A.1**) No assumption is made with regards to the 3D representation techniques of spatial entities.

(**A.2**) Complex spatial entities and configurations can be represented to any specified resolution.

(**A.3**) Determining the intersection of the boundaries, interiors and exteriors of two spatial entities is extremely fast and independent of the complexity of the spatial entities.

The remainder of this paper is organized as follows. Section 2 introduces some preliminaries on mereotopological relationship theories. Next, the related literature on mereotopological qualification of spatial configurations in $\mathbb{R}^3$ is discussed in Sec. 3. Section 4 explains the concept of signed distance fields. We propose our novel qualification algorithm using signed distance fields in Sec. 5, and report on implementation and performance results in Secs. 6 and 7. We conclude with summary in Sec. 8.

2. Mereotopological Relationship Theories

As a framework for the definition of mereotopological relations between spatial entities embedded in $\mathbb{R}^n$, some prominent QSR theories specify a finite number of jointly exhaustive and pairwise disjoint (JEPD) binary relations in terms of the topological operators *interior* $\cdot^o$, *boundary* $\partial \cdot$ and *exterior* $\cdot^e$ along with the intersection operation from set theory.[4]

---

*Corresponding author.

Fig. 1. RCC-8 qualification of *Joe's Ice Cream* (left) results in *externally Connected (Joe, Ice)*, *disconnected (Joe, Bucket)* and *disconnected (Ice, Bucket)*. RCC-8 qualification of *Armadillos* (right) results in *partiallyOverlapping(A,B)*.

For example, the 4-Intersection Model (4IM)[5] defines the possible relations between two spatial entities $\mathcal{P}_1, \mathcal{P}_2 \subset \mathbb{R}^n$ in terms of the intersections of $\mathcal{P}_1$'s boundary $(\partial\mathcal{P}_1)$ and interior $(\mathcal{P}_1^o)$ with the boundary $(\partial\mathcal{P}_2)$ and interior $(\mathcal{P}_2^o)$ of B. The existence of intersections is registered in $4\text{IM}(\mathcal{P}_1, \mathcal{P}_2) \in \mathbb{B}^{2\times2}$, a matrix that precisely describes the topological relation between $\mathcal{P}_1$ and $\mathcal{P}_2$:

$$4\text{IM}(\mathcal{P}_1, \mathcal{P}_2) = \begin{bmatrix} \mathcal{P}_1^o \cap \mathcal{P}_2^o & \mathcal{P}_1^o \cap \partial\mathcal{P}_2 \\ \partial\mathcal{P}_1 \cap \mathcal{P}_2^o & \partial\mathcal{P}_1 \cap \partial\mathcal{P}_2 \end{bmatrix}.$$

The 9-Intersection Model (9IM)[6] extends the 4IM by considering interior and boundary with respect to the other object's exterior:

$$9\text{IM}(\mathcal{P}_1, \mathcal{P}_2) = \begin{bmatrix} \mathcal{P}_1^o \cap \mathcal{P}_2^o & \mathcal{P}_1^o \cap \partial\mathcal{P}_2 & \mathcal{P}_1^o \cap \mathcal{P}_2^e \\ \partial\mathcal{P}_1 \cap \mathcal{P}_2^o & \partial\mathcal{P}_1 \cap \partial\mathcal{P}_2 & \partial\mathcal{P}_1 \cap \mathcal{P}_2^e \\ \mathcal{P}_1^e \cap \mathcal{P}_2^o & \mathcal{P}_1^e \cap \partial\mathcal{P}_2 & \mathcal{P}_1^e \cap \mathcal{P}_2^e \end{bmatrix}.$$

Based on $9\text{IM}(\mathcal{P}_1, \mathcal{P}_2) \in \mathbb{B}^{3\times3}$, eight qualitative spatial relations between $\mathcal{P}_1$ and $\mathcal{P}_2$ can be distinguished.[7,8]

These relations (cf. Table 1) are equivalent to the relations defined by the RCC-8 calculus,[9] a first-order logical language for formalizing topological relationships between abstract spatial regions (cf. Fig. 2).

Table 1. Basic relations of the RCC-8 calculus in terms of the 9-Intersection Model.

	$\mathcal{P}_1^o \cap \partial\mathcal{P}_2$	$\mathcal{P}_1^e \cap \partial\mathcal{P}_2$	$\partial\mathcal{P}_1 \cap \partial\mathcal{P}_2$	$\mathcal{P}_2^o \cap \partial\mathcal{P}_1$
$\text{NTPP}(\mathcal{P}_1, \mathcal{P}_2)$	0	1	0	1
$\text{NTPP}_i(\mathcal{P}_1, \mathcal{P}_2)$	1	0	0	0
$\text{DC}(\mathcal{P}_1, \mathcal{P}_2)$	0	1	0	0
$\text{TPP}(\mathcal{P}_1, \mathcal{P}_2)$	0	1	1	1
$\text{TPP}_i(\mathcal{P}_1, \mathcal{P}_2)$	1	0	1	0
$\text{EC}(\mathcal{P}_1, \mathcal{P}_2)$	0	1	1	0
$\text{PO}(\mathcal{P}_1, \mathcal{P}_2)$	1	1	1	1
$\text{EQ}(\mathcal{P}_1, \mathcal{P}_2)$	0	0	1	0

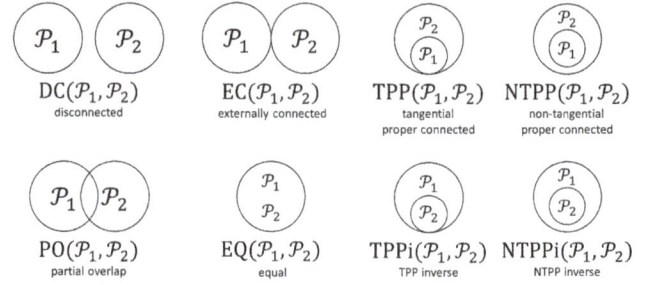

Fig. 2. The eight basic relations of the RCC-8 calculus.

3. Qualification of Mereotopological Relations

Although mereotopological relationship theories, e.g., RCC-8 and 9IM, and their qualification[10] have been extensively studied in $\mathbb{R}^2$, the qualification of mereotopological relations in $\mathbb{R}^3$ remains challenging. This is due to

(C.1) the limited availability of topological operators interior $\cdot^o$, boundary $\partial\cdot$ and exterior $\cdot^e$; and

(C.2) the computational costs of the boundary intersection tests $\mathcal{P}_1^o \cap \partial\mathcal{P}_2$, $\mathcal{P}_1^e \cap \partial\mathcal{P}_2$, $\partial\mathcal{P}_1 \cap \partial\mathcal{P}_2$ and $\mathcal{P}_2^o \cap \partial\mathcal{P}_1$,

with respect to a specific representation technique for three-dimensional spatial entities. In what follows, we briefly discuss the related literature on mereotopological qualification in this respect.

Surface models. Albath *et al.* introduce RCC-3D,[11] a QSR system based on the Generalized 2D Region Connection Calculus. RCC-3D originally relies on polygonal boundary information determined for each three-dimensional spatial entity prior to the reasoning process. In later work,[12] a qualification algorithm to determine the mereotopological relation between two three-dimensional spatial entities without *a priori* spatial knowledge is presented. The qualification algorithm determines pairwise triangle–triangle intersections[13] in worst-case time complexity of $O(n^5)$ with n being the maximum number of faces or vertices used to define the spatial domains under consideration.

In a series of papers,[14–16] Sabharwal *et al.* present VRCC-3D+, an extension of RCC-3D considering binary relationships for nonocclusion, partial occlusion and complete occlusion. Subsequent works[17–19] devise a decision framework for more efficient calculations of boundary intersections using either triangle–triangle intersection tests or intersection tests with AABB trees.[20]

Solid modeling. Borrmann and Rank[21] introduce a QSR system that allows for the spatial analysis of 3D CAD building models. Spatial domains are individually encoded to any specified resolution in octrees.[22] Each octree node corresponds to a subregion of $\mathbb{R}^3$ and specifies if it lies completely inside, outside or on the boundary of the encoded spatial domain. In order to determine the topological

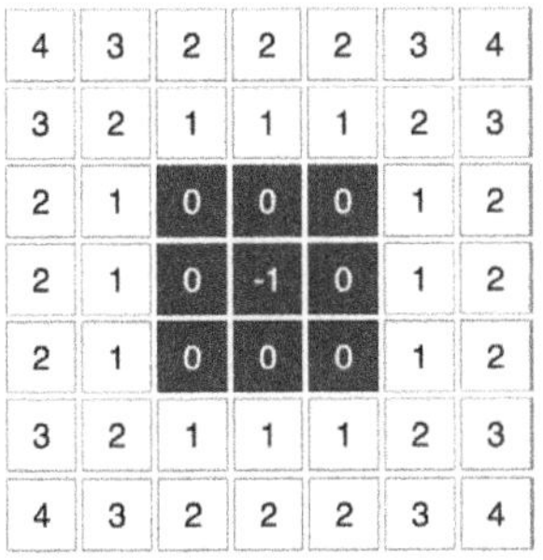

Fig. 3. Signed distance field in $\mathbb{R}^2$ using Manhattan distance.

relationship between two given spatial domains, a synchronized breadth-first traversal of both octree encodings is performed and 21 decision rules are applied on each level of the traversal.

Hmida *et al.*[23–25] present a method to automatically compute 9IM topological relations between Nef polyhedra.[26] The procedure is based on the generation of Selective Nef Complexes (SNC), a representation of Nef polyhedra providing binary Boolean operators and unary operators such as interior, closure and boundary, from standard polyhedra, and the redefinition of the 9IM matrix using solely SNC operators.

4. Signed Distance Fields in Brief

In the following, we give some background on signed distance fields and refer the interested reader to related works with respect to the efficient calculation of such.

Distance fields are a popular data structure in the fields of computer graphics, geometric modeling and robotics and serve a vast range of applications including surface reconstruction,[27] shape representation,[28] collision detection,[29] path planning and navigation.[30]

Given a three-dimensional spatial entity $\mathcal{P} \subset \mathbb{R}^3$ with distance metric $d : \mathbb{R}^3 \times \mathbb{R}^3 \to \mathbb{R}$, we denote with $\partial \mathcal{P} \subset \mathcal{P}$ the surface (or boundary) of $\mathcal{P}$. A *distance field* of $\mathcal{P}$ is a scalar field $D' : \mathbb{R}^3 \to \mathbb{R}$ that specifies the minimum distance from any $\boldsymbol{q} \in \mathbb{R}^3$ to the closest point $\boldsymbol{p} \in \partial \mathcal{P}$ such that

$$D'(\boldsymbol{q}) = \min_{\boldsymbol{p} \in \partial \mathcal{P}} \{d(\boldsymbol{p}, \boldsymbol{q})\}.$$

If $\partial \mathcal{P}$ is a closed surface, i.e., $\partial \mathcal{P}$ is compact and without boundary, we can define a sign function $\sigma' : \mathbb{R}^3 \to \{-1, 1\}$ such that

$$\sigma'(\boldsymbol{q}) = \begin{cases} -1 & \text{if } \boldsymbol{q} \in \mathcal{P}, \\ 1 & \text{if } \boldsymbol{q} \notin \mathcal{P}, \end{cases}$$

for any $\boldsymbol{q} \in \mathbb{R}^3$. If $\partial \mathcal{P}$ is oriented, i.e., a face normal $\boldsymbol{n} \in \mathbb{R}^3$ is known for every $\boldsymbol{p} \in \partial \mathcal{P}$, we can define a sign function $\sigma : \mathbb{R}^3 \to \{-1, 1\}$ such that

$$\sigma(\boldsymbol{q}) = \begin{cases} -1 & \text{if } (\boldsymbol{q} - \boldsymbol{p}) \cdot \boldsymbol{n} < 0, \\ 1 & \text{if } (\boldsymbol{q} - \boldsymbol{p}) \cdot \boldsymbol{n} \geq 0. \end{cases}$$

As noted by Xu and Barbič,[31] we can find for sufficiently small $\Delta \in \mathbb{R}$ a point $\boldsymbol{q} \in \mathbb{R}^3$ for every $\boldsymbol{p} \in \partial \mathcal{P}$ within the offset surface

$$\partial_\Delta \mathcal{P} = \{\boldsymbol{q} \in \mathbb{R}^3 \| D'(\boldsymbol{q}) = \Delta\},$$

such that $\sigma(\boldsymbol{q}) = \sigma'(\boldsymbol{q})$. Hence, we can define a *signed distance field* $D : \mathbb{R}^3 \to \mathbb{R}$ with

$$D(\boldsymbol{q}) = \sigma(\boldsymbol{q}) \cdot D'(\boldsymbol{q}),$$

for any oriented, not necessarily boundary-free, three-dimensional spatial entity $\mathcal{P} \subset \mathbb{R}^3$.

The problem of computing a distance field is fairly well understood, and there is a wide array of algorithms for triangle meshes, triangle soups, implicit surfaces and parametric surfaces.[32]

5. Qualification using Signed Distance Fields

Given two spatial entities $\mathcal{P}_1, \mathcal{P}_2 \subset \mathbb{R}^3$ with body-fixed coordinate systems $\mathcal{C}_1$ and $\mathcal{C}_2$ and a six-degree-of-freedom transformation $\boldsymbol{P}$ between $\mathcal{C}_1$ and $\mathcal{C}_2$, our objective is to turn this quantitative geometric description into a qualitative description composed of RCC-8 base relations.

We propose to solve this qualification problem using pre-computed signed distance fields as lookup data structures for boundary intersection tests. For both spatial entities $\mathcal{P}_1$ and $\mathcal{P}_2$, we assume the availability of

(R.1) signed distance fields $D_{\mathcal{P}_1}$ and $D_{\mathcal{P}_2}$ precomputed with one of the methods mentioned in Sec. 4; as well as

(R.2) point-based representations of boundaries $\partial \mathcal{P}_1$ and $\partial \mathcal{P}_2$ obtained either directly from three-dimensional polygonal models or by point sampling[33] of $\mathcal{P}_1$ and $\mathcal{P}_2$.

We consider assumptions **(R.1)** and **(R.2)** to be justified for nondeformable spatial entities $\mathcal{P}_1$ and $\mathcal{P}_2$, and, therefore, in conformity with related work.

Our qualification technique relies on the fast computation of histograms specified by the following indicator functions:

$$\mathbf{1}_\partial(\mathcal{P}_i, \boldsymbol{x}) = \begin{cases} 1 & \text{if } D_{\mathcal{P}_i}(\boldsymbol{x}) = 0, \\ 0 & \text{otherwise}, \end{cases}$$

$$\mathbf{1}_o(\mathcal{P}_i, \boldsymbol{x}) = \begin{cases} 1 & \text{if } D_{\mathcal{P}_i}(\boldsymbol{x}) < 0, \\ 0 & \text{otherwise}, \end{cases}$$

$$\mathbf{1}_e(\mathcal{P}_i, \boldsymbol{x}) = \begin{cases} 1 & \text{if } D_{\mathcal{P}_i}(\boldsymbol{x}) > 0, \\ 0 & \text{otherwise}. \end{cases}$$

The indicator functions $\mathbf{1}_\partial(\mathcal{P}_i, \boldsymbol{x})$, $\mathbf{1}_o(\mathcal{P}_i, \boldsymbol{x})$ and $\mathbf{1}_e(\mathcal{P}_i, \boldsymbol{x})$ employ a precomputed signed distance field $D_{\mathcal{P}_i}$ in order to determine if a given point $\boldsymbol{x} \in \mathbb{R}^3$ lies completely inside, outside or on the boundary of the spatial domain $\mathcal{P}_i$.

BuildHistogram (cf. Algorithm 1) effectively computes the boundary intersection tests $\mathcal{P}_1^o \cap \partial \mathcal{P}_2$, $\mathcal{P}_1^e \cap \partial \mathcal{P}_2$ and $\partial \mathcal{P}_1 \cap \partial \mathcal{P}_2$ in worst-case time complexity of $O(n)$ with n

Algorithm 1. Histogram computation

```
 1: procedure BUILDHISTOGRAM(P₁, P₂, P)
 2:     c∂, cₒ, cₑ ← 0
 3:     for all x ∈ ∂P₂ do
 4:         c∂ ← c∂ + 1∂(P₁, Px)
 5:         cₒ ← cₒ + 1ₒ(P₁, Px)
 6:         cₑ ← cₑ + 1ₑ(P₁, Px)
 7:         if cₑ > 0 and cₒ > 0 then
 8:             break
 9:         end if
10:     end for
11:     return c∂, cₒ, cₑ
12: end procedure
```

Algorithm 2. Determine if $\mathcal{P}_1$ is enclosed by $\mathcal{P}_2$

```
procedure INSIDE(P₁, P₂, P)
    for all x ∈ ∂P₁ do
        if 1ₒ(P₂, Px) then
            return True
        else if 1ₑ(P₂, Px) then
            return False
        end if
    end for
end procedure
```

Algorithm 3. RCC-8 qualification algorithm

```
 1: procedure QUALIFY(P₁, P₂, P)
 2:     if |∂P₁| < |∂P₂| then
 3:         QUALIFY(P₂, P₁, P⁻¹)
 4:     end if
 5:     c∂, cₒ, cₑ ← BUILDHISTOGRAM(P₁, P₂, P)
 6:     if cₑ = 0 then
 7:         if cₒ = 0 then
 8:             return EQ(P₁, P₂)
 9:         else if cₒ > 0 and c∂ = 0 then
10:             return NTPPᵢ(P₁, P₂)
11:         else if cₒ > 0 and c∂ > 0 then
12:             return TPPᵢ(P₁, P₂)
13:         end if
14:     else if cₑ > 0 then
15:         if cₒ = 0 then
16:             inside ← INSIDE (P₁, P₂, P)
17:             if c∂ = 0 and inside then
18:                 return NTPP(P₁, P₂)
19:             else if c∂ = 0 and ¬inside then
20:                 return DC(P₁, P₂)
21:             else if c∂ > 0 and inside then
22:                 return TPP(P₁, P₂)
23:             else
24:                 return EC(P₁, P₂)
25:             end if
26:         else if cₒ > 0 then
27:             return PO(P₁, P₂)
28:         end if
29:     end if
30: end procedure
```

being the size of the point-based representation of $\partial \mathcal{P}_2$. The computation of BUILDHISTOGRAM allows for straight-forward optimization:

(O.1) Stop BUILDHISTOGRAM as soon as buckets c_o and c_e contain at least one point as we have found a partial overlap (PO) (cf. Algorithm 1, lines 7–9; cf. Fig. 4).

Further performance optimizations of BUILDHISTOGRAM are achievable by exploiting obvious loop-level parallelism. We detail on this in Sec. 6.

QUALIFY (cf. Algorithm 3), our overall qualification procedure, evaluates a binary decision diagram as illustrated in Fig. 4.

QUALIFY relies on BUILDHISTOGRAM in order to perform the boundary intersection tests $\mathcal{P}_1^e \cap \partial \mathcal{P}_2$, $\mathcal{P}_1^o \cap \partial \mathcal{P}_2$ and $\partial \mathcal{P}_1 \cap \partial \mathcal{P}_2$ (cf. Algorithm 3, line 5). Since QUALIFY's execution time is dominated by BUILDHISTOGRAM, we choose to

(O.2) Evaluate QUALIFY against the smaller point-based representation and swap arguments accordingly (cf. Algorithm 3, lines 2–4).

Note that optimization **(O.2)** will not affect the correctness of qualification. This can be easily seen from the following

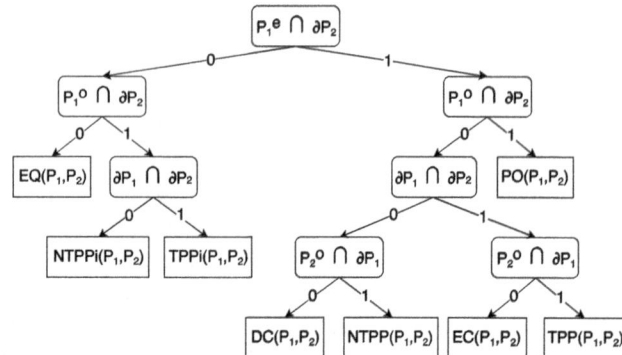

Fig. 4. Binary decision diagram evaluated by QUALIFY.

equivalences:

$$
\begin{aligned}
\text{EC}(x, y) &\equiv \text{EC}(y, x), & \text{DC}(x, y) &\equiv \text{DC}(y, x), \\
\text{EQ}(x, y) &\equiv \text{EQ}(y, x), & \text{PO}(x, y) &\equiv \text{PO}(y, x), \\
\text{TPP}(x, y) &\equiv \text{TPP}_i(y, x), & \text{NTPP}(x, y) &\equiv \text{NTPP}_i(y, x).
\end{aligned}
$$

For the remaining boundary intersection test $\mathcal{P}_2^o \cap \partial\mathcal{P}_1$ (cf. Algorithm 3, lines 15–25), we refrain from computing BUILDHISTOGRAM($\mathcal{P}_2$, $\mathcal{P}_1$, $\mathbf{P}^{-1}$). Instead we use INSIDE (cf. Algorithm 2) to determine if $\mathcal{P}_1$ is enclosed by or completely outside of $\mathcal{P}_2$ by

(O.3) Search for a single boundary point $x \in \partial\mathcal{P}_1$ such that $x \in \mathcal{P}_2^o$ or $x \in \mathcal{P}_2^e$, respectively (cf. Algorithm 3, line 16).

6. Implementation

We implement a single-threaded variant of our approach in C++ using the Vega FEM framework[34] that provides a rich feature-set and tools for distance fields. For each spatial entity, we precompute a signed distance field based on the techniques presented by Xu and Barbič.[31] Furthermore, we rely on MeshLab,[35] an open-source 3D mesh processing software system, for precomputing a point-based surface representation for each spatial entity.

As pointed out in Sec. 5, the execution time of our overall qualification procedure QUALIFY is dominated by BUILDHISTOGRAM.

At the core of BUILDHISTOGRAM is a loop over all boundary points in $\partial\mathcal{P}_2$ (cf. Algorithm 1, lines 3–10). We take advantage of OpenMP, an API specification for parallel programming, for loop parallelization (cf. Algorithm 4, lines 3–7). Since optimization (O.1) (cf. Algorithm 1, lines 7–9) hinders OpenMP's ability to parallelize BUILD-HISTOGRAM's loop, we remove the respective BREAK statement in the MULTITHREADED procedure.

Instead, we suggest to use BUILDHISTOGRAM on a randomly sampled subset $\mathcal{P}_R \subset \mathcal{P}_2$ with $|\mathcal{P}_R| = \mathcal{N}$ and $\mathcal{N} \ll |\mathcal{P}_2|$ trying to detect a partial overlap [cf. (O.1)] in BUILDHISTO-GRAMMT (cf. Algorithm 5, lines 2 and 3).

In case we do not detect a partial overlap, we use MULTI-THREADED for parallelized computation of $\mathcal{P}_1^o \cap \partial\mathcal{P}_2$, $\mathcal{P}_1^e \cap \partial\mathcal{P}_2$ and $\partial\mathcal{P}_1 \cap \partial\mathcal{P}_2$.

Algorithm 4. Loop parallelization

1: **procedure** MULTITHREADED($\mathcal{P}_1$, $\mathcal{P}_2$, $\mathbf{P}$)
2:     $c_\partial, c_o, c_e \leftarrow 0$
3:     **parfor** $x \in \partial\mathcal{P}_2$ **do**
4:         $c_\partial \leftarrow c_\partial + \mathbf{1}_\partial(\mathcal{P}_1, \mathbf{P}x)$
5:         $c_o \leftarrow c_o + \mathbf{1}_o(\mathcal{P}_1, \mathbf{P}x)$
6:         $c_e \leftarrow c_e + \mathbf{1}_e(\mathcal{P}_1, \mathbf{P}x)$
7:     **end parfor**
8:     **return** c_∂, c_o, c_e
9: **end procedure**

Algorithm 5. Multithreaded histogram computation

1: **procedure** BUILDHISTOGRAMMT($\mathcal{P}_1$, $\mathcal{P}_2$, $\mathbf{P}$, $\mathcal{N}$)
2:     $\mathcal{P}_R \leftarrow$ RANDOMPOINTS($\mathcal{P}_2, \mathcal{N}$)
3:     $c_\partial, c_o, c_e \leftarrow$ BUILDHISTOGRAM($\mathcal{P}_1$, $\mathcal{P}_R$, $\mathbf{P}$)
4:     **if** $c_e = 0$ or $c_o = 0$ **then**
5:         $c_\partial, c_o, c_e \leftarrow$ MULTITHREADED($\mathcal{P}_1$, $\mathcal{P}_2$, $\mathbf{P}$)
6:     **end if**
7:     **return** c_∂, c_o, c_e
8: **end procedure**

---

7. Evaluation

We apply our approach to a number of spatial configurations and report on the measured execution times of computing qualitative descriptions composed of basic RCC-8 relations.

7.1. Spatial configurations

Joe's Ice Cream [cf. Fig. 1 (left)] is composed of three spatial entities, i.e., *Joe* (6054 vertices, 11,988 faces, 16 connected components, two manifolds, six holes, genus 9), *Ice Lolly* (247 vertices, 481 faces, three connected components, two manifolds, three holes, genus 0) and *Bucket* (476 vertices, 896 faces, three connected components, two manifolds, two holes, genus 0). RCC-8 qualification using our approach correctly results in EC(Joe, Ice), DC(Joe, Bucket) and DC (Ice, Bucket).

Life in a Glasshouse [cf. Fig. 5 (left)] is composed of four spatial entities, i.e., *Joe, Ice Lolly, Baseplate* (2804 vertices, 5360 faces, 62 connected components, two manifolds, zero holes, genus 0) and *Glasshouse* (26 vertices, 48 faces, one connected component, two manifolds, zero holes, genus 0). RCC-8 qualification using our approach correctly results in EC(Joe, Ice), TPP(Ice, Glasshouse), TPP (Joe, Glasshouse), EC(Joe, Baseplate), NTPP(Baseplate, Glasshouse).

Fig. 5. RCC-8 qualification of *Life in a Glasshouse* (left) results in EC(Joe, Ice), NTPP(Ice, Glasshouse), NTPP(Joe, Glasshouse), EC (Joe, Baseplate) and TPP(Baseplate, Glasshouse). RCC-8 qualification of *Armadillo* (right) results in EQ(A, B).

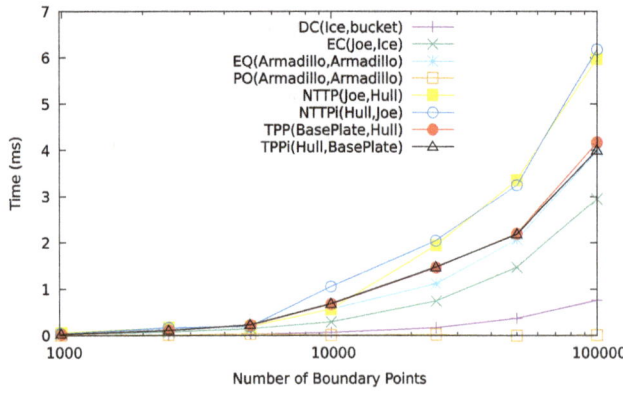

Fig. 6. Runtime performance of QUALIFY($\mathcal{P}_1$, $\mathcal{P}_2$) using *single-threaded* BUILDHISTOGRAM with fixed SDF resolution ($256 \times 256 \times 256$) and point-based boundary representations of varying sizes.

Armadillos [cf. Fig. 1 (right)] is composed of two partially overlapping instances of the *Stanford Armadillo*[36] (172,974 vertices, 345,944 faces, one connected component, two manifolds, zero holes, genus 0). RCC-8 qualification using our approach correctly results in PO($\mathcal{P}_1$, $\mathcal{P}_2$).

Armadillo [cf. Fig. 5 (right)] is composed of two perfectly overlapping instances of the *Stanford Armadillo*. RCC-8 qualification using our approach correctly results in EQ ($\mathcal{P}_1$, $\mathcal{P}_2$).

7.2. *Performance*

We report on the performance (Fig. 6) and scaling behavior (cf. Fig. 7) of our single-threaded implementation. Next, we provide comparison between the single-threaded and multi-threaded implementations in Fig. 8.

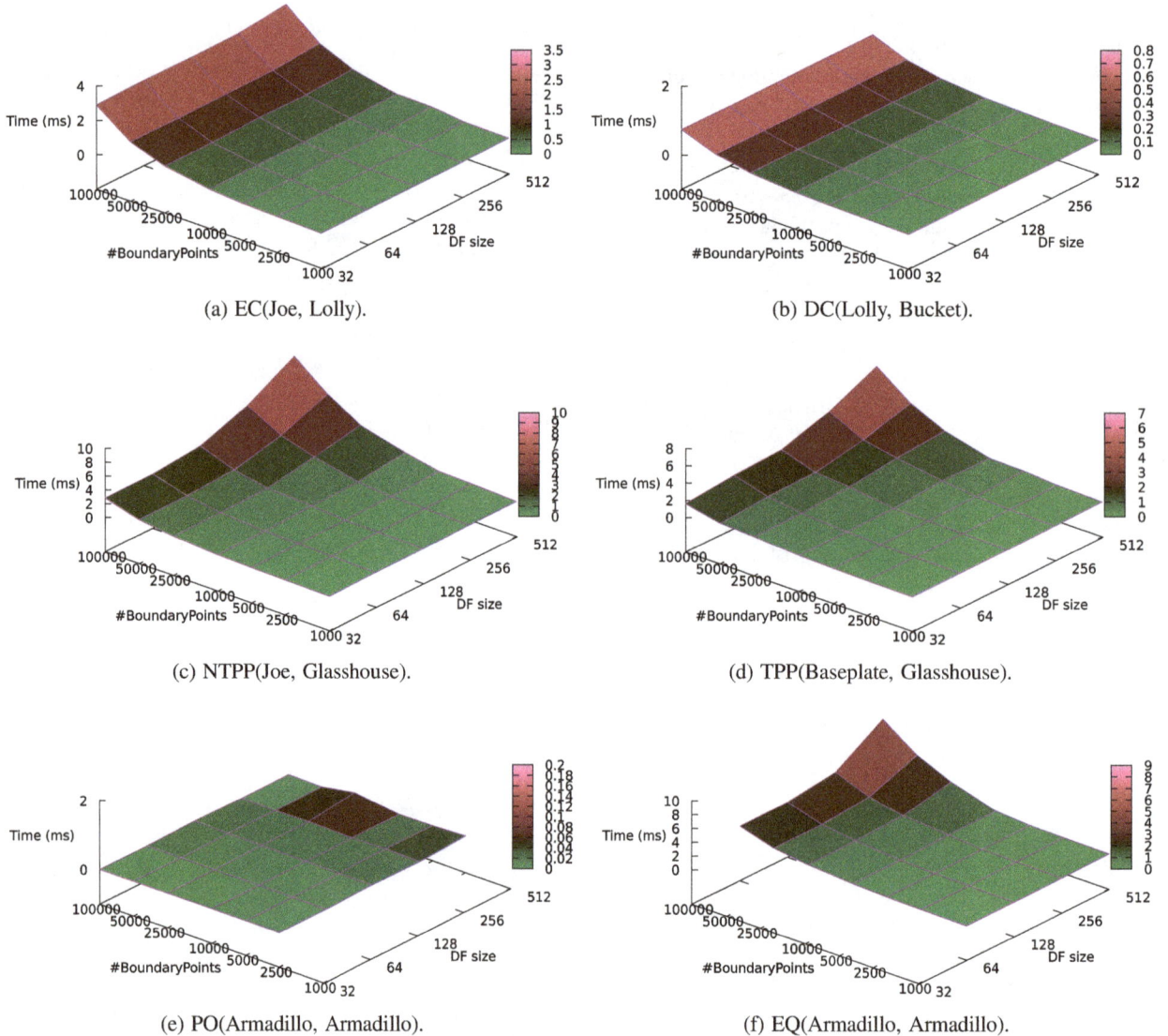

(a) EC(Joe, Lolly).

(b) DC(Lolly, Bucket).

(c) NTPP(Joe, Glasshouse).

(d) TPP(Baseplate, Glasshouse).

(e) PO(Armadillo, Armadillo).

(f) EQ(Armadillo, Armadillo).

Fig. 7. Scaling behavior of QUALIFY($\mathcal{P}_1$, $\mathcal{P}_2$) with respect to varying SDF resolutions and varying point-based boundary representations.

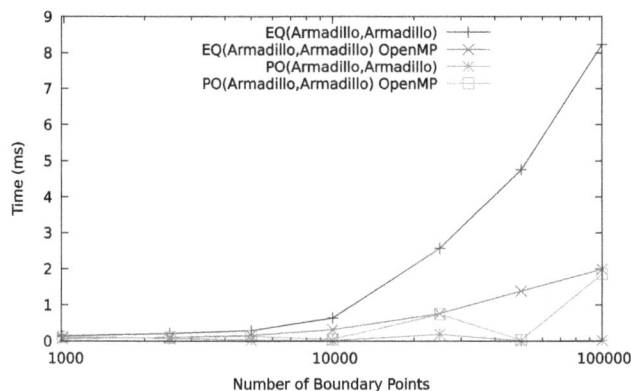

Fig. 8. Comparing runtime performances of QUALIFY($\mathcal{P}_1,\mathcal{P}_2$) using *single-threaded* and *multithreaded* BUILDHISTOGRAM with fixed SDF resolution ($512 \times 512 \times 512$) and point-based boundary representations of varying sizes.

All timings presented in this paper are generated on a MacBook Pro 2.8 GHz Intel Core i7 with four cores, 16 GB 1600 MHz DDR3, running MacOS Sierra 10.12.6.

Single-threaded performance

We discover the scaling behavior of our single-threaded implementation to be nearly invariant with respect to the resolution of distance fields and to be linear in the size of point-based boundary representations.

Single-threaded versus multi-threaded performance

We observe the runtime performance of our multi-threaded implementation to scale with the number of available cores. As discussed in Sec. 6, BUILDHISTOGRAMMT cannot always compensate the performance gain of optimization (**O.1**) for the special case of a partial overlap.

8. Conlusions

In this paper, we propose a novel qualification technique for mereotopological relations in $\mathbb{R}^3$. Our approach rapidly computes RCC-8 base relations using precomputed signed distance fields, and makes no assumptions with regards to complexity or 3D representation method of the spatial entities under consideration. Conducted performance evaluations suggest significant improvements over relevant related works as well as favorable scaling behavior.

Acknowledgments

This work is supported by the Federal Ministry of Education and Research of Germany in the Project Hybr-iT (Förderkennzeichen 01IS16026A).

References

[1] B. Kuipers and Y.-T. Byun, A robot exploration and mapping strategy based on a semantic hierachy of spatial representations, *Robot. Auton. Syst.* **8**, 47 (1991).

[2] A. Gupta, A. A. Efros and M. Hebert, Blocks world revisited: Image understanding using qualitative geometry and mechanics, *Computer Vision: ECCV 2010*, Lecture Notes in Computer Science, Vol. 6314 (Springer, Berlin, 2010), pp. 482–496.

[3] A. U. Frank, Qualitative spatial reasoning about distances and directions in geographic space, *J. Vis. Lang. Comput.* **3**, 343 (1992).

[4] M. J. Egenhofer and J. Herring, A mathematical framework for the definition of topological relationships, *Proc. Fourth Int. Symp. Spatial Data Handling* (1990), pp. 803–813.

[5] M. Egenhofer and R. D. Franzosa, Point-set topological spatial relations, *Int. J. Geogr. Inf. Sys.* **5**, 161 (1991).

[6] M. Egenhofer and J. Herring, Categorizing binary topological relations between regions, lines and points in geographic databases, the 9-intersection: Formalism and its use for natural-allanguage spatial predicates, Technical Report, No. 94-1, Santa Barbara CA National Center for Geographic Information and Analysis (1990).

[7] C. L. Sabharwal and J. L. Leopold, Reducing 9-intersection to 4-intersection for identifying relations in region connection calculus, *Proc. 24th Int. Conf. Computer Applications in Industry and Engineering* (2011).

[8] C. L. Sabharwal and J. L. Leopold, Identification of relations in region connection calculus: 9-Intersection reduced to 3^+-Intersection predicates, *Advances in Soft Computing and Its Applications: 12th Mexican Int. Conf. Artificial Intelligence Proc. Part II*, eds. F. Castro, A. Gelbukh and M. González, Lecture Notes in Computer Science, Vol. 8266 (Springer, Berlin, 2013), pp. 362–375.

[9] A. Izadi, K. M. Stock and H. W. Guesgen, Multidimensional region connection calculus, Unpublished Paper, Association for the Advancement of Artifical Intelligence (2017).

[10] F. Dylla, L. Frommberger, J. O. Wallgrün and D. Wolter, SparQ: A toolbox for qualitative spatial representation and reasoning, *Proc. Qualitative Constraint Calculi: Application and Integration: Workshop at KI* (2006), pp. 79–90.

[11] J. Albath, J. L. Leopold, C. L. Sabharwal and A. M. Maglia, RCC-3D: Qualitative spatial reasoning in 3D, *Proc. 23nd Int. Conf. Computer Applications in Industry and Engineering* (2010), pp. 74–79.

[12] J. Albath, J. L. Leopold, C. L. Sabharwal and K. Perry, Efficient reasoning with RCC-3D, *Knowledge Science, Engineering and Management*, Lecture Notes in Computer Science, Vol. 6291 (Springer, Berlin, 2010), pp. 470–481.

[13] T. Möller, A fast triangle-triangle intersection test, *J. Graph. Tools* **2**, 25 (1997).

[14] C. L. Sabharwal, J. L. Leopold and N. Eloe, A more expressive 3D region connection calculus, *Proc. 2011 Int. Workshop Visual Languages and Computing* (2011), pp. 307–311.

[15] C. L. Sabharwal and J. L. Leopold, Evolution of region connection calculus to VRCC-3D+, *New Math. Nat. Comput.* **10**, 103 (2014).

[16] N. Eloe, VRCC-3D+: Qualitative spatial and temporal reasoning in 3 dimensions, Ph.D. thesis, Missouri University of Science and Technology, Rolla, Missouri (2015).

[17] N. Eloe, J. L. Leopold, C. L. Sabharwal and Z. Yin, Efficient computation of object boundary intersection and error tolerance in VRCC-3D+, *Proc. 18th Int. Conf. Distributed Multimedia Systems* (2012), pp. 67–70.

18 C. L. Sabharwal and J. L. Leopold, A fast intersection detection algorithm for qualitative spatial reasoning, *Proc. 2013 Int. Workshop Visual Languages and Computing* (2013), pp. 145–149.

19 N. Eloe, J. Leopold, C. Sabharwal and D. McGeehan, Efficient determination of spatial relations using composition tables and decision trees, *Proc. 2013 IEEE Symp. Computational Intelligence for Multimedia, Signal and Vision Processing (CIMSIVP)* (2013), pp. 1–7.

20 G. van den Bergen, Efficient collision detection of complex deformable models using AABB trees, *J. Graph. Tools* **2**, 1 (1997).

21 A. Borrmann and E. Rank, Topological analysis of 3D building models using a spatial query language, *Adv. Eng. Inf.* **23**, 370 (2009).

22 D. Meagher, Geometric modeling using octree encoding, *Comput. Graph. Image Process.* **19**, 129 (1982).

23 H. B. Hmida, F. Boochs, C. Cruz and C. Nicolle, From quantitative spatial operator to qualitative spatial relation using Constructive Solid Geometry, logic rules and optimized 9-IM model: A semantic based approach, *Proc. 2012 IEEE Int. Conf. Computer Science and Automation Engineering (CSAE)* (2012), pp. 453–458.

24 H. B. Hmida, C. Cruz, F. Boochs and C. Nicolle, From 9-IM topological operators to qualitative spatial relations using 3d selective Nef complexes and logic rules for bodies, arXiv:1301.4992 [cs.AI].

25 H. B. Hmida, C. Cruz, F. Boochs and C. Nicolle, From 9-IM topological operators to qualitative spatial relations using 3D selective Nef complexes and logic rules for bodies, *Proc. Int. Conf. Knowledge Engineering and Ontology Development (KEOD 2012)* (2012), pp. 208–213.

26 W. Nef, *Beiträge zur Theorie der Polyeder: mit Anwendungen in der Computergraphik* (Herbert Lang, Bern, 1978).

27 G. Taubin, Smooth signed distance surface reconstruction and applications, *Progress in Pattern Recognition, Image Analysis,* *Computer Vision, and Applications*, Lecture Notes in Computer Science, Vol. 7441 (Springer, Berlin, 2012), pp. 38–45.

28 S. F. Frisken, R. N. Perry, A. P. Rockwood and T. R. Jones, Adaptively sampled distance fields: A general representation of shape for computer graphics, *Proc. 27th Annual. Conf. on Computer Graphics and Interactive Techniques* (ACM Press/Addison-Wesley Publishing Co., New York, 2000), pp. 249–254.

29 A. Fuhrmann, G. Sobotka and C. Gross, Distance fields for rapid collision detection in physically based modeling, *Proc. GraphiCon 2003* (2003), pp. 58–65.

30 H. Oleynikova, A. Millane, Z. Taylor, E. Galceran, J. Nieto and R. Siegwart, Signed distance fields: A natural representation for both mapping and planning, *Proc.: RSS 2016 Workshop: Geometry and Beyond Representations, Physics, and Scene Understanding for Robotics* (University of Michigan, 2016).

31 H. Xu and J. Barbič, Signed distance fields for polygon soup meshes, *Proc. Graphics Interface 2014* (Canadian Information Processing Society, Toronto, 2014), pp. 35–41.

32 M. W. Jones, J. A. Baerentzen and M. Sramek, 3D distance fields: A survey of techniques and applications, *IEEE Trans. Vis. Comput. Graph.* **12**, 581 (2006).

33 M. Alexa, J. Behr, D. Cohen-Or, S. Fleishman, D. Levin and C. T. Silva, Computing and rendering point set surfaces, *IEEE Trans. Vis. Comput. Graph.* **9**, 3 (2003).

34 J. Barbič, F. S. Sin and D. Schroeder, Vega FEM Library (2012), http://www.jernejbarbic.com/vega.

35 P. Cignoni, M. Callieri, M. Corsini, M. Dellepiane, F. Ganovelli and G. Ranzuglia, MeshLab: An open-source mesh processing tool, *Proc. Eurographics Italian Chapter Conf.*, eds. V. Scarano, R. D. Chiara and U. Erra (The Eurographics Association, 2008).

36 V. Krishnamurthy and M. Levoy, Fitting smooth surfaces to dense polygon meshes, *Proc. 23rd Annu. Conf. Computer Graphics and Interactive Techniques* (ACM, New York, 1996), pp. 313–324.

Part 5

Robotic Intelligence

A learning from demonstration framework for implementation of a feeding task

Nabil Ettehadi[*,‡] and Aman Behal[*,†]

*Department of Electrical and Computer Engineering
University of Central Florida, Orlando, FL 32816 USA

†NanoScience Technology Center
University of Central Florida, Orlando, FL 32826 USA

‡nabil.ettehadi@knights.ucf.edu

In this paper, a learning from demonstration (LFD) approach is used to design an autonomous meal-assistance agent. The feeding task is modeled as a mixture of Gaussian distributions. Using the data collected via kinesthetic teaching, the parameters of the Gaussian mixture model (GMM) are learnt using Gaussian mixture regression (GMR) and expectation maximization (EM) algorithm. Reproduction of feeding trajectories for different environments is obtained by solving a constrained optimization problem. In this method we show that obstacles can be avoided by robot's end-effector by adding a set of extra constraints to the optimization problem. Finally, the performance of the designed meal assistant is evaluated in two feeding scenario experiments: one considering obstacles in the path between the bowl and the mouth and the other without.

Keywords: Activities of daily living tasks (ADLs); feeding task; meal assistant; Gaussian mixture model (GMM); learning by imitation; human–robot interaction (HRI); learning from demonstration (LFD); programming by demonstration (PBD); probabilistic motion encoding.

1. Introduction

Almost 56.7 million people living in the US have some sort of disabilities according to a comprehensive report released in 2010 by Census Bureau.[1] This number is nearly 20% of the US civilian noninstitutional population. Currently, the number of disabled people in the world is around 650 million (nearly 10% of the world's total population).[2] The disabled community needs assistance with performing activities of daily living (ADLs) such as cleaning, cooking, eating, dressing, etc. Assisting these individuals to perform ADLs requires very rigorous labor and often is not convenient for the individuals themselves. Hence, the need for a smart personal robot acting safely and naturally as a human caregiver is at its peak nowadays.

One of the most important ADLs is eating as it is the most basic human need. Therefore, developing meal-assistance robots for disabled individuals is a challenging interest of the robotics research community. Previously developed commercialized meal-assistance robots such as "Handy 1"[3] and "MANUS" (Exact Dynamics Co.) are used to feed individuals. However, due to their lack of manipulation flexibility, these robots are not satisfactory for general dining assistance. Recently developed meal-assistance robots such as those in Refs. 4 and 5 have more potential toward being a suitable, general meal-assistance agents than "Handy 1" and "MANUS." Nevertheless, these robots often tend to use a set

of pregenerated trajectories, have limited adaptation skills or lack natural human-like motions. Thus, in order create a more general, human-like meal-assistance agents, a different approach should be taken to generate adaptable trajectories for the task of feeding.

Learning from demonstrations (LFD), also referred to as programming by demonstration (PBD) or learning by imitation, has picked out a number of major issues to develop an appropriate approach to learn and transfer new skills across different environments, situations and agents. Such methods are highly effective to learn context-free task constraints from a set of demonstrations. The key to extract a feature-based representation for a task is to process the local trajectories in the coordinate frames of all the objects involved in the task and look for the variations happening from one demonstration to another. Hence, probabilistic models are used to achieve a general context-free model for a task in each coordinate frame. In Ref. 6, a PBD approach is proposed to extract the constraints of a task and reproduce it in new environments. Principle components analysis (PCA)[7] is used to reduce the dimensionality of the data and Gaussian mixture model/regression (GMM/GMR)[8] is used for having a probabilistic representation. In Refs. 9–11 the authors present a demonstration-guided motion planning (DGMP) to generate motion plans in the joint space of the robot. Direct mean and covariance matrix of the data is used in Ref. 9, hidden Markov model (HMM) is used in Ref. 10 and multivariate Gaussian

---

‡Corresponding author.

distribution (MGD) is utilized in Ref. 11 to achieve a probabilistic model for the task. These frameworks are based on constrained motion planning which incrementally computes a motion plan while avoiding obstacles and optimizing a learnt cost metric based on the probabilistic model to maintain the task features. Implementation of a pizza dough rolling task is done in Ref. 12 using automatic task segmentation and action primitive discovery algorithm. An extension of beta process hidden Markov model (BP-HMM) is used for segmentation and action constraints are modeled and learnt via GMM. Finally, the pizza dough rolling task is implemented using a Cartesian impedance controller. In Ref. 13, a novel automatic task segmentation algorithm is presented to generalize a task to different environments by rotating and/or scaling corresponding segments of the task. The challenging problem of catching in-flight objects is addressed in Ref. 14. The authors use support vector regression (SVR) to model the dynamics of the in-flight object while using GMM to model and predict the catching configuration of the robot. This work shows how GMM/GMR is a powerful probabilistic tool for modeling difficult task constraints. In a different approach, dynamic movement primitives (DMPs) are used in Ref. 15 with generalization to different start and goal points for the trajectory of the learnt task. Robust learning and adaptation of robot manipulation tasks is investigated in Ref. 16 using semi-tied GMM. Instead of estimating the full covariance matrix, the authors tie frequent synergistic basis vectors with the covariance matrix. This approach has the advantage of reusing the synergistic directions for different parts of the task sharing similar coordination structure. For the purpose of autonomous adaptation to new environments, the authors use task-parameterized and hidden semi-Markov models. In Ref. 17, leveraged Gaussian process regression is used to integrate demonstrations from both casual/naive and skilled users for the training set. In Ref. 18, a task is modeled as a nonlinear time-independent dynamical system (DS) from a set of demonstrations. Stable estimator of dynamical systems (SEDS) is proposed to learn the parameters of the DS. Sufficient conditions for ensuring global asymptotic stability are derived by authors. A rich survey of robot LFD is presented in Ref. 19.

In this paper, we follow a PBD approach to implement the feeding task using GMM for modeling the task and a convex constrained optimization to generate proper trajectories for new environments. The important constraints of the feeding task such as reaching for the food, scooping, keeping the loaded spoon level, etc. are learnt using GMR. While most PBD approaches focus mostly on learning merely the position trajectories, here, by adopting a novel approach from Ref. 14, the orientations are learnt too as they are very important for the task of feeding, i.e., the loaded spoon should always be kept level. We also consider simple task space obstacle avoidance by adding additional constraints to the optimization problem. Using the framework, the generated trajectories of feeding task are more human-like and the meal-assistance robot is capable of adapting itself to new environments, i.e., different positions/orientations for the bowl of food and the user's mouth. The performance of the framework on the implementation of the feeding task is validated through a set of experiments conducted with the Baxter robot.

The organization of the rest of the paper is as follows. Notations and definition are presented in Sec. 2. The problem definition is presented in Sec. 3. The proposed approach is fully detailed in Sec. 4. Section 4.1 covers modeling, learning approach and task constraints derivation method and Sec. 4.2 presents the imitation part in which the methodology for generating trajectories for new environments is discussed. Section 5 presents the experimental results. Finally, conclusions are presented in Sec. 6.

2. Notations and Definition

The set of real n vectors is denoted by $\mathbb{R}^n$ and the set of real $m \times n$ matrices is denoted by $\mathbb{R}^{m \times n}$. Matrices and vectors are denoted by capital and lower-case bold letters, respectively. Rotation matrix of a coordinate frame p with respect to a coordinate frame q is denoted by $\mathbf{R}_p^q = [\mathbf{r}_1, \mathbf{r}_2, \mathbf{r}_3] \in \mathbb{R}^{3 \times 3}$ where $\mathbf{r}_k \in \mathbb{R}^3$ is the kth column vector of the rotation matrix. Trajectory of the end-effector (or of any other objects attached to it) in jth demonstration with respect to ith coordinate frame evaluated at each timestamp t is denoted by $\eta_i^j(t) = [\eta_{\text{pos}}; \eta_{\text{ori}}] \in \mathbb{R}^9$ where $\eta_{\text{pos}} \in \mathbb{R}^3$ denotes the Cartesian position and $\eta_{\text{ori}} = [\mathbf{r}_1; \mathbf{r}_2] \in \mathbb{R}^6$ denotes the orientation.[a]

Definition 1 (Landmark). Any relevant object involved in a task is called a landmark. Note that since the robot itself is involved in the task, the base of the robot is always considered as a landmark. A body-fixed coordinate frame is attached to each landmark. Landmark i is denoted by l_i. As an example, in the task of feeding, both the bowl and the individual's mouth along with the base of the robot are considered as landmarks, i.e., $l_1 = $ bowl, $l_2 = $ mouth and $l_3 = $ base.

3. Problem Statement

We consider a set of N demonstrations of the feeding task performed in different environments (i.e., different

---

[a]The representation of the orientation as $\eta_{\text{ori}} = [\mathbf{r}_1; \mathbf{r}_2]$, which is motivated by Ref. [14], increases the dimensionality of the dataset compared to other representations but has several benefits. First, since the rotation matrix representation is unambiguous and singularity-free, we do not encounter the problem of gimbal lock occurring in Euler angles representation, Second, representation of the orientation in different coordinate frames is easily addressed by merely multiplying the orientation by the corresponding rotation matrix between the two coordinate frames. Third, the similarity between orientations is represented more accurately than Euler angle or quaternion representation As an example, the Euclidean distance between the roll angles ϕ and $-\phi$ in Euler representation is large while the actual rotation is the same.

positions/orientations of bowl and mouth with respect to base). These demonstrations are recorded with respect to different coordinate frames, i.e., the coordinate frames of all three landmarks. It is assumed that the length of all trajectories in the demonstrations are equal or are rescaled to a fixed value of T (via interpolation or other similar techniques). The training set recorded in the coordinate frame of the ith landmark is formed as $\mathbf{X}_i(t) = [\eta_i^1; \eta_i^2; \ldots; \eta_i^N] \in \mathbb{R}^{(N \cdot T) \times 9}$.

We assume the positions of the three landmarks (bowl, mouth and base) are fully known by the robot. Given a set of demonstrations of feeding task, i.e., $\mathbf{X}_1(t)$, $\mathbf{X}_2(t)$ and $\mathbf{X}_3(t)$, we seek to extract important relevant features and constraints of this task in order to have a general context-free representation being utilized to generate trajectories accomplishing the task successfully in new environments.

In order to address this challenging problem, our approach consists of two main parts: (1) learning a general probabilistic model for feeding task from a set of demonstrations and (2) defining a metric of imitation to generate new trajectories of feeding task for environments in which the landmarks may be in different locations. In the next section the proposed approach is fully detailed, namely Sec. 4.1 discusses the first part and Sec. 4.2 discusses the second part.

4. Method Overview

4.1. *Determining task constraints using probabilistic modeling*

One of the most famous approaches for probabilistic modeling and density function approximation of a dataset is mixture modeling.[8] A probabilistic mixture model of K components is defined as the following probability density function:

$$p(\mathbf{x}(q)) = \sum_{k=1}^{K} p(k)p(\mathbf{x}(q)|k),$$

where $\mathbf{x}(q)$ is a vector datapoint at time index q (the query point in general), $p(k)$ is the prior probability and $p(\mathbf{x}(q)|k)$ is the conditional probability density function conditioning on which component k the datapoint $\mathbf{x}(q)$ is drawn from. In the case of a D-dimensional GMM, the conditional probability density function follows a Gaussian distribution:

$$p(\mathbf{x}(q)|k) = \frac{1}{\sqrt{(2\pi)^D|\mathbf{\Sigma}_k|}} e^{\left(\frac{-(\mathbf{x}(q)-\mu_k)'\mathbf{\Sigma}_k^{-1}(\mathbf{x}(q)-\mu_k)}{2}\right)},$$

where μ_k is the mean and $\mathbf{\Sigma}_k$ is the full covariance matrix of the kth Gaussian component. These parameters are calculated using expectation maximization (EM) algorithm.[20] One of the drawbacks of EM algorithm is that the optimal total number of Gaussian components K is not known in advance. Hence, a trade-off is needed to be reached between maximizing the log-likelihood of the probabilistic model and minimizing the number of components and their corresponding

parameters. Several criteria have been proposed and used in literature such as: Akaike information criteria[21] and Bayesian information criteria.[22] In this work, Bayesian information criteria (BIC) are used to determine the total number of components mixed together.

In order to have a probabilistic model of the feeding task, GMR[23] is used. However, since the trajectories $\eta_i^j(t)$ are not temporally aligned with each other, a temporal alignment technique is required to be used prior to GMR in order to compare the datapoints of the correct timestamps with each other.[6] Hence, dynamical time warping (DTW)[24] is used to temporally align the trajectories $\eta_i^j(t)$ of different demonstrations. K-means clustering algorithm[25] is applied to the timely-aligned data to avoid ending up in a poor local minima when EM algorithm is being used.[6] Using the timely-aligned, clustered dataset with the consecutive temporal values as the query points $\{\mathbf{X}_i(t), t\}$, a general form for the trajectories of the feeding task is achieved. The temporal and spatial variables are separated to form the parameters of each Gaussian component k as follows[6]:

$$\mu_k = \{\mu_{t,k}, \mu_{s,k}\}, \quad \mathbf{\Sigma}_k = \begin{pmatrix} \mathbf{\Sigma}_{tt,k} & \mathbf{\Sigma}_{ts,k} \\ \mathbf{\Sigma}_{st,k} & \mathbf{\Sigma}_{ss,k} \end{pmatrix},$$

where $\mu_{t,k}$ shows the mean of consecutive temporal values, $\mu_{s,k}$ denotes the mean of spatial values $\mathbf{X}_i(t)$ at each time stamp, $\mathbf{\Sigma}_{ts,k}$ and $\mathbf{\Sigma}_{st,k}$ show the spatiotemporal covariance matrices, $\mathbf{\Sigma}_{tt,k}$ denotes the temporal covariance matrix and $\mathbf{\Sigma}_{ss,k}$ shows the spatial covariance matrix of the kth Gaussian component. The conditional expectation $\hat{\mu}_{s,k}$ and the covariance matrix $\hat{\mathbf{\Sigma}}_{s,k}$ of each Gaussian component are estimated as follows[6]:

$$\hat{\mu}_{s,k} = \mu_{s,k} + \mathbf{\Sigma}_{st,k}(\mathbf{\Sigma}_{tt,k})^{-1}(t - \mu_{t,k}),$$

$$\hat{\mathbf{\Sigma}}_{s,k} = \mathbf{\Sigma}_{ss,k} - \mathbf{\Sigma}_{st,k}(\mathbf{\Sigma}_{tt,k})^{-1}\mathbf{\Sigma}_{ts,k}.$$

The components are mixed together as follows, using $\omega_k = \frac{p(t|k)}{\sum_{k'=1}^{K} p(t|k')}$ as the mixing coefficient:

$$\hat{\mu}_s = \sum_{k=1}^{K} \omega_k \hat{\mu}_{s,k}, \quad \hat{\mathbf{\Sigma}}_s = \sum_{k=1}^{K} \omega_k^2 \hat{\mathbf{\Sigma}}_{s,k}.$$

$M = \{\hat{\mu}_s, \hat{\mathbf{\Sigma}}_s\}$ forms a general probabilistic model for feeding task. Evaluating $\hat{\mu}_s = [\hat{\mu}_{pos}; \hat{\mu}_{ori}]$ at each timestamp gives a general trajectory of the task of feeding, where $\hat{\mu}_{pos} \in \mathbb{R}^3$ captures the important features of the positions while $\hat{\mu}_{ori} = [\hat{\mathbf{r}}_1; \hat{\mathbf{r}}_2] \in \mathbb{R}^6$ captures the important orientation constraints. In addition, by evaluating $\hat{\mathbf{\Sigma}}_s$ at each timestamp, the corresponding correlation as well as the allowed variations are achieved.

Remark 2. In order to have an accurate and valid model $M = \{\hat{\mu}_s, \hat{\mathbf{\Sigma}}_s\}$, $\hat{\mathbf{r}}_1$ and $\hat{\mathbf{r}}_2$ should be manually normalized, as the column vectors in any rotation matrix are unit vectors.

4.2. *Metric of imitation and optimization problem*

4.2.1. *Imitation without obstacle avoidance*

After using GMR, three probabilistic models for the task of feeding are achieved, each represented in the coordinate frame of one landmark, i.e., $M_{l_1} = \{\hat{\mu}_{s1}, \hat{\boldsymbol{\Sigma}}_{s1}\}$ in bowl's coordinate frame, $M_{l_2} = \{\hat{\mu}_{s2}, \hat{\boldsymbol{\Sigma}}_{s2}\}$ in mouth's coordinate frame and $M_{l_3} = \{\hat{\mu}_{s3}, \hat{\boldsymbol{\Sigma}}_{s3}\}$ in the coordinate frame of the base. For the purpose of reproduction of the feeding task, a metric of imitation needs to be defined measuring the similarity between the candidate trajectories $\{\eta_1(t), \eta_2(t), \eta_3(t)\}$ and the generalized trajectories $\{\hat{\mu}_{s1}, \hat{\mu}_{s2}, \hat{\mu}_{s3}\}$. Hence, a weighted Euclidean distance measure is defined as follows:

$$H(\eta_1, \eta_2, \eta_3) = \sum_{i=1}^{3} ((\eta_i(t) - \hat{\mu}_{si}(t))'(\hat{\boldsymbol{\Sigma}}_{si}(t))^{-1}$$
$$\times (\eta_i(t) - \hat{\mu}_{si}(t))).$$

The weights of the time-dependent metric of imitation H are chosen as the inverse of the full covariance matrices. This way, the correlations across different variables are taken into account while ensuring that the variables with less variances are more dominant to the cost function than those with high variances. In order to reproduce a proper trajectory for the feeding task in the coordinate frame of landmark l_3 (base of the robot), a constraint-based optimization problem can be set up using the quadratic function H as the cost function:

$$\min \quad H(\eta_1(t), \eta_2(t), \eta_3(t))$$
$$\text{subject to } \eta_1(t) - \mathbf{R}_1(\eta_3(t) - \mathbf{a}_1) = 0,$$
$$\eta_2(t) - \mathbf{R}_2(\eta_3(t) - \mathbf{a}_2) = 0,$$
$$\frac{1}{2}(\eta_3(t)'\mathbf{Q}\eta_3(t)) - 1 = 0,$$
$$\frac{1}{2}(\eta_3(t)'\mathbf{P}\eta_3(t)) - 1 = 0,$$
$$\frac{1}{2}(\eta_3(t)'\mathbf{S}\eta_3(t)) = 0,$$

where $\mathbf{R}_1 = \text{diag}(\mathbf{R}_3^1, (\mathbf{R}_3^1)', (\mathbf{R}_3^1)') \in \mathbb{R}^{9\times9}$, $\mathbf{R}_2 = \text{diag}(\mathbf{R}_3^2, (\mathbf{R}_3^2)', (\mathbf{R}_3^2)') \in \mathbb{R}^{9\times9}$, $\mathbf{a}_1 = [\mathbf{p}_1; \mathbf{0}_{6\times1}] \in \mathbb{R}^9$, $\mathbf{a}_2 = [\mathbf{p}_2; \mathbf{0}_{6\times1}] \in \mathbb{R}^9$, $\mathbf{Q} = \text{diag}(0,0,0,2,2,2,0,0,0) \in \mathbb{R}^{9\times9}$, $\mathbf{P} = \text{diag}(0, 0,0,0,0,0,2,2,2) \in \mathbb{R}^{9\times9}$ and $\mathbf{S} = [\mathbf{0}_{3\times3}, \mathbf{0}_{3\times3}, \mathbf{0}_{3\times3}; \mathbf{0}_{3\times3}, \mathbf{0}_{3\times3}, \mathbf{I}_{3\times3}; \mathbf{0}_{3\times3}, \mathbf{I}_{3\times3}, \mathbf{0}_{3\times3}] \in \mathbb{R}^{9\times9}$. Here, $\mathbf{p}_1 \in \mathbb{R}^3$ is the position vector of landmark l_1 (bowl) with respect to landmark l_3 (base of the robot) represented in l_3 coordinate frame and $\mathbf{p}_2 \in \mathbb{R}^3$ is the position vector of landmark l_2 (mouth) with respect to landmark l_3 (base of the robot) represented in l_3 coordinate frame. The first two constraints of this optimization problem take into account the relationship between the trajectories of different coordinate systems. The third and fourth constraints ensure that the solution of the optimization provides unit vectors for $\mathbf{r}_1$ and $\mathbf{r}_2$ in η_{ori}. Finally, the last constraint ensures that $\mathbf{r}_1$ and $\mathbf{r}_2$ are orthogonal. Since, to the best of our knowledge, there is no closed-form solution to this constrained optimization problem, a numerical method for

convex constrained minimization, called the interior point algorithm,[26] is used to find minimum of the cost function H at each time instance. The solution $\eta_3^*(t)$ that minimizes H is achieved by implementing the interior point algorithm with MATLAB optimization toolbox.[27] $\eta_3^*(t) = [\eta_{\text{pos}}^*; \eta_{\text{ori}}^*]$ provides the appropriate trajectory of feeding task for the given environment (defined by the first two constraints) in the base frame of the robot. η_{pos}^* provides the suitable positions and $\eta_{\text{ori}}^* = [\mathbf{r}_1^*; \mathbf{r}_2^*]$ gives the first two columns of the rotation matrix representing the spoon's orientation while the third column can be achieved using the cross product of these two columns, i.e., $\mathbf{r}_3^* = \mathbf{r}_1^* \times \mathbf{r}_2^*$.

4.2.2. *Imitation with obstacle avoidance*

In this sub-subsection, we consider the problem of imitation with simple task space obstacle avoidance. The goal of the obstacle avoidance problem discussed in here is to merely make the end-effector (or the spoon attached to it) avoid obstacles. We assume the positions, sizes and shapes of the obstacles are fully known to the robot. In order to tackle this problem, a set of additional constraints (dependent on the number of obstacles in the task space environment) is added to the optimization problem causing the end-effector/spoon to keep its distance from obstacles above a predefined safety margin ϵ. In general, the obstacles are modeled as ellipsoids with three semi-principal axes lengths a_1, a_2 and a_3 whose values are depending on the size and shape of the obstacles. Let $\mathbf{c}_3 \in \mathbb{R}^3$ be the Cartesian position vector of the center of an obstacle with respect to landmark l_3, i.e., the base frame of the robot. Using the most general matrix representation of an ellipsoid in a Cartesian coordinate system, the additional constraint appended to the previous optimization problem for avoiding this obstacle is given as

$$(\eta_{\text{pos}_3} - \mathbf{c}_3)'\mathbf{R}_e'\mathbf{A}\mathbf{R}_e(\eta_{\text{pos}_3} - \mathbf{c}_3) \geq 1 + \epsilon,$$

where $\mathbf{A} = \text{diag}((\frac{1}{a_1})^2, (\frac{1}{a_2})^2, (\frac{1}{a_3})^2) \in \mathbb{R}^{3\times3}$, $\mathbf{R}_e \in \mathbb{R}^{3\times3}$ is a rotation matrix showing the rotation of the ellipsoid from the standard form in base coordinate frame and ϵ is the safety margin. As we only consider collision avoidance of the end-effector/spoon, note that this constraint only forces the end-effector/spoon to avoid the obstacle while there is no guarantee that the other links of the robot's arm do not collide with the obstacle.

5. Experimental Results

In order to evaluate the performance of the proposed system, we implemented the task of feeding using the Baxter robot.[28] The experiment environment consisted of a bowl full of cereal and a symbolic mannequin representing the user's mouth, both placed on a table (see Fig. 1). A spoon was rigidly attached to the gripper of the right hand of the robot as

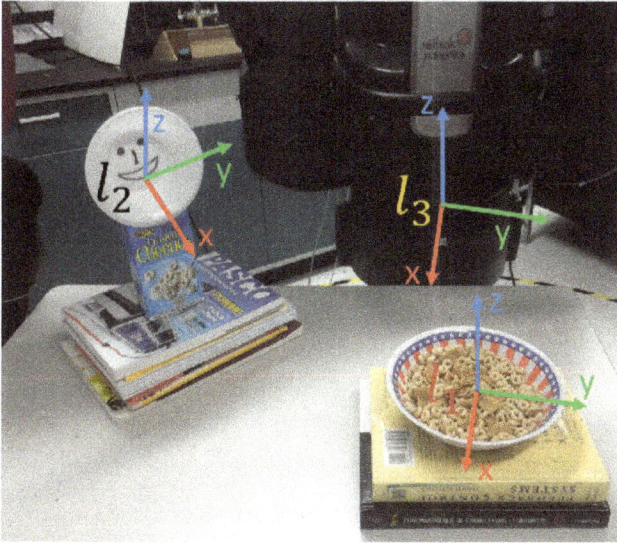

Fig. 1. Experimental setup for feeding task.

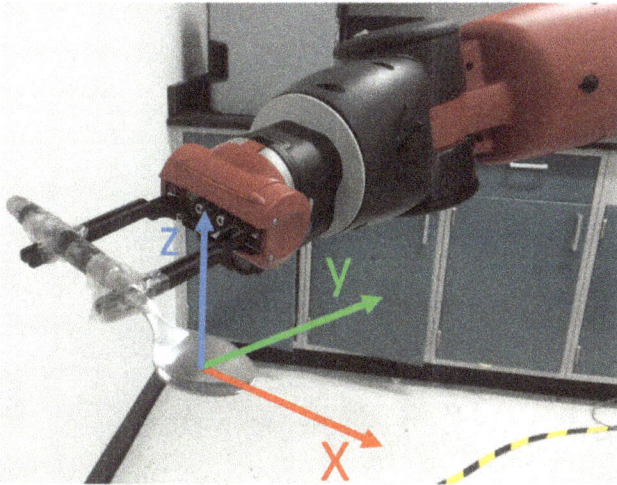

Fig. 2. The spoon and its coordinate frame. The x-axis denotes the approach direction of the spoon and the z-axis is normal to the tip of the spoon.

shown in Fig. 2. The feeding task we considered here consisted of reaching for the bowl, scooping with the spoon, moving the loaded spoon toward the user's mouth, waiting for user to hypothetically unload the spoon, coming back toward the bowl, unloading the cereal back in the bowl and moving away from the bowl. Throughout the experiment it was assumed there is no specific action taking place by the robot or the user's mouth at the location of the mouth, only that the spoon stopped moving for an amount of time, i.e., there was no actual desire for the act of eating by individual's mouth. We conducted 61 demonstrations of the feeding task in 21 different locations and/or orientations for the bowl and the mouth. These demonstrations were conducted by demonstrator manually moving the end-effector to perform the feeding task, a process called *kinesthetic teaching*. The lengths of all recorded trajectories were rescaled to 570

points via linear interpolation using *interp1* function in MATLAB. Figures 3(a)–3(c) show the BIC score graph for finding the optimal numbers of GMM components for bowl, mouth and base, respectively. Based on the BIC score analysis, the optimal numbers of GMM components for bowl, mouth and the base were 25, 27 and 25, respectively. By applying GMR with the chosen optimal number of Gaussian components, three models for feeding task in the coordinate frame of bowl, mouth and the base were achieved. The code used for learning the parameters of GMM is available online.[29] The generalized (mean) position trajectories and the corresponding standard deviations in the coordinate frame of the bowl and the mouth are depicted in Figs. 4 and 5, respectively. As shown in Fig. 4, the three position components x, y and z of the spoon are highly constrained when the spoon is reaching the mouth allowing for less variations as it is expected. Similar observations are made for x, y and z of the spoon in the bowl's coordinate frame as the spoon was initially reaching the bowl for scooping and on the way back when unloading the cereals. Figures 6 and 7 show the generalized first and second columns of the rotation matrix representing the orientation of the spoon with respect to the base frame, along with the corresponding standard deviations. For a better understanding and visualization, the Euler angle representations of the orientations of the spoon in the base frame are depicted in Fig. 8. The Euler sequence used in here is *ZYX*. The first two angles θ_x and θ_y in Fig. 8 are the ones affecting the level adjustment of the spoon while θ_z is responsible for turning the spoon around the z-direction, which is normal to the table and does not affect the level adjustment of the spoon. As shown in Fig. 8, θ_x and θ_y are highly constrained when the loaded spoon was moving from the bowl toward the mouth and back again to the bowl keeping its orientation level in order to prevent the cereal from falling on the ground. Since in the demonstrations, the positions of the bowl and the mouth varied with respect to the base, θ_z allows for high variations as expected. As shown in Figs. 4–8, the most important features of the feeding task, i.e., reaching the bowl, scooping properly, keeping the spoon level between the bowl and the mouth and reaching the individual's mouth, are successfully captured and learnt using GMM/GMR framework.

5.1. *Optimization results without obstacle avoidance*

In order to illustrate the results of the optimization part, we have conducted four experiments of the feeding task each having a different position/orientation for the bowl and the mouth with respect to the base. All experiments were successfully completed and the video of the full experiments is available online.[30] To keep this subsection short, here we show the results of one of the experiments. For this case, the position and orientation of the bowl with respect to the base were $\mathbf{p}_1 = [0.952; -0.234; 0.085]$ m and $\mathbf{R}_1^3 = \mathbf{I}_{3\times3}$, respectively, and the position and orientation of the mouth with respect to the base were $\mathbf{p}_2 = [0.500; -0.317; 0.095]$ m and

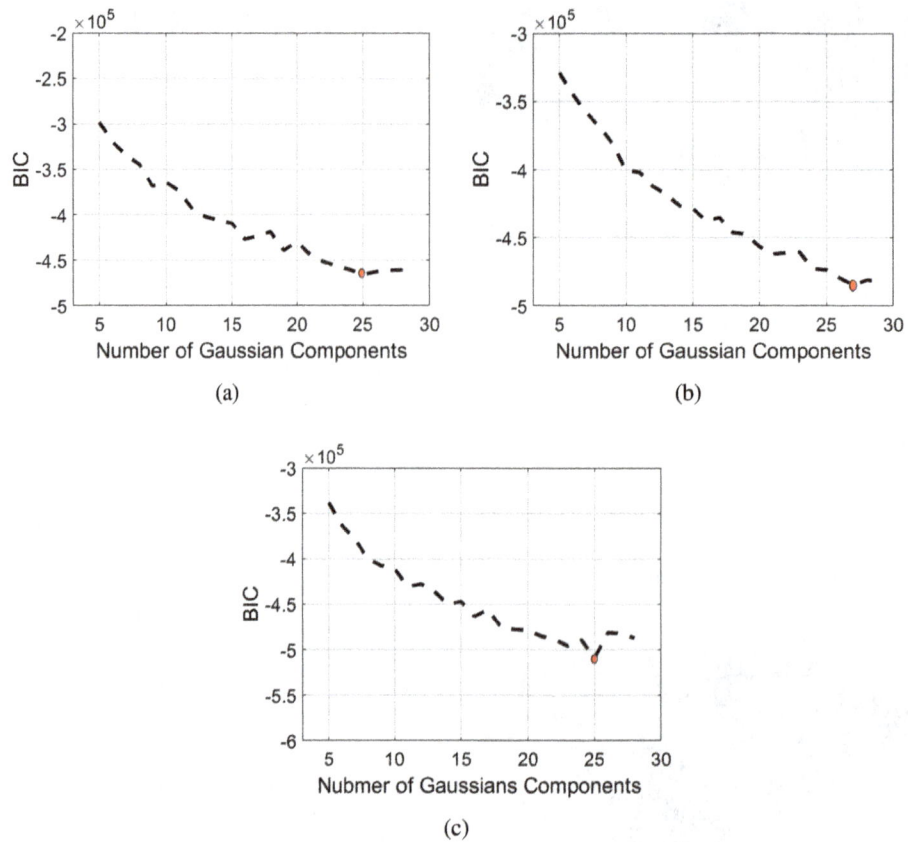

Fig. 3. Optimal numbers of Gaussian components for (a) M_{l_1} (bowl), (b) M_{l_2} (mouth) and (c) M_{l_3} (base).

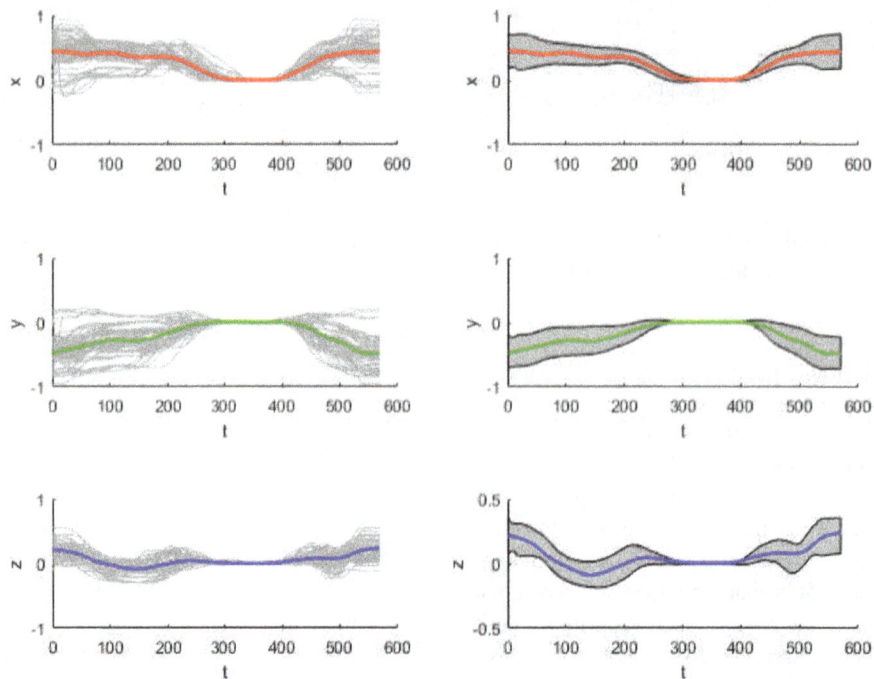

Fig. 4. The generalized (mean) position trajectories of the spoon in the coordinate frame of the mouth and the corresponding standard deviations around them. (Left column) Recorded demonstrations and the generalized trajectories for x, y and z are depicted in gray and bold lines, respectively. (Right column) The generalized trajectories of the spoon with the allowed variations are shown. As depicted, the position trajectories are highly constrained in the time index interval of 280–400 as the spoon reaches the mouth and stands still for an amount of time.

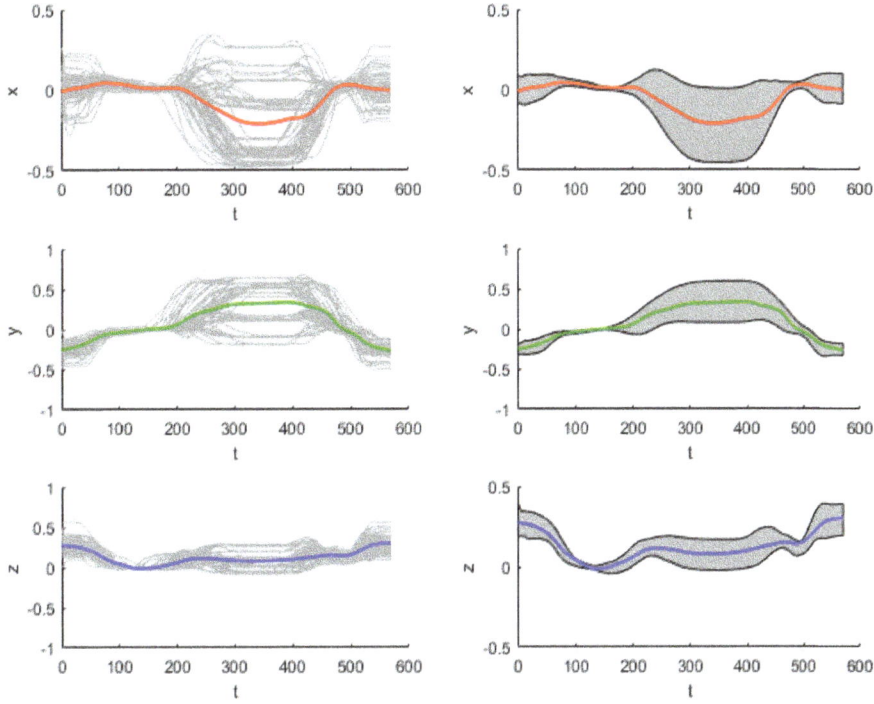

Fig. 5. The generalized position trajectories of the spoon in the coordinate frame of the bowl and the corresponding standard deviations around them. (Left column) Recorded demonstrations and the generalized trajectories for x, y and z are depicted in gray and bold lines, respectively. (Right column) The generalized trajectories of the spoon with the allowed variations are shown. As depicted, the position trajectories are highly constrained in the time index interval of 90–200 as the spoon reaches the bowl while scooping takes place and in the time index interval of 480–510 as the spoon again reaches the bowl to unload the cereal.

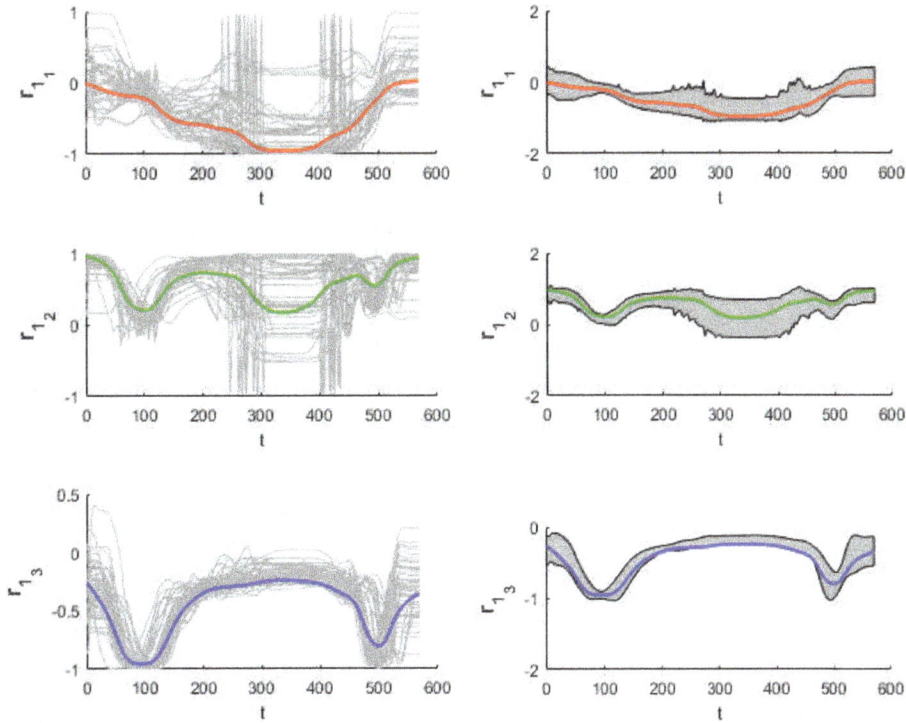

Fig. 6. The generalized three components of the first column of the rotation matrix representing the orientation of the spoon in the coordinate frame of the base along with the allowed variations. (Left column) Recorded demonstrations and the generalized trajectories for r_{1_1}, r_{1_2} and r_{1_3} are depicted in gray and bold lines, respectively. (Right column) The generalized trajectories of the three components of the first column of the rotation matrix with the allowed variations are shown.

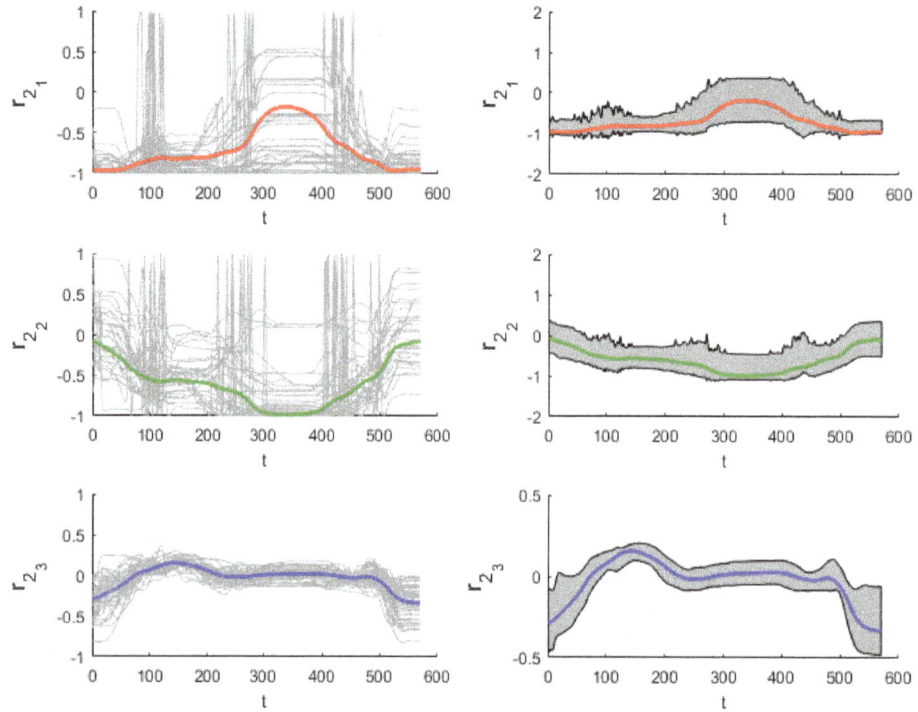

Fig. 7. The generalized three components of the second column of the rotation matrix representing the orientation of the spoon in the coordinate frame of the base along with the allowed variations. (Left column) Recorded demonstrations and the generalized trajectories for r_{2_1}, r_{2_2} and r_{2_3} are depicted in gray and bold lines, respectively. (Right column) The generalized trajectories of the three components of the second column of the rotation matrix with the allowed variations are shown.

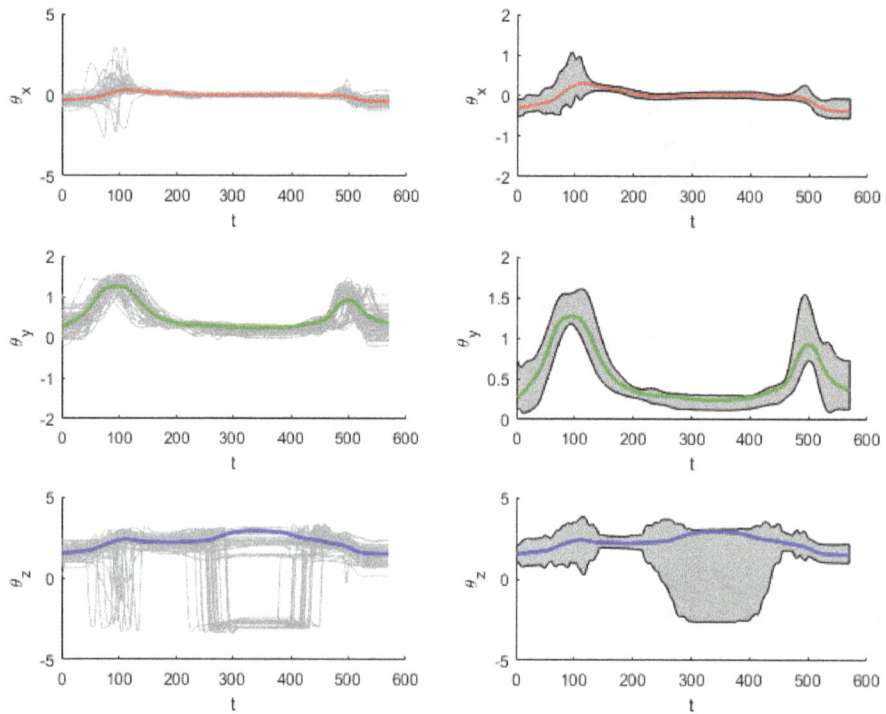

Fig. 8. The Euler angle representations of the generalized orientations with the corresponding standard deviations. (Left column) Recorded demonstrations and the generalized trajectories for θ_x, θ_y and θ_z are depicted in gray and bold lines, respectively. (Right column) The generalized trajectories of the orientation of the spoon with the allowed variations are shown. In the time index interval of 200–480, θ_x and θ_y are highly constrained keeping the spoon level as it moves from the bowl to the mouth and moves back to the bowl again. θ_z allows for high variations as the positions of the bowl and the mouth with respect to the base vary in our demonstrations.

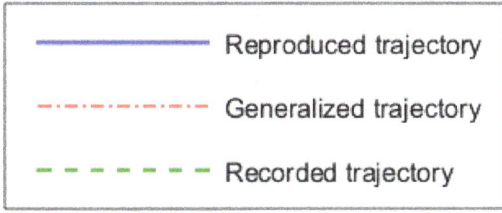

Fig. 9. Legend for Figs. 10(a)–10(i).

$$\mathbf{R}_2^3 = \begin{bmatrix} 0.986 & -0.169 & 0 \\ 0.169 & 0.986 & 0 \\ 0 & 0 & 1 \end{bmatrix}, \text{ respectively. Figure 10 shows}$$

the output of the optimization (the reproduced trajectory η^*), the generalized (mean) trajectory $\hat{\mu}_s$ and a recorded trajectory of the feeding task for this position with respect to the base frame, achieved by manually performing the task with the robot using kinesthetic teaching. As shown in Fig. 10, the spoon was inside the bowl during the time index interval of 125–180, then it moved toward the mouth and was at the mouth during the time index interval of 300–380 and finally, it moved back toward the bowl again and reached the bowl

during the time index interval of 450–500. Legend of Figs. 10(a)–10(i) is presented in Fig. 9.

5.2. *Optimization results with obstacle avoidance*

Different obstacles with different sizes and shapes were placed between the bowl and the mouth for the same four positions as in the previous subsection. All the four experiments were successfully completed. Here, the results of the optimization with obstacle avoidance for the same position of the landmarks as in the last subsection and for a single obstacle are depicted in Fig. 12. The legend of Figs. 12(a)–12(c) is presented in Fig. 11. The position of the center of the obstacle with respect to the base frame was $\mathbf{c} = [0.723; -0.251; 0.158]$ m. The rotation of the ellipsoid, representing the obstacle, from the standard form in the base frame is considered to be $\mathbf{R}_e = \mathbf{I}_{3 \times 3}$. The three semi-principal axes lengths a_1, a_2 and a_3 for the chosen obstacle were 0.022 m, 0.022 m and 0.043 m, respectively. The safety margin ϵ was chosen to be $\epsilon = 0.020$ m. Since the additional constraint of avoiding the obstacle only affects the position components and not the orientations, only positions are shown in Fig. 12 (orientation components are identical to the results of the previous subsection shown in Fig. 10). As shown in Fig. 12, the spoon successfully went around the obstacle avoiding

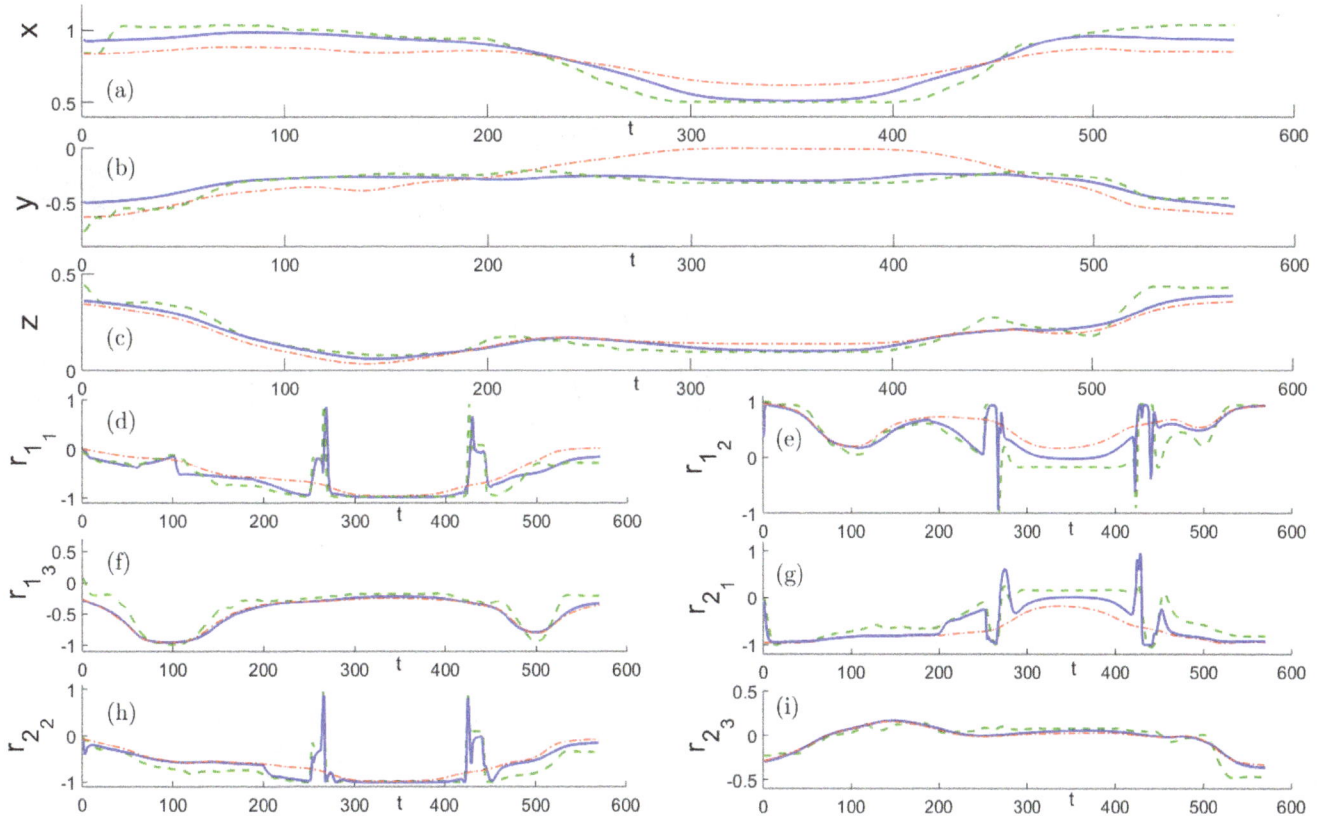

Fig. 10. The reproduced position and orientation trajectories in the experiment considering no obstacle.

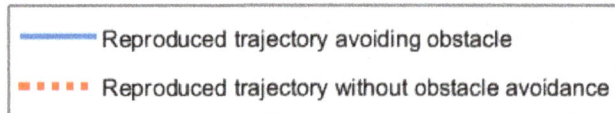

Fig. 11. Legend for Figs. 12(a)–12(c).

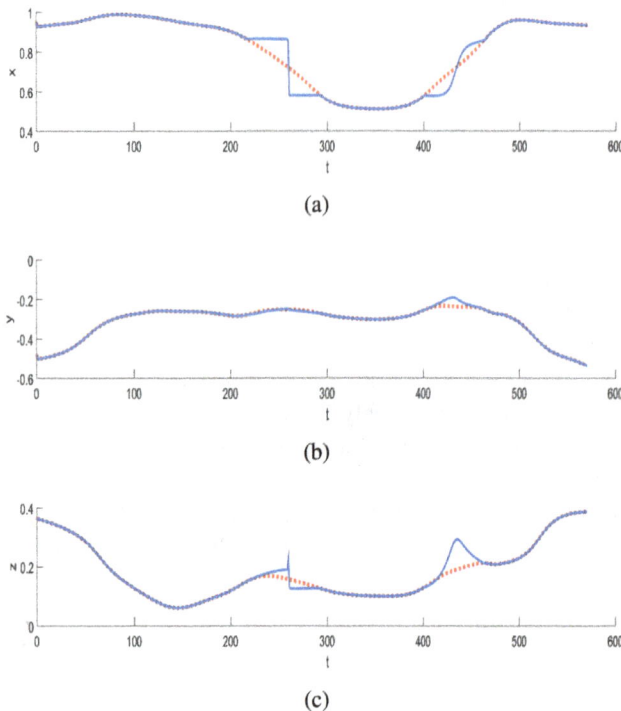

(a)

(b)

(c)

Fig. 12. The reproduced position trajectories in the experiment considering obstacle avoidance.

collision while maintaining its highly constrained orientation in this segment, i.e., keeping the spoon level. It is worthwhile to mention again that by using the inverse of the covariance matrices for the weights of the cost function in the optimization problem, there is no need to add an extra explicit condition to keep the orientation of the spoon level. This information has been already captured in the covariance matrix.

6. Conclusion and Future Work

In this paper, a LFD (also known as PBD) framework is adopted and developed to teach a robot how to feed a person using a bowl and a spoon rigidly attached to its end-effector. The framework consists of two main parts: (1) a data-driven probabilistic model of the feeding task which provides the generalized spoon trajectory and the corresponding variations and correlations across the variables; and (2) a convex constrained optimization process generating proper feeding trajectories for new environments which also deals with simple end-effector collision avoidance. In order to demonstrate the

performance of the proposed framework for feeding task, four implementations of the task were successfully conducted placing the bowl and the mouth in different positions/orientations.

The reproduced trajectories of feeding task are in the task space of the robot. While this may be viewed as a disadvantage resulting in dealing with inverse kinematics issues, it might be seen as an advantage allowing us to transfer the learnt task from one robot to another robot with similar physical characteristics, i.e., similar links and joints. In the current work, the process of learning was done under the assumption of having solely successful demonstrations of the task in the dataset. It is interesting to investigate what can be learnt from partial or completely failed demonstrations of a task. Moreover, in the current work, a batch learning process has been used to find the constraints of the task. In order to be able to customize the autonomous implementation of a task by the robot for each user, an active learning process may be preferred. Although the current method deals with simple end-effector collision avoidance, but there is no guarantee for collision avoidance of other parts of the robot. In future work, we plan to address the full collision avoidance, the interesting problem of learning from both successful and failed demonstrations, the active learning problem as well as the challenging problem of transferring the feeding task to a similar robot.

Acknowledgments

This study was funded in part by NIDILRR Grant No. H133G120275 and in part by NSF Grant Nos. IIS-1409823 and IIS-1527794. However, these contents do not necessarily represent the policy of the aforementioned funding agencies, and one should not assume endorsement by the Federal Government.

References

[1] M. W. Brault, Americans with disabilities: 2010, Current Population Reports No. p70-131, United States Census Bureau (2012).
[2] Disabled World, Disability statistics: Facts on disabilities and disability issues (2017), https://www.disabled-world.com/disability/statistics/.
[3] M. J. Topping and J. K. Smith, The development of Handy 1: A robotic system to assist the severely disabled, *Technol. Disability* **10**, 95 (1999).
[4] I. Sumio, Meal-assistance robot "My Spoon", *J. Robot. Soc. Jpn.* **21**, 378 (2003).
[5] H. Kawamoto *et al.*, Meal-assistance by robot suit HAL using detection of food position with camera, *Proc. 2011 IEEE Int. Conf. Robotics and Biomimetics (ROBIO)* (IEEE, 2011).
[6] S. Calinon, F. Guenter and A. Billard, On learning, representing, and generalizing a task in a humanoid robot, *IEEE Trans. Syst. Man Cybern. B, Cybern.* **37**, 286 (2007).
[7] I. Jolliffe, *Principal Component Analysis* (John Wiley & Sons, 2002).
[8] G. McLachlan and D. Peel, *Finite Mixture Models* (John Wiley & Sons, 2004).

[9]G. Ye and R. Alterovitz, Demonstration-guided motion planning, *Proc. Int. Symp. Robotics Research (ISRR)*, Vol. 5 (2011).

[10]C. Bowen and R. Alterovitz, Closed-loop global motion planning for reactive execution of learned tasks, *Proc. 2014 IEEE/RSJ Int. Conf. Intelligent Robots and Systems* (2014), pp. 1754–1760.

[11]C. Bowen, G. Ye and R. Alterovitz, Asymptotically optimal motion planning for learned tasks using time-dependent cost maps, *IEEE Trans. Autom. Sci. Eng.* **12**, 171 (2015).

[12]N. Figueroa, A. L. P. Ureche and A. Billard, Learning complex sequential tasks from demonstration: A pizza dough rolling case study, *Proc. Eleventh ACM/IEEE Int. Conf. Human Robot Interaction* (IEEE, 2016).

[13]N. Ettehadi, S. Manaffam and A. Behal, Learning from demonstration: Generalization via task segmentation, *IOP Conf. Ser., Mater. Sci. Eng.* **261**, 012001 (2017).

[14]S. Kim, A. Shukla and A. Billard, Catching objects in flight, *IEEE Trans. Robot.* **30**, 1049 (2014).

[15]E. Ghalamzan *et al.*, An incremental approach to learning generalizable robot tasks from human demonstration, *Proc. 2015 Int. Conf. Robotics and Automation* (2015).

[16]A. K. Tanwani and S. Calinon, Learning robot manipulation tasks with task-parameterized semitied hidden semi-Markov model, *IEEE Robot. Autom. Lett.* **1**, 235 (2016).

[17]S. Choi, K. Lee and S. Oh, Robust learning from demonstration using leveraged Gaussian processes and sparse-constrained optimization, *Proc. 2016 IEEE Int. Conf. Robotics and Automation (ICRA)* (2016), pp. 470–475.

[18]S. M. Khansari-Zadeh and A. Billard, Learning stable nonlinear dynamical systems with Gaussian mixture models, *IEEE Trans. Robot.* **27**, 943 (2011).

[19]B. D. Argall *et al.*, A survey of robot learning from demonstration, *Robot. Auton. Syst.* **57**, 469 (2009).

[20]A. P. Dempster, N. M. Laird and D. B. Rubin, Maximum likelihood from incomplete data via the EM algorithm, *J. R. Stat. Soc. B, Methodol.* **39**, 1 (1977).

[21]H. Akaike, On entropy maximization principle, in *Application of Statistics* (North-Holland, 1977), pp. 27–41.

[22]G. Schwarz, Estimating the dimension of a model, *Ann. Stat.* **6**, 461 (1978).

[23]D. A. Cohn, Z. Ghahramani and M. I. Jordan, Active learning with statistical models, *J. Artif. Intell. Res.* **4**, 129 (1996).

[24]C.-Y. Chiu *et al.*, Content-based retrieval for human motion data, *J. Vis. Commun. Image Represent.* **15**, 446 (2004).

[25]J. A. Hartigan and M. A. Wong, Algorithm AS 136: A k-means clustering algorithm, *J. R. Stat. Soc. C, 100 Appl. Stat.* **28**, 100 (1979).

[26]Y. Nesterov and A. Nemirovskii, *Interior-Point Polynomial Algorithms in Convex Programming* (Society for Industrial and Applied Mathematics, 1994).

[27]M. A. Branch and A. Grace, *Optimization Toolbox: For Use with MATLAB: User's Guide: Version 1* (MathWorks, 1998).

[28]Rethink Robotics, Baxter research robot (2013), http://cdn-staging.rethinkrobotics.com/wp-content/uploads/2014/08/BRR [Vol. 9, p. 13].

[29]S. Calinon, A tutorial on task-parameterized movement learning and retrieval, *Intell. Serv. Robot.* **9**, 1 (2016).

[30]UCF Assistive Robotics Lab, Experiment: Implementation of feeding task via learning from demonstration (2017), https://youtube/R0wCXjVi73U.

Cooperative obstacle avoidance for heterogeneous unmanned systems during search mission

K. Harikumar[*,‡], Titas Bera[*,§], Rajarshi Bardhan[*,¶] and Suresh Sundaram[*,†,‖]

*ST Engineering-NTU Corporate Laboratory
School of Electrical and Electronic Engineering
Nanyang Technological University
Singapore 639798, Singapore
†School of Computer Science and Engineering
Nanyang Technological University
Singapore 639798, Singapore
‡kharikumar@ntu.edu.sg
§btitas@ntu.edu.sg
¶rbardhan@ntu.edu.sg
‖ssundaram@ntu.edu.sg

Accepted 6 May 2018; Published 13 June 2018

The problem of cooperative obstacle avoidance by a group of unmanned ground vehicles (UGVs) and unmanned air vehicles (UAVs) during a typical search mission is addressed in this paper. The group of UAVs and UGVs are performing a search operation in the designated area. All the UAVs and UGVs are equipped with a vision sensor/LIDAR to identify the possible obstacles in the search space. Due to their operation on the ground, UGVs are more likely to encounter obstacles. The obstacle avoidance for UGV under the event of sensor failure is done with environment information from the nearest UAV. The UAV plans its trajectory according to the UGV's expected future trajectory, leading towards the base station. The UGV replans its trajectory to avoid obstacles after obtaining the information from its nearest UAV. A simulation study is performed with 10 UAVs and five UGVs performing a search mission in 1 km × 1 km area. The proposed obstacle avoidance method is experimentally validated in the outdoor environment with an autonomous UAV equipped with a camera and an autonomous UGV navigating based on GPS localization and environment information from the UAV.

Keywords: Heterogeneous system; cooperation; obstacle avoidance; vision-based detection.

1. Introduction

The use of a heterogeneous system of unmanned air vehicles (UAVs) and unmanned ground vehicles (UGVs) for various missions provides additional advantages when compared to the homogeneous system due to their accessibility to both land and airspace.[1] The cooperation between UAV and UGV for removal of landmines is reported in Ref. 2. To locate a moving target in a given area, an algorithm is developed that uses the cooperation between a group of UAVs and UGVs.[3] A UAV supporting a team of UGVs to transport objects in an industrial area is presented in Ref. 4. The detection and avoidance of obstacle is a challenging problem for UGVs, when operating in an unknown environment. Usually, UGVs are equipped with sensors like camera, LIDAR, SONAR, etc.,[5,6] for obstacle detection. When a sudden sensor failure occurs, the UGV needs to make a decision based on any prior information and it increases the risk of collision with obstacles.

Use of multiple UGVs and UAVs to search and explore an unknown environment is efficient than using a single agent. If one of the UGVs encounters a sensor failure, then other UGVs share some information about the environment and thus enable the UGV to navigate with a low risk of collision.[7] Exploration and obstacle detection by a team of UAVs and UGVs is an efficient approach when compared to homogeneous agents performing a mission. The cooperation between heterogeneous agents (UAV and UGV) is utilized in this paper, in the event of a sensor failure for UGV. As UAV can fly at different altitudes, it is easier for a UAV equipped with a camera and gimbal to focus on the region of interest. A navigation system is developed for a UGV operating in an indoor environment through visual feedback provided by a miniature UAV.[8] The navigation of UGV depends only on the visual feedback by UAV and hence limits the UAV motion to be in the close proximity of UGV. Similarly, use of UAV to assist UGV for performing a mission is considered in Ref. 9.

In this paper, a team of UAVs and UGVs perform the search operation in a given area. The waypoints are selected from a uniform probability distribution and each UAV/UGV navigates to the respectively generated waypoint. When a

---

[‡]Corresponding author.

UGV encounters a sensor failure, it communicates the information to its nearest UAV. It is assumed here that the sensor fault occurs only for UGVs. The nearest UAV then terminates search operation temporarily and moves along the trajectory to be followed by UGV. Using the images procured from camera, the UAV tries to estimate the position and size of the obstacles present and communicates to the UGV in real time. UGV then decides to continue with its original trajectory or plans a new trajectory based on the information received from UAV. UGV is autonomous with GPS localization. Stable trajectory tracking control laws are used for both UAV and UGV navigation. After UGV reaches the base station, the UAV resumes the search mission.

A preliminary version of this paper is presented in Ref. 10, where the concept was obstacle avoidance for a single UGV and UAV. The present paper includes multi-UGV and multi-UAV search and a coordination between a UGV and its nearest UAV in the event of sensor failure. A simulation study is presented for a search mission involving five UGVs and 10 UAVs, where sensor failure occurs for all the five UGVs during the mission. Experimental results are presented for vision-based obstacle detection and avoidance for one UGV and one UAV.

This paper is organized as follows. Section 2 describes the search and obstacle avoidance algorithm, including trajectory tracking control law for UAV and UGV. It also explains the obstacle detection methodology using a camera. Simulation results for five UGV and 10 UAV are given in Sec. 3. Experimental results are presented in Sec. 4, followed by conclusions in Sec. 5.

2.   Search and Cooperative Obstacle Avoidance

The heterogeneous team consisting of UAVs and UGVs is performing the search operation in the given area. It is assumed that each UAV is equipped with a vision sensor to detect any ground obstacle or target. It is also assumed that each UAV flies at different altitude to avoid collision between UAVs. Similarly, each UGV is assumed to be having a vision sensor and LIDAR for detecting ground obstacles or targets. Both UAVs and UGVs generate waypoints in the given search space following a uniform probability distribution $[U()]$. Let the current position of ith UAV be denoted by $[x_i^u(t), y_i^u(t)]$ and for UGV be denoted by $[x_i^g(t), y_i^g(t)]$. Let the kth waypoint be denoted by $W_k(x, y)$. Then

$$W_k(x, y) \in [U(x_a, x_b), U(y_a, y_b)], \tag{1}$$

where $[(x_a, y_a), (x_a, y_b), (x_b, y_b), (x_b, y_a)]$ are the boundary coordinates of the search region. Upon reaching a waypoint, the UAV/UGV moves to the next waypoint generated using (1). An ith UAV is nearest to jth UGV if

$$d_{ji} = \min\{d_{j1}, d_{j2}, \ldots, d_{jn}\}, \tag{2}$$

where d_{ji} denotes the Euclidean distance between jth UGV and ith UAV. Let jth UGV encounter a sensor fault at time $t = t_j^i$. Then the UGV remains stationary till its nearest UAV (let it be ith UAV) acknowledges the fault information and comes near to it. Once the ith UAV receives the fault information from jth UGV, it terminates its search mission temporarily. The ith UAV's next waypoint is the current location of jth UGV. Upon reaching near jth UGV, the ith UAV flies along the straight line path connecting the jth UGV's current position and base station. The jth UGV follows the ith UAV, receiving the obstacle information from the UAV. All the obstacles are approximated by a circle. The kth obstacle is shown in blue color in Fig. 1 with radius r_{ok} and center at O_k. The path followed by UGV to avoid the obstacle is shown as black dotted line. The avoidance radius r_k is selected as given below:

$$r_k > r_{ok} + 2 * r_{\min} + O_{ke}, \tag{3}$$

where $r_{\min}$ is the minimum turn radius of the UGV and O_{ke} is the worst case error in estimating O_k. The accuracy in estimation of O_k depends upon the camera specifications and GPS positioning accuracy of UAV. The UAV resumes the search mission once the UGV reaches the base station.

2.1.   *UAV trajectory tracking control law*

Each UAV navigates from one waypoint to other till it receives fault information from a UGV. After a UAV receives fault information from a UGV, it follows a trajectory that connects the current UGV's position to the base station. Each UAV is assumed to have inner loop velocity control. The UAV is expected to follow the nominal desired trajectory of UGV in the near future. Let ith UAV be the nearest to the jth UGV under fault. The desired reference trajectory for ith

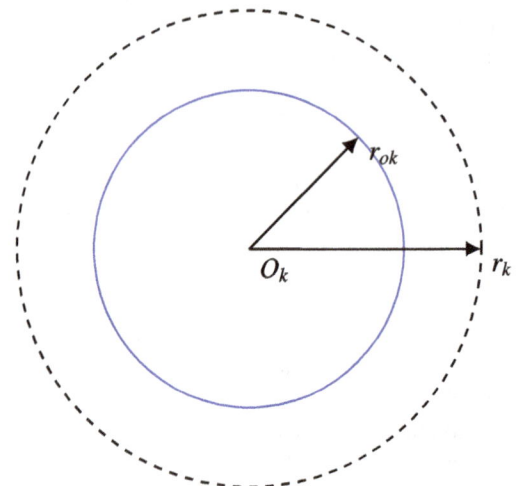

Fig. 1. (Color online) Obstacle shown in blue and path taken by UGV in black.

UAV is given by

$$x_{ri}^u(t) = x_j^g(t) + \delta_x, \tag{4}$$

$$y_{ri}^u(t) = y_j^g(t) + \delta_y. \tag{5}$$

The parameters δ_x and δ_y are constants that depend on the region around the path of the UGV that needs to be monitored. The outer loop velocity command for trajectory tracking is given by

$$vx_{ri}^u(t) = vx_j^g(t) + \frac{V_m(x_{ri}^u(t) - x^u(t))}{\epsilon + \|e_i(t)\|_2}, \tag{6}$$

$$vy_{ri}^u(t) = vy_j^g(t) + \frac{V_m(y_{ri}^u(t) - y^u(t))}{\epsilon + \|e_i(t)\|_2}, \tag{7}$$

where $e_i(t) = [x_{ri}^u(t) - x^u(t), y_{ri}^u(t) - y^u(t)]^T = [e_{xi}(t), e_{yi}(t)]^T$ and the parameters ϵ and V_m are determined as given below.

Define a Lyapunov function

$$V = \frac{\|e_i(t)\|_2^2}{2}. \tag{8}$$

The derivative of V can be written as

$$\dot{V} = e_{xi}(t)(vx_j^g(t) - vx_{ri}^u(t)) + e_{yi}(t)(vy_j^g(t) - vy_{ri}^u(t)). \tag{9}$$

Using (6) and (7), the above equation can be rewritten as shown below:

$$\dot{V} = -\frac{V_m(\|e_i(t)\|_2^2)}{\epsilon + \|e_i(t)\|_2}. \tag{10}$$

Therefore, $\dot{V} < 0$ for $\epsilon > 0$ and $V_m > 0$, and hence the tracking law given in (6) and (7) is asymptotically stable.[11] For waypoint following, $[x_{ri}^u(t), y_{ri}^u(t)]^T = W_k(x, y)$, $vx_j^g(t) = 0$ and $vy_j^g(t) = 0$.

2.2. Vision-based obstacle detection by UAV

During the obstacle detection phase, the UAV is flying with a downward pointing camera at a constant altitude of h from ground level. The UGV is assumed to be moving along a level surface. A gimbal is used to stabilize camera from disturbances arising due to UAV motion. A diagram explaining obstacle information extraction through image processing by UAV and subsequent information transmission to UGV is shown in Fig. 2. The images captured by UAV are corrected for distortion like wide angle, noise, etc. The corrected image is processed in OPENCV for extraction of features like lines and contours. The approximate size of the obstacles are assumed to be known *a priori*. The extracted contours are checked for convexity. Thresholding is applied for the convex contours to identify the obstacle and to eliminate the outliers. A dedicated Wi-Fi network between UAV and UGV serves as the medium of information exchange between them. Let the position of the obstacle be inertial frame be denoted by $(x_{pi}(t), y_{pi}(t))$.

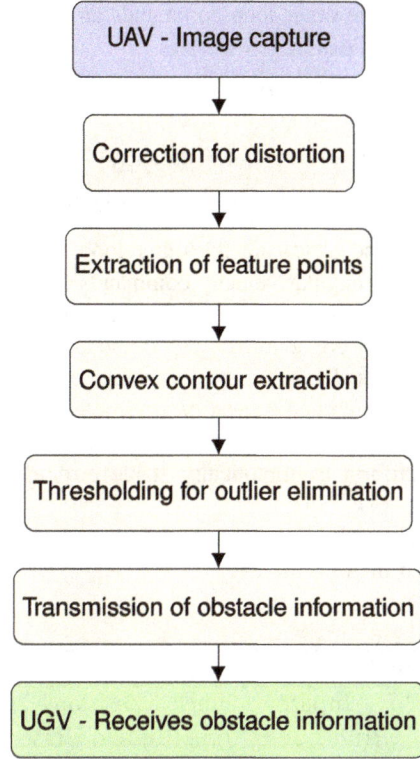

Fig. 2. Diagram explaining obstacle information extraction through image processing by UAV.

The relation between a pixel location in the image plane and its position in the inertial frame is given below:

$$\begin{pmatrix} x_{pi}(t) \\ y_{pi}(t) \\ 0 \end{pmatrix} = \begin{pmatrix} x^u(t) \\ y^u(t) \\ h \end{pmatrix} + d_c({}^IR_B \quad {}^BR_G \quad {}^GR_C)S_{J_p}, \tag{11}$$

where d_c denotes the distance between the target feature point and the center of the image plane and $({}^GR_C)$, $({}^BR_G)$ and $({}^IR_B)$ are the rotation matrices from camera frame to gimbal frame, from gimbal frame to the body fixed frame of UAV and from body fixed frame of UAV to the local inertial frame, respectively. J_p denotes the feature point and S_{J_p} is the unit vector along the projection line of the feature point from the center of the image plane.

2.3. UGV trajectory tracking control law

The kinematic model of UGV is given below is considered for controller design:

$$\dot{x}(t) = v(t)\cos\psi(t), \tag{12}$$

$$\dot{y}(t) = v(t)\sin\psi(t), \tag{13}$$

$$\dot{\psi}(t) = \omega(t), \tag{14}$$

where the heading angle is denoted by $\psi(t)$, the angular velocity by $\omega(t)$ and the magnitude of linear velocity by $v(t)$. The variables $v(t)$ and $\omega(t)$ are controlled independently by

PID control. The outer loop commands are generated using the below equations[12]:

$$v^*(t) = v_r(t)\cos(\psi_r(t) - \psi(t)) + k_1(y_r(t) - y(t)), \quad (15)$$

$$\omega^*(t) = \omega_r(t) + v_r(t)(k_2(x_r(t) - x(t)) \\ + k_3\sin(\psi_r(t) - \psi(t))), \quad (16)$$

where $v^*(t)$ and $\omega^*(t)$ are the outer loop's linear velocity magnitude and angular velocity commands to the inner loop of UGV. The linear velocity and angular velocity of the reference trajectory are denoted by $v_r(t)$ and $\omega_r(t)$, respectively. The parameters k_1, k_2, k_3 are constants and the heading angle of the reference trajectory is denoted by $\psi_r(t)$. It is shown in Ref. 12 that the control law stated in (15) and (16) provides uniform asymptotically stable error dynamics for $k_1 > 0, k_2 > 0, k_3 > 0$.

The main steps of the search and obstacle avoidance are summarized in Algorithm 1.

Algorithm 1

- STEP I: Input: Launch positions of UAVs $[x_1^u(0), y_1^g(0), \dots, x_n^u(0), y_n^g(0)]$ and UGVs $[x_1^g(0), y_1^g(0), \dots, x_m^g(0), y_m^g(0)]$.
- STEP II: Select waypoints for UAVs and UGVs using (1).
- STEP III: Perform search mission by moving from one waypoint to another.
- STEP IV: IF sensor fault for jth UGV, THEN move to STEP V ELSE move to STEP II.
- STEP V: Find the nearest UAV (ith) for jth UGV using (2).
- STEP VI: The ith UAV performs trajectory following as given in (6), (7) and jth UGV navigates towards base station.
- STEP VII: IF obstacle is detected by the vision sensor of ith UAV, obstacle avoidance path is followed by jth UGV using (3).
- STEP VIII: After jth UGV reached base station, ith UAV resumes search mission following STEP II.

---

3. Numerical Simulation Results

A numerical simulation study is conducted with 10 UAVs and five UGVs performing a search mission. The search area is having dimensions of $1000\,\text{m} \times 1000\,\text{m}$. Four obstacles are placed in the search area at locations (200 m, 200 m), (300 m, 400 m), (600 m, 200 m) and (900 m, 700 m). The base station location is placed outside the search region at $(-50\,\text{m}, -50\,\text{m})$. The time at which sensor fault occurs for each UGV is randomly selected. Table 1 shows the nearest UAVs for the

Table 1. UGV–UAV mapping.

UGV	1	2	3	4	5
UAV	6	7	2	4	10

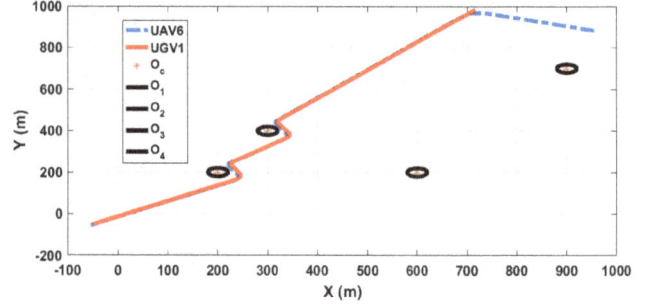

Fig. 3. Trajectories followed by UGV 1 and UAV 6.

five UGVs when sensor fault occurs. The trajectories followed by UGV 1 and UAV 6 are shown in Fig. 3. The red line represents the trajectory of UGV and the blue line represents the trajectory followed by UAV. It is clear from Fig. 3 that UAV 6 goes to the position where sensor fault occurred for UGV 1 and later follows the trajectory leading to the base station. Two obstacles located at (200 m, 200 m) and (300 m, 400 m), respectively, are avoided by UGV 1 with the obstacle information from UAV 6. Similarly, the trajectories followed by UGV 2 and UAV 7, UGV 3 and UAV 2, UGV 4 and UAV 4 and UGV 5 and UAV 10 are shown in Figs. 4–7,

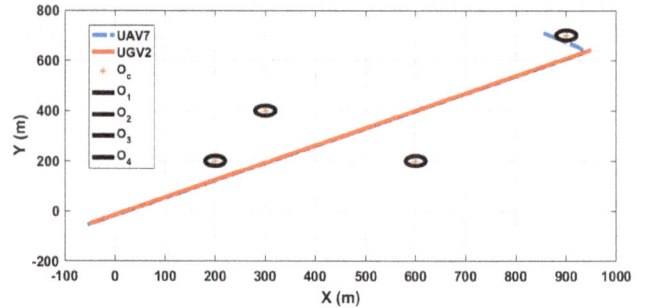

Fig. 4. Trajectories followed by UGV 2 and UAV 7.

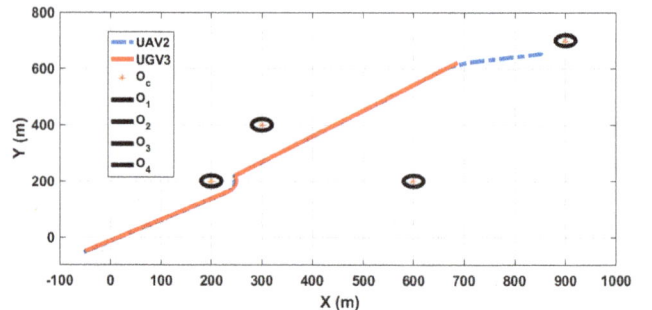

Fig. 5. Trajectories followed by UGV 3 and UAV 2.

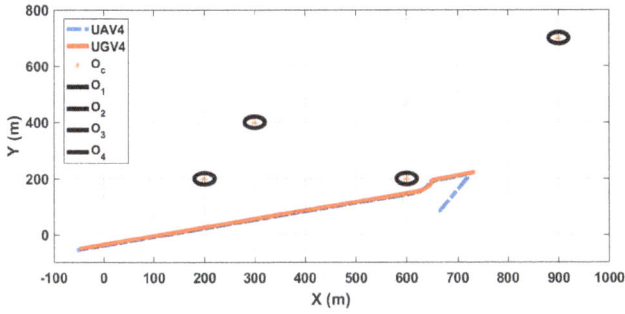

Fig. 6. Trajectories followed by UGV 4 and UAV 4.

Fig. 7. Trajectories followed by UGV 5 and UAV 10.

respectively. Figures 3–7 indicate that all the five UGVs navigate safely back to base station avoiding obstacles with the information from nearest UAVs.

4. Experimental Results for Obstacle Avoidance by UGV

The hardware experiments are conducted using SOLO-3DR quadrotor[13] and a customized UGV. The UGV and UAV used in experiments are shown in Figs. 8 and 9, respectively.

Fig. 8. UGV used in the experiment.

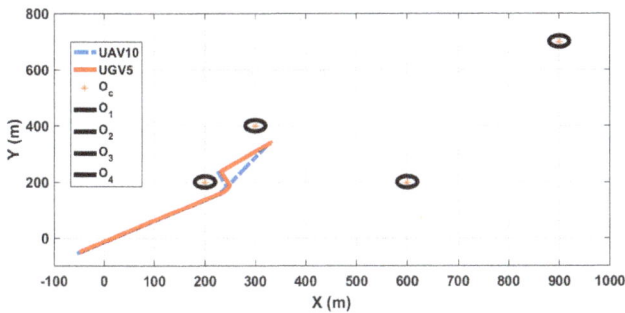

Fig. 9. UAV used in the experiment.

The UAV is equipped with a GIT2-Pro wide angle camera with $120°$ diagonal field of view (FOV).[14] The images are processed using a Raspberry-Pi3 computer mounted on UAV. The images are captured at a rate of 15 frames per second (fps). The captured images are corrected for wide angle distortion and processed as described in Fig. 2. A sample captured image and the convex contours obtained after processing it are shown in Figs. 10 and 11, respectively.

The position and size information of the obstacle are transformed into a global frame using Fig. 11. As a consequence of the GPS position variations of the UAV, a locus of adjacent points corresponding to the obstacle is generated. The UGV computes the mean value of the possible location of obstacles from UAV and considers it as the true position of the obstacle. A GPS receiver module is used for UGV localization. The algorithms for UGV navigation are implemented in a Raspberry-Pi3 computer. The GPS receiver is a

Fig. 10. Sample captured image using camera.

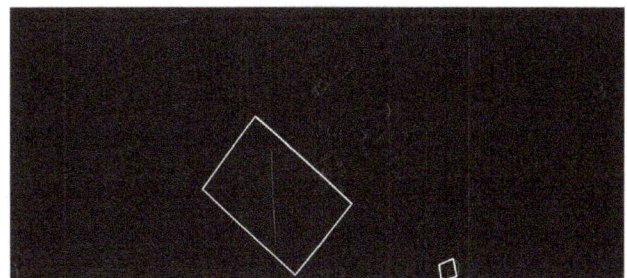

Fig. 11. Convex contours obtained after processing the image.

96   *K. Harikumar et al.*

Real-Time Kinematic (RTK) module that gives accurate position and velocity measurements when compared to conventional single-point GPS module. A servo motor is used for angular velocity control and a DC motor for linear velocity control. A POLULU servo control board driven by a Raspberry-Pi3 computer generates pulse width-modulated (PWM) signals to the motor. The inner loop PID controller maps the desired angular velocity and linear velocity commands to the PWM signals.

The position of UGV is approximately $(-10\,\text{m}, 7.5\,\text{m})$, when the sensor failure has occurred. The next two waypoints of UGV are at $(-7\,\text{m}, -7\,\text{m})$ and $(-2.5\,\text{m}, -19.5\,\text{m})$, respectively. The obstacle is a rectangular-shaped object of dimensions $3\,\text{m} \times 3\,\text{m}$ and is positioned between the first and second waypoints of UGV. Since the navigation is along a straight line, the reference commands for UGV are $v_r(t) = 0.5\,\text{m/s}$ and $\omega_r(t) = 0\,\text{rad/s}$. The line of sight (LOS) angle connecting the current UGV position and the next destination coordinate is the reference heading command $\psi_r(t)$. The position of UAV is around $(-12\,\text{m}, 7\,\text{m})$ when the information about sensor failure is received from UGV. The UAV flies along the waypoints of UGV with a constant velocity of 1 m/s and a constant altitude of 5.5 m from ground level. The communication between UAV and UGV is through a dedicated 2.4-GHz Wi-Fi network. Figure 12 shows the set of points indicating the obstacle position estimated by UAV from images. Due to the error in GPS position of UAV, the obstacle position is scattered.

The three-dimensional trajectories of UAV and UGV are given in Fig. 13. The trajectory of UAV is shown in blue color and the portion marked in red color indicates the region of the flight where UAV is detecting the target. UGV's initial position, first and second waypoints are joined by red colored line. The trajectory followed by UGV is shown in green. A radial distance of 4 m from a waypoint is considered as waypoint attained and the UGV moves to the next waypoint. Black colored rectangle represents the obstacle. A more clear description of the UGV trajectory is given in Fig. 14. UGV receives information about obstacle from UAV and after reaching first waypoint, the UGV takes an alternate path to reach the second destination waypoint. The intermediate

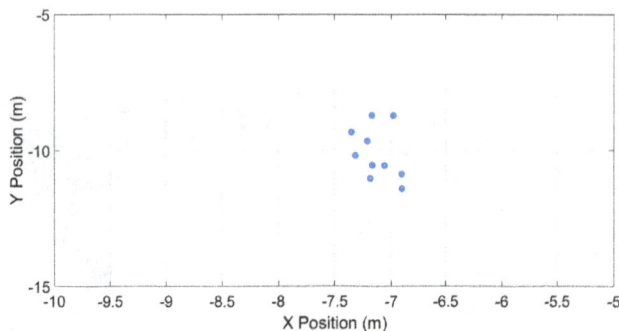

Fig. 13. Three-dimensional trajectories of UAV and UGV with detected obstacle shown as rectangle.

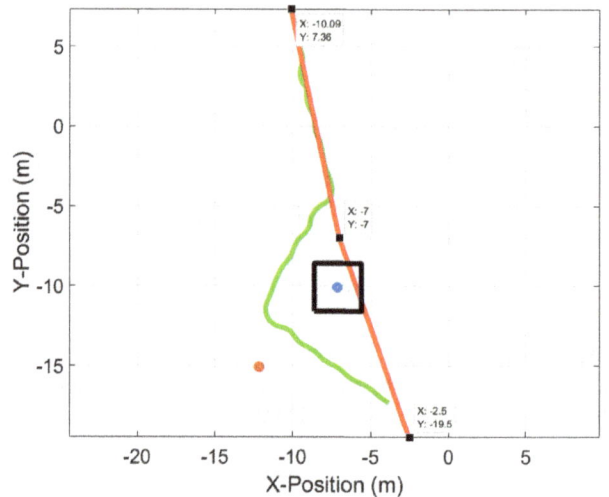

Fig. 14. Two-dimensional trajectory of UGV with detected obstacle shown as rectangle.

waypoint is fixed 5 m away from the obstacle's center coordinates to facilitate obstacle-free transition to second waypoint.

Figure 15 shows the distance (error) to reach the next waypoint for UGV as a function of time. The first waypoint is switched to intermediate waypoint at around 20 s and later to second waypoint at around 32 s. From Fig. 15, it is clear that UAV detects the obstacle before 20 s, facilitating the UGV to

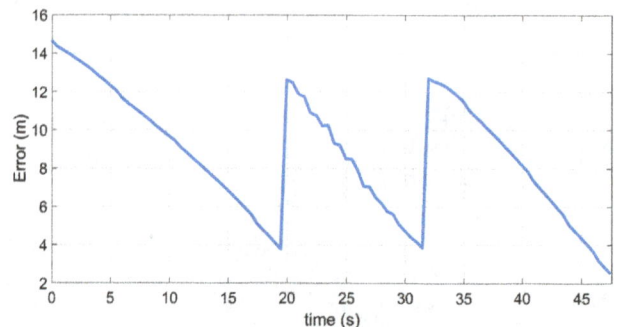

Fig. 12. Points showing the position of obstacle estimated by UAV from images.

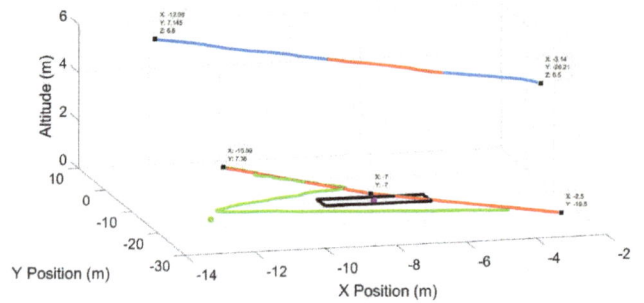

Fig. 15. Distance to reach the next waypoint for UGV as a function of time.

Fig. 16. UGV approaching the base station along the path with obstacle, while UAV is flying ahead of UGV.

Fig. 17. UGV taking deviation after obtaining the obstacle location information from UAV.

Fig. 18. UGV approaching the final destination after deviating from the original path.

alter its trajectory to reach the second waypoint by avoiding obstacles.

The UGV heading towards the first waypoint along the line of the obstacle is shown in Fig. 16. The UGV deviating from its original path to avoid the obstacle is shown in Fig. 17. UGV approaching the final destination waypoint after avoiding the obstacle is shown in Fig. 18. In Figs. 16–18, we can see that the UAV flies ahead of UGV monitoring the designated path of the UGV.

5. Conclusion

A method for search and cooperative obstacle avoidance for UGVs is developed in this paper. UGVs encountering a sudden sensor failure while performing the mission increases the possibility of a collision with obstacles. The cooperation between the UGV having sensor failure and its nearest UAV enables the UGV to safely navigate towards the base station. UAVs incorporated with a vision sensor can detect the obstacles ahead of UGV and this helps in efficient replan of the future trajectory for UGV. Two asymptotically stable trajectory tracking algorithms are presented for UAV and UGV. The numerical simulation and experimental results show the effectiveness of the proposed method as the UGV avoids the obstacle and reaches the destination point without an obstacle detection sensor. They also indicate that the trajectory tracking algorithms are stable.

Acknowledgments

This work is fully funded and supported by ST Engineering-NTU Corporate Laboratory under the Project CRP3-P2P.

References

[1] S. Minaeian, J. Liu and Y. J. Son, Vision-based target detection and localization via a team of cooperative UAV and UGVs, *IEEE Trans. Syst. Man Cybern., Syst.* **46**, 1005 (2016).

[2] L. Cantelli, M. Mangiameli, C. D. Melita and G. Muscato, UAV/ UGV cooperation for surveying operations in humanitarian demining, *Proc. IEEE Int. Symp. Safety, Security and Rescue Robotics* (2013), pp. 1–6.

[3] H. G. Tanner, Switched UAV-UGV cooperation scheme for target detection, *Proc. IEEE Int. Conf. Robotics and Automation* (2007), pp. 3457–3462.

[4] E. H. C. Harik, F. Guerin, F. Guinand, J. F. Brethe and H. Pelvillian, UAV-UGV cooperation for objects transportation in an industrial area, *Proc. IEEE Int. Conf. Industrial Technology* (2015), pp. 547–552.

[5] H. C. Moon, H. C. Lee and J. H. Kim, Obstacle detecting system of unmanned ground vehicle, *Proc. SICE-ICASE Int. Joint Conf.* (2006), pp. 1295–1299.

[6] J. Larson and M. Trivedi, Lidar based off-road negative obstacle detection and analysis, *Proc. 14th Int. IEEE Conf. Intelligent Transportation Systems* (2011), pp. 192–197.

[7]M. T. Long, R. R. Murphy and L. E. Parker, Distributed multi-agent diagnosis and recovery from sensor failures, *Proc. Int. Conf. Intelligent Robots and Systems* (2003), pp. 2506–2513.

[8]M. Garn, J. Valente, D. Zapata, R. Chil and A. Barrientos, Towards a ground navigation system based in visual feedback provided by a mini UAV, *Proc. IEEE Intelligent Vehicles Symp. Workshops* (2012), pp. 1–6.

[9]L. Klodt, S. Khodaverdian and V. Willert, Motion control for UAV-UGV cooperation with visibility constraint, *Proc. IEEE Multi-Conf. Systems and Control* (2015), pp. 1379–1385.

[10]K. Harikumar, T. Bera, R. Bardhan and S. Sundaram, Autonomous navigation and sensorless obstacle avoidance for UGV with environment information from UAV, *Proc. IEEE Int. Conf. Robotic Computing* (2018), pp. 266–269.

[11]J. J. E. Slotine and W. Li, *Applied Nonlinear Control*, Chapter 3 (Prentice Hall, USA, 1991), pp. 58–62.

[12]Y. Kanayama, Y. Kimura, F. Miyazaki and T. Noguchi, A stable tracking control method for an autonomous mobile robot, *Proc. IEEE Int. Conf. Robotics and Automation* (1990), pp. 384–389.

[13]3D Robotics, Inc., *SOLO User Manual V8* (3D Robotics, 2015), pp. 1–81.

[14]K. Harikumar, T. Bera, R. Bardhan and S. Sundaram, State estimation of an agile target using discrete sliding mode observer, *5th IEEE Int. Conf. Control, Decision and Information Technologies* (2018).

Improving code quality in ROS packages using a temporal extension of first-order logic

David Come, Julien Brunel* and David Doose

DTIM, ONERA, 2 Avenue Edouard Belin,
31400 Toulouse, France
*Julien.brunel@onera.fr

Robots are given more and more challenging tasks in domains such as transport and delivery, farming or health. Software is a key component for robots, and Robot Operating System (ROS) is a popular open-source middleware for writing robotics applications. Code quality matters a lot because a poorly written software is much more likely to contain bugs and will be harder to maintain over time. Within a code base, finding faulty patterns takes a lot of time and money. We propose a framework to search automatically user-provided faulty code patterns. This framework is based on FO^{++}, a temporal extension of first-order logic, and Pangolin, a verification engine for C++ programs. We formalized with FO^{++} five faulty patterns related to ROS and embedded systems. We analyzed with Pangolin 25 ROS packages looking for occurrences of these patterns and found a total of 218 defects. To prevent the faulty patterns from arising in new ROS packages, we propose a design pattern, and we show how Pangolin can be used to enforce it.

Keywords: ROS; C++; temporal logic; formal methods; code query.

1. Introduction

Robotics is a growing field, and in future, robots are expected to be deployed in open and unconstrained environments with human presence. Traditionally, robots were confined within controlled environments which ensured safety. However, in a public environment with human interaction, safety is crucial to protect people and ensure public trust.

Robot Operating System (ROS)[1] is a popular open-source middleware for robotics. It provides tools and abstractions to ease the creation of a robotic application and has bindings for several programming languages (C++, Lisp and Python). A typical ROS application consists of several basic processing units (called nodes) deployed in a certain manner and exchanging data through ROS. There are many existing components from various sources (the official website lists more than 1600 packages for the latest LTS release).

ROS is widely used in an academic context and spreads in the industry with initiatives like ROS-Industrial.[2] ROS-Industrial aims for extending ROS to provide solutions needed by the industry (such as interoperability with standard hardware or ready-to-use industrial applications). ROS's other drawbacks for industrial applications are the lack of real-time control and the lack of insurance on the code quality. Code quality matters a lot because a poorly written software is much more likely to contain bugs and will be harder to maintain over time.[3] This is all the more critical as robots often have a long service life.

Our goal is to find patterns in robotics software that do not respect good programming practices. Fixing these patterns (which are not necessarily bugs) will improve code quality. Finding such faulty patterns can be done manually by peer-review, but this is time and money consuming as it requires to divert one (or several) programmer(s) from their current task to perform the review. Instead, we propose an approach in which (1) each pattern is specified in a formal language, having thus an unambiguous meaning, and (2) the detection of the pattern within the code relies on a formal technique (model checking) which is fully automatic and provides an exhaustive exploration of the code.

The specification language for patterns, which we call FO^{++}, addresses two concerns: it allows reasoning over the structure of the source code (classes, functions, inheritance relationship, etc.) and over properties related to the execution paths within the control flow graph (CFG) of functions (such as a constraint on the ordering of statements). We define this language as an extension of first-order logic with temporal logic. Indeed, first-order logic is suited to reasoning over structural aspects and temporal logic allows expressing properties related to the ordering of events. Our framework also includes the (automatic) detection engine for C++ code, Pangolin, which takes a pattern specification in FO^{++} and the target C++ code as inputs.

The contributions of this paper are as follows:

- We present a framework to detect patterns in source code, which includes a formal specification language, FO^{++}, and Pangolin, an engine based on model checking, which

---

*Corresponding author.

detects automatically the occurrences of patterns in source code.

- We study five suspicious code patterns that reduce code quality and apply Pangolin to detect occurrences of these patterns in 25 common ROS packages.
- We propose a design pattern for the development of new ROS packages, which prevents the problems that come with the previously identified patterns and show how to enforce this design pattern using FO^{++} and Pangolin.

The rest of the paper is organized as follows. In Sec. 2, we develop the code quality in robotics. In Secs. 3 and 4, we present Pangolin and FO^{++}. In Sec. 5, we study five suspicious patterns and apply Pangolin to detect their occurrences. In Sec. 6, we propose a design pattern for the development of new ROS packages and show how to enforce it using FO^{++} and Pangolin.

2. Robotics and Code Quality

2.1. *ROS overview*

ROS is a popular open-source framework providing a collection of tools to ease the development of robots, including: a set of language- and platform-independent tools used for building and distributing ROS-based software; capabilities with many state-of-the-art algorithms ready to be integrated; library implementation for C++, Python and Lisp for developing new features; a worldwide community.

A package is the smallest unit of development and release for ROS-related projects. It usually provides either access to some physical devices (such as LIDAR or IMU) or the implementation of standard algorithms for robots (navigation, guidance, control, localization, etc.). A typical ROS application consists of nodes (processes executing different tasks) connected through a software bus provided by ROS. The nodes use a publish–subscribe architecture to exchange data. This design is flexible, promotes code reuse and simplifies complex programming.

2.2. *Code quality on critical systems*

On critical systems, safety is of the utmost importance given that lives and public trust are at stake. Thus, it is important to ensure that the code embedded in these systems is correct with respect to a specification. Two different aspects can be checked for correctness: its semantics (what it does) or its "style" (how it is designed and written).

There are several ways to ensure that a piece of code computes the right result. Tests are commonly used to find errors and help to gain confidence in the code correctness. The code can be correct-by-construction if it is generated from models by tools that are both proven correct.[4] Finally, the code could be proven to be correct, with the help of some static analysis tools. These tools rely on methods such

as abstract interpretation or deductive methods based on Hoare logic. For instance, Frama-C with Jessie[5] is a well-known tool to verify properties in a deductive manner on a C program.

However, correctness is often not enough for software. It has to be understandable, readable and easy to update: it should be *well written*, and one should also have confidence in it. This translates into software metrics, but also into idioms of the language to follow, domain-specific constraints, project guidelines and coding rules. For C++, see MISRA C++,[6] JSF++ (Ref. 7) or HIC++.[8] They often focus on banning or restricting C++ constructs to produce a safer subset of C++ and are widely enforced in embedded systems. The goal of enforcing them is to improve the safety and reliability of such systems. Many static analysis tools (such as Coverity or Klocwork) can do it.

.QL[9] proposes to see the code as data. It enables to run queries over a special database which contains a representation of the program, which was built using a language-specific extractor. .QL is underpinned by an untyped variant of Datalog, and unlike FO^{++}, it cannot express temporal properties over the CFG of a function.

CppDepend is another code exploration software. It allows to analyze the code structure and specify design rules to compute technical debt estimations and find code smells. It provides Code query over Linq[10] (CQLinq) to write custom queries. Again, unlike FO^{++}, it cannot express temporal properties over the CFG of a function. Besides, it does not come with a formal semantics.

Coccinelle[11] performs advanced program matching on C-code. It is underpinned by CTL-VW, a variant of Computation Tree Logic (CTL; a temporal logic) with the ability to use quantified variables over the set of expressions within a function and record their values. Its user specification language (SPatch) is close to the one used by the *diff* utility and translated into CTL-VW formula. It was successfully applied on several open-source projects such as the Linux kernel[12] and allows not only detecting patterns but also transforming the source code. Unlike Coccinelle, Pangolin targets C++ programs, and because of the object-oriented paradigm of C++, reasons about the structural aspects of the code. It also offers CTL and linear temporal logic (LTL), while Coccinelle only supports CTL-VW.

2.3. *ROS code quality*

The need for asserting the quality of ROS packages has been recently acknowledged. In Ref. 13, the authors propose *HAROS*, a framework assessing the quality of ROS packages. The goal of this framework is to ease the computation of metrics and conformance to coding standards by leveraging ROS unique features. In Ref. 14, the authors extract usage patterns and statistics on how ROS nodes are effectively programmed (e.g., hard-coded queue sizes, topic names, etc.). In Ref. 15, an analysis on physical unit manipulation is performed, in order to find inconsistencies in the source code of ROS packages.

Enforcing methods from embedded systems on ROS packages will undoubtedly improve code quality. However, this will only fix issues that are related to C++ and that could be found in other fields (for instance, using invalid pointers). ROS has some distinctive features, and thus ROS packages should also follow some specific design rules and patterns. Currently, the ROS community has issued a coding guide[16] based on Google's C++ Style Guide and set of code metrics.[17] The coding guide mostly provides stylistic guidelines regarding naming conventions and some basic informal rules regarding code design. roslint[18] is a linter integrated within ROS, which can perform some checks.

2.4. *Motivation*

As a motivation example, we consider the use of free functions as callbacks in ROS applications. A typical ROS application is made of nodes which use a publish–subscribe mechanism to exchange data. ROS delivers the messages to the nodes with callbacks. Since a callback is a function that is not supposed to be called directly by the developer, it should be hidden as much as possible.

Technically, within a C++ ROS program, a callback is a function that appears as the third argument of a call to `subscribe` on a `NodeHandle` variable. ROS allows both free functions and member functions to be used as callbacks. A free function can never be totally hidden. At best, it can be static or within an anonymous namespace, which prevents the developer from calling it from another file. Nevertheless, it does not prevent the developer from calling it in the current file. Moreover, a free function often couples the reception of the data with the algorithm that treats them, hindering code reusability. Thus, we want to ensure that *all callbacks are private member functions*.

Specifying a pattern that corresponds to the satisfaction (or the violation) of this rule requires reasoning about all functions, specifying that a function is member of a class, it is private, the third argument of a certain call is of type `NodeHandle`, etc. This is what we call the structural aspects of a pattern. Besides, if we consider for instance the violation of the rule, we also need to express that in the CFG of a certain function, there is an execution path in which at some point there is a call to a function on a variable of type `NodeHandle`, such that the third argument is not a private member function. Existing methods do not address well this type of analysis which combines structural aspects and properties related to execution paths within a CFG.

In this paper, we propose a framework to address this issue. It relies on FO^{++}, a temporal extension of first-order logic, and comes with Pangolin, a prototype which implements the detection of patterns through a model checking algorithm for FO^{++} formula.

3. Pangolin

Pangolin[a] is a verification engine for C++ programs. The user provides an FO^{++} specification (corresponding to the code pattern), the source code to analyze and Pangolin checks the code compliance with respect to the specification. Figure 1 illustrates the whole process.

Pangolin parses the C++ code source with Clang and its API libtooling.[19] Pangolin evaluates the formulas by rewriting them until it reaches true or false. It prints the result with a complete trace of the evaluation process which can be reviewed. It deals internally with most of the first-order logic. It performs first-order quantifiers' elimination, and evaluates structural predicates and functions. For evaluating temporal predicates, it uses the model checker nuXmv[20] by reducing the evaluation of such predicates to a model checking problem.

Pangolin implements two modes for code queries, which differ in the first-order logic evaluation: a *fast mode* and a *complete mode*. With the fast mode, Pangolin stops the evaluation of the formula as soon as it finds a counterexample. With the complete mode, Pangolin continues to evaluate the formula in order to generate all possible values of each first-order variable for which there is a counter-example.

Clang provides an up-to-date and complete support for C++ and direct access to the code abstract syntax tree (AST). If Clang is also used for compilation, it strengthens the analysis as the choices made about unspecified elements within the C++ standard will be the same for Pangolin and the actual execution of the program. Using Clang implies that Pangolin relies on Clang building process,[b] meaning it can only analyze one file at a time. As Clang expects the code to compile, Pangolin cannot analyze a piece of code if it does not compile or a simple snippet of code taken out of its context.

In the following subsections, the complete construction of FO^{++} is detailed. Section 4.2 defines FO^{++} syntax and then Sec. 4.3 defines its semantics. Since FO^{++} is a temporal

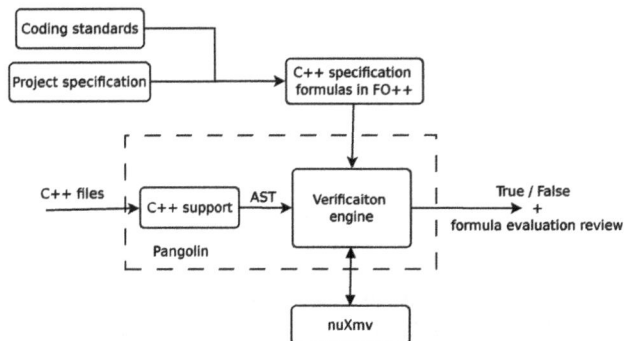

Fig. 1. Pangolin overview.

---

[a]Pangolin is available at https://gitlab.com/Davidbrcz/Pangolin.
[b]It is the standard C++ build process.

extension of first-order logic, most of its syntax and semantics is identical to one of the first-order logics. The extension is done through a process close to parametrization. It consists in adding two temporal predicates to the atoms of first-order logic. The semantics of these two temporal predicates follows the usual semantics of temporal logics. The details of the temporal part of FO^{++} are in Sec. 4.4. Finally, Sec. 4.5 shows how FO^{++} is instantiated for C++ so that it can be used as a specification language for Pangolin.

4. Specification Formalism: FO^{++}

4.1. *Overview*

FO^{++} is a new logic defined as a temporal extension of first-order logic through a process called *parametrization*.[21] It has a well-defined semantics and is not tied to any programming language in its generic form. Once instantiated for a specific programming language, we use it as a specification formalism to describe faulty patterns. The first-order part of the logic is used to express structural properties.

In FO^{++}, temporal logics are used to describe a sequence of events within the CFG of a function. Traditionally, temporal logic formulas use temporal operators that allow describing sequences of events over time. In FO^{++}, an event represents a statement that can be found in the CFG of a function. Moreover, they are not sequenced with respect to time but with respect to the order in which the statements occur within the CFG.

The first-order logic is parameterized with temporal logics through two special first-order predicates models$_{CTL}$ and models$_{LTL}$. Intuitively, if f is a first-order variable and φ is an LTL formula (respectively, a CTL formula) then models$_{LTL}(f, \varphi)$ [respectively, models$_{CTL}(f, \varphi)$] is true if f denotes a function and if the CFG of f satisfies the formula φ according to LTL (respectively, CTL) semantics rules.

4.2. *Syntax*

Terms. Let V be a finite set of variables, and F^{++} a set of function symbols, each having its own arity.

We define inductively the set T^{++} of FO^{++} terms as follows:

- if x is a variable then x is a term in T^{++};
- if $f \in F^{++}$ has arity n and for each $i \in 1, \ldots, n, t_i \in T^{++}$ then $f(t_1, \ldots, t_n)$ is a term in T^{++}.

Atoms. An atom consists of a call to a predicate (a predicate symbol together with a list of terms that is consistent with the profile of the predicate). The set ATOMS^{++} of FO^{++} atoms is built from the set of predicate symbols, which consists of:

- a set P^{++} of predicate symbols (each having its own arity);
- two special binary predicate symbols: models$_{LTL}$ and models$_{CTL}$. The first argument of both predicates is a term

in T^{++} term. The second argument of models$_{LTL}$ (respectively, models$_{CTL}$) is an LTL (respectively, CTL) formula as defined in Sec. 4.4.

Formulas. FO^{++} formulas are defined as follows:

- $\top, \bot$ are formulas;
- if $a \in$ ATOMS^{++} then a is also a formula;
- if Q is a formula, then $\neg Q, Q \vee Q, Q \wedge Q, \forall x\, Q, \exists x\, Q,$ $Q \Longleftrightarrow Q, Q \Rightarrow Q$ are also formulas.

A sentence is an FO^{++} formula is a formula without free variable. In the following subsections, all formulas are sentences.

4.3. *Semantics*

Interpretation structure. An FO^{++} formula is interpreted over a structure $M = (D, \text{CFGs}, \text{has\_cfg}, \text{cfg}, I, P)$ such that domain is a domain in which terms are interpreted. CFGs is a set of transition systems (as defined in Sec. 4.4.2) that represent the dynamic behaviors, i.e., in the context of program analysis, the CFGs of some elements in the domain (functions or methods). They are used to interpret temporal formulas. has_cfg $: D \rightarrow \{\text{true}, \text{false}\}$ is a total function to indicate whether a value in the domain is associated with a dynamic behavior, i.e., a CFG. cfg $: D \rightarrow$ CFGs is a partial function, which maps some values in the domain to a CFG. $I : F^{++} \rightarrow (D^n \rightarrow D)$ defines an interpretation for functions in F^{++}. $P : P^{++} \rightarrow (D^n \rightarrow \{\text{true}, \text{false}\})$ defines an interpretation for predicates in P^{++}.

Notice that a structure M has to satisfy the following constraint: for every element $d \in D$, has_cfg(d) if and only if d is in the domain of cfg. F^{++} and P^{++} are specific to the programming language used for the project under analysis, and D and CFGs are even specific to the program itself.

Environment. An environment is a partial function from the set V of variables to the domain D. If σ is an environment, x a variable in V and d a value in D, then $\sigma[x \leftarrow d]$ denotes the environment σ_1 where $\sigma_1(x) = d$ and for every $x \neq y$, $\sigma_1(y) = \sigma(y)$.

From an environment σ and an interpretation I for functions, we define an interpretation $K_\sigma : T^{++} \rightarrow D$ for terms in the following way:

$$\text{for each variable } x \text{ in } V, \quad K_\sigma(x) = \sigma(x), \quad (1)$$

$$K_\sigma(f(t_1, \ldots, t_n)) = I(f)(K_\sigma(t_1), \ldots, K_\sigma(t_n)). \quad (2)$$

Satisfaction rules. Let M be a model, σ an environment and K_σ an interpretation according to this environment. We define the satisfaction relation of FO^{++} as follows[c]:

$$M, K_\sigma \models \neg Q \text{ if and only if } M, K_\sigma \nvDash Q,$$

---

[c]Due to space constraints, we only provide the semantics of a minimal set of logical connectives, which are sufficient to define the others (disjunction, implication, universal quantification, etc.).

$M,K_\sigma \models Q_1 \wedge Q_2$ if and only if $M,K_\sigma \models Q_1$ and $M,K_\sigma \models Q_2$,

$M,K_\sigma \models \exists x\, Q$ if and only if there is an $a \in D$ such that $M,K_{\sigma[x \leftarrow a]} \models Q$,

$M,K_\sigma \models p(t_1,\ldots,t_1)$ if and only if $P(p)(K_\sigma(t_1),\ldots,K_\sigma(t_n)) = \text{true}$,

$M,K_\sigma \models \text{models}_{\text{CTL}}(x,\psi)$ if and only if $M,K_\sigma \models \text{has\_cfg}(x)$ and $\text{cfg}(x), K_\sigma \models_{\text{CTL}} \psi$,

$M,K_\sigma \models \text{models}_{\text{LTL}}(x,\psi)$ if and only if $M,K_\sigma \models \text{has\_cfg}(x)$ and $\text{cfg}(x), K_\sigma \models_{\text{LTL}} \psi$.

4.4. *Temporal formulas*

Temporal logics are formalisms usually used to describe the flow of time and how some events are ordered. Within FO^{++}, a temporal formula is evaluated over a function CFG. Thus, events are sequenced with respect to the order in which the statements occur within the CFG. Two temporal logics are available within FO^{++}: CTL and LTL. LTL is discrete linear time logic, only one path is considered at a time whereas CTL is a discrete branching time. At each moment, all paths leaving the current state are considered and it offers path quantifications for temporal operators. Both are included because they are well-established logics, and tackle different issues: CTL is more natural over a CFG as it is branching time whereas LTL offers a path-sensitive analysis and its ability to refer to the past of an event is convenient.

4.4.1. *Syntax*

Temporal logics are usually built on top of some fixed set of atoms (called here temporal atoms to distinguish them from FO^{++} atoms). But here, temporal atoms are not taken among a fixed set. Instead, they consist of calls to predicates from a set PREDCFG (disjoint from P^{++}). PREDCFG is a set of predicates which describe a fragment of AST that can be found within a function. The parameters (if any) of these predicates are terms in T^{++} term, built from variables that are quantified outside the temporal predicates (we do not allow quantification in temporal formulas). Thus temporal atoms have the following shape $q(t_1,\ldots,t_n)$ with $q \in \text{PREDCFG}, t_1,\ldots,t_n \in T^{++}$ term. We may omit parenthesis when there is no arguments.

We inductively construct LTL formulas as follows:

- temporal atoms are valid LTL formula;
- if ψ_1, ψ_2 are valid LTL formulas, so are $\psi_1 \circ \psi_2, \neg\psi_1, \mathbf{X}\psi_1$, $\mathbf{G}\psi_1, \mathbf{F}\psi_1, \psi_1\mathbf{U}\psi_2, \mathbf{Y}\psi_1, \mathbf{O}\psi_1, \mathbf{H}\psi_1$ and $\psi_1\mathbf{S}\psi_2$ where $\circ$ is a binary Boolean logic operator.

CTL construction is similar to LTL, expect that the temporal operators are $\mathbf{A}\circ-$, $\mathbf{E}\circ-$, $\mathbf{A}[-U-]$, $\mathbf{E}[-U-]$ with $\circ \in \{\mathbf{X}, \mathbf{F}, \mathbf{G}\}$.

We also define CTL *weak until* operator for all paths, defined as $\mathbf{A}[x\mathbf{W}y] = \neg\mathbf{E}[\neg yU(\neg x \wedge \neg y)]$. Intuitively, $\mathbf{A}[x\mathbf{W}y]$

is true if x is true until y becomes true (which is not required to occur).

Finally, for n CTL formulas p_i we inductively define a sequence $\mathbf{seq}(p_1,\ldots,p_n)$ operator such that

$$\mathbf{seq}_n(p_1,\ldots,p_n) = \mathbf{A}\left[\left(\bigwedge_{i\leq n}\neg p_i\right)\mathbf{U}p_n\right],$$

$$\text{for } k < n \; \mathbf{seq}_k(p_1,\ldots,p_n)$$

$$= \mathbf{A}\left[\left(\bigwedge_{i\leq n}\neg p_i\right)\mathbf{U}(p_k \wedge \mathbf{seq}_{k+1}(p_1,\ldots,p_n))\right],$$

$$\mathbf{seq}(p_1,\ldots,p_n) = \mathbf{seq}_0(p_1,\ldots,p_n).$$

Intuitively, operator $\mathbf{seq}$ describes a sequence of n events p_i. The events must occur in order (i.e., p_1 occurs, then p_2 and so on) and between two events, everything (except one of the p_i) can happen.

4.4.2. *Semantics*

CFG formal definition. Let $B = (S, \rightarrow, I_{\text{CFG}}, [[\,]])$ be a CFG. S is a set of states, $\rightarrow \subseteq S \times S$ is the transition relation between states (written $\circ \rightarrow \circ$), $I_{\text{CFG}} \subseteq S$ is the set of initial states and $[[\,]] : \text{PRED}CFG \times S \rightarrow P(D^n)$ associates a predicate p of arity n with its valuation in a state s, denoted $[[p]]_s$ (i.e., the list of all tuples of concrete values for which the predicate is true).

Interpretation rules for temporal formulas. The satisfaction of LTL and CTL formulas are defined in the standard way. For the precise rules, refer for instance to Ref. 22. In our context, the only particular satisfaction rule is for the atoms of temporal logic formulas, relying on predicates in PREDCFG. Given an environment σ, an interpretation K_σ, a predicate $p \in \text{PREDCFG}$, a state s and some terms $v_1,\ldots,v_n$, the satisfaction relation for atoms is defined as follows [d]:

$$s, K_\sigma \models p(v_1,\ldots,v_n) \text{ if and only if}$$
$$(K_\sigma(v_1),\ldots,K_\sigma(v_n)) \in [[p]]_s.$$

4.5. FO^{++} *instantiation for C++*

FO^{++} construction does not mention any particular programming language and cannot be used *as it is* to formalize some properties. FO^{++} needs to be instantiated for a specific programming language to be concretely used as a specification language. As the rules aim to improve the architecture of C++ ROS package, FO^{++} will be specified for C++. This process requires to instantiate the set of functions and predicates and to give them a semantics. We also have to specify precisely the concrete domain over which the formula is evaluated.

4.5.1. *Domain of discourse*

The domain of discourse contains the values over which FO^{++} formulas are interpreted. It is unique to each file under

---

[d]It applies to both $\models_{\text{LTL}}$ and $\models_{\text{CTL}}$ and is thus simply denoted with $\models$.

Table 1. Small subset of structural predicates in the FO^{++} instantiation for C++.

Predicate	Informal semantics
isClass(c)	true if and only if c refers to struct, class or union
isAttribute(c)	true if and only if c refers to a class/struct/union field
parent(c, p)	true if and only if p is the father of c
isprivate(f)	true if and only if f is private within the class

analysis. For a given C++ program, it contains all classes (including template ones), their attributes, member functions, constructors and destructor, all free functions, all global variables, all function arguments and all defined types.

4.5.2. *Predicates and functions*

Nontemporal predicates deal with the most fundamental structural properties. In C++, structural properties are nature inquiry (*is it a class, an attribute etc.*), parenthood relationship between elements, visibility, inheritance relationship, types and their qualification. For conciseness, the full list of functions and predicates is not shown here.

The meaning of FO^{++} functions and predicates corresponds to the one described in the C++ standard.[23] Table 1 lists some functions and predicates needed to formally express some properties in the subsequent sections and describes an informal semantics for them.

5. Measuring ROS Packages Conformance to Coding Rules

In this section, we define a set of five coding rules that are relevant to increase the code quality within ROS packages. We then formalize with FO^{++} code patterns corresponding to the violation of the rules. Finally, we analyze 25 ROS packages with Pangolin to see if there are any occurrences of these patterns.

5.1. *Suspicious patterns*

This subsection shows five code patterns that can be considered bad practices for robotics systems. The first three are generic (they apply to any C++ projects on embedded systems) whereas the last two are specific to ROS. These patterns are not, strictly speaking, bugs but they do not convey confidence in the code. They must be at least detected, at best be fixed.

We focus on these patterns to exhibit Pangolin abilities and do not try to check the compliance of the packages with a more generic set of rules such as MISRA C++ or JSF++.

5.1.1. *Generic patterns*

The generic patterns address the use of global variables, variables with a scope too wide, and the use of inadequate logging mechanisms.

R1: *All user-provided global variables must be constant.* Global variables make the code harder to read and reason about for the programmer. Developers have to track the use of each variable across many lines of code to know what the code does. This is why many guidelines either reject or strictly supervise the use of global variables.

R2: *There should be no local nonconstant variable passed to a function and never used again.* Variable with a scope too wide hinder code readability and can be misused in future code evolution. This is particularly visible for variables created in one scope, passed to an object or a function and never used again in the original scope. This is especially true for ROS NodeHandle as they are entry points for ROS functionality, their scope should be narrowed as much as possible. In Listing 1 variables p and pnh should be attributes of that class rather than passed as arguments. Even if it is not a bug, it represents a defect in the design;

R3: *There should be not call to* `std::cout<<`, `std::cerr<<` *in any function. No std :: ofstream variables should be created.* On embedded systems, accessing resources usually follows some strict constraints in order to have guarantees on execution time and scheduling. For inputs and outputs, one should not directly write on standard output and error or open custom files, all the more there is usually a dedicated mechanism for logging.

```
int main(int argc, char **argv){
ros::init(argc, argv,
"depthimage_to_laserscan");
ros::NodeHandle n;
ros::NodeHandle pnh("~");
depthimage_to_laserscan::
  DepthImageToLaserScanROS dtl(n, pnh);
ros::spin();
}
```

5.1.2. *ROS-specific patterns*

For ROS-specific patterns, we focus on the correct use of publishers and on callbacks.

Each ROS publisher should be advertised and be published on

(a) R4a: *If the publisher is local to a function, then there is a call to publish within that function.*
(b) R4b: *If the publisher is an attribute, then there is a member function in which there is a call to publish on it.*
(c) R5: *All callbacks are private member functions.*

The motivation for the rule was exposed in Sec. 2.4.

5.2. Patterns formalization

In this subsection, we show how to formally express the rules described previously. For the sake of conciseness, we exclude the formal translation of the two simplest rules, i.e., R1 and R2.

Rule R3. The formal translation for this property is straightforward as demonstrated by Eq. (3). We look in all functions for a call to `std::cout<<` or `std::cerr<<`. For states with such calls, buseCout and useCerr are true. Also, the property looks for a local variable whose type is `std::ofstream`:

$$\exists f (\text{isFunction}(f) \wedge ((\exists s (\text{locallyDeclared}(s,f)$$
$$\wedge \text{ hasType}(s, ofstream)))$$
$$\vee \text{ models}_{\text{CTL}}(f, \textbf{EFuseCout} \vee \textbf{EFuseCerr}))). \quad (3)$$

Rules R4a and R4b. Equation (4) shows the formal translation of rule R4a. The property looks first for a function f in which there is a variable p whose type is *Publisher*. For such variables, there is one path within f's CFG in which pub(p) is true:

$$\forall f (\text{isFunction}(f) \Rightarrow \forall p (\text{locallyDeclared}(p,f)$$
$$\wedge \text{ hasType}(p, Publisher)$$
$$\Rightarrow \text{ models}_{\text{CTL}}(f, \textbf{EF}\text{pub}(p)))). \quad (4)$$

The formal translation for rule R4b is similar to one the one for R4a.

Rule R5. We are looking for pieces of code that do not respect the rule, thus we will formally describe a code pattern that is the negation of the rule. The pattern is formally expressed with the following formula:

$$\exists m (\text{isFreeFunction}(m))$$
$$\wedge \exists n (\text{locallyDeclared}(n,m)$$
$$\wedge \text{ hasType}(n, NodeHandle)$$
$$\wedge \exists c (\text{allFunctions}(c)$$
$$\wedge \text{ models}_{\text{CTL}}(m, \textbf{EF}\text{sub}(n,c))$$
$$\wedge \neg \text{isprivate}(c)))). \quad (5)$$

The first part of the formula consists of finding a free function m, and then a local variable n whose type is `NodeHandle`. Then, the rest of the formula looks for a function c (that could be a free or member) and checks if it is not private. Notice that, as a free function cannot be private, the predicate isPrivate will be false if c denotes such a function. Finally, models$_{\text{CTL}}$ will be true if sub(n,c) becomes finally true on, at least, one execution path in the CFG of m. sub is an element of PREDCFG such that sub(n,c) is true on states where there is a call of the following shape `n.subscribe(_,_,c)` (c is used as the third argument of a call to `subscribe` on n).

5.3. Experiments

We analyze a set of ROS packets looking for violations of the previously defined rules.

5.3.1. Corpus

We analyze 25 ROS packages for a total of 173 files and 50k lines of code. We chose these packages because they are well-known components and are likely to be used off the shelf for new robotics platforms. Thus, it is important to make sure their code does not contain suspicious code patterns.

The packages are sorted in three main groups:

- Navigation: All packages from ROS Navigation, extended with teb_local_planner and ros_dso (ROS Wrapper around Direct Sparse Odometry).
- Perception: A subset of ROS Perception (depthimage_to_laserscan, imu_pipeline, laser_filters, gmapping).
- LIDAR: rplidar, urg_node, loam_velodyne.

5.3.2. Results

Pangolin output and analysis. For each formula and each file, Pangolin yields true if the formula holds on the file and false otherwise with the current value of quantified variables. To reach the end-goal of code quality improvement, when a formula is false, the user needs to review the code as there are two cases:

- a legitimate code turns out to be a counter-example for the formula. It may be because the formula was not well designed (unforeseen cases, not the intended meaning) or due to Pangolin's limitations as detailed in Sec. 3;
- the code is truly suspicious.

Although Pangolin could have analyzed all the packages from ROS's last LTS release, the manual review of the results (in an effort to improve code quality) prevents us from doing so.

Violations found. A summary of the results is shown in Table 2. For each rule, the number of files in which a counter-example was found in shown in the first column. This number is also equal to the number of counter-examples for the *fast mode*. The next two columns represent the minimal and

Table 2. ROS quality measures.

Rule	No. of files	Min	Max	Total	Nav.	Percep.	LIDAR
R1	20	1	50	179	10	5	5
R2	3	1	2	4	1	2	0
R3	3	1	2	4	1	2	0
R4a	0	0	0	0	0	0	0
R4b	6	1	2	9	5	0	1
R5	8	1	6	22	1	2	5

maximal numbers of violations found in one of the reported file. The total column is the total number of violations in all reported files (which is also the total number of counter-examples for the *complete mode*). The number of violations was iteratively computed as Pangolin stops on the first counter-example it finds. The other three columns show the distribution of the files between the different groups (navigation, perception or LIDAR).

The first three rules were generic rules and not focused on ROS. Rule R1 deals with global variables and is one of the simplest properties. Yet, Pangolin found 20 files in which there is at least one user-provided nonconstant global variable (that number drops to 10 files if we exclude test files). In the 10 remaining files, the number of global variables varies from 1 to 50. Three files have more than 40 global variables and belong to the same package: this indicates a serious design issue. This is surprising as we expected the problem of global variables to be well known and therefore their use to be limited or even absent.

Rule R2 looks for local variables used at most once and Pangolin found three files where a node handle had a scope too wide. The first one is shown Listing 1. Even if the code works well, a better design would have been to make the node handlers attribute of the class. Indeed, these variables can be abused in a future development in order to quickly integrate a functionality in the code at the expense of the quality of the code. In the other two files, the node handler was created in main, and never used afterward. Thus, they pose the exact same issue.

Finally, rule R3 targets the use of general I/O instead of dedicated ones. Pangolin found three files where `std::cerr` or `std::cout` was used. Two of them were in files unrelated to ROS (main file for the Google Test Framework and one in gmapping). The last one used `std::cout` and `std::cert` before running `ros::init` to print help for command line options and report errors when parsing them. This is an example where legitimate code turns out to be a counter-example because of an unforeseen case. To take this into account, we would have to change the rule (and its corresponding pattern) to authorize the use of generic I/O in main until `ros::init` is called.

The last three look for suspicious code with ROS-specific features. Even if Pangolin found no counter-examples of the property R4a, this rule might be useful during the development process to prevent bugs from copy and paste for instance.

The rule R4b is similar to rule R4a but deals with publishers which are attributes of a class. Pangolin reported six files in which it found a counter-example. All the publishers were actually published but not the way it was specified in the formula. These counter-examples are legitimate code that could be eliminated by lifting Pangolin limitations (such as interprocedural or multi-file analysis).

Finally, rule R5 looks for the use of free functions as callbacks. With Pangolin, we found eight files in which at

Table 3. Experiments timing.

Rule	T_f (s)	Stddev (s)	T_c (s)	Stdev (s)	$\frac{T_c}{T_f}$
R1	467	2.5	473	4.50	1.01
R2	4400	60.2	4488	85.5	1.02
R3	583	3.9	968	2.2	1.66
R4a	493	1.44	533	3.9	1.08
R4b	630	2.3	712	2.8	1.13
R6	534	2.0	600	3.4	1.12

least one free function is used as a callback. Six files reported by Pangolin for this rule partly overlap with those reported for the use of global variables, suggesting a design issue. Among these six, there are the three files concentrating more than 40 global variables each.

A total of 218 defects were found (including 11 false positives), resulting in a false positive ratio of 5%. The details for each defect (rule formulation, package and file) can be found in Pangolin's repository.

Influence of search mode. Table 3 shows a summary of the results. For each rule, T_f (respectively, T_c) is the computation time of the analysis in *fast mode* (respectively, in *complete mode*). Columns 3 and 5 show the standard deviations for T_f and T_c respectively. All timings were averaged over 10 runs.

Specifications with temporal properties take longer to evaluate than fully structural properties, regardless of the analysis mode. Indeed, there is a cost related to the communication between Pangolin and nuXmv, in addition to the nuXmv execution, which solves an LTL/CTL model checking problem.

The additional cost for the *complete mode* essentially depends on the number of files in which there is a counter-example and of the number of counter-examples within each file. Indeed, for the files where there is no counter-example, both modes perform very similar calculations. *A posteriori*, the violations found by Pangolin are each time located in a limited number of files, which have a small number of counter-examples. This explains why the ratio $\frac{T_c}{T_f}$ is close to 1. Yet, the *fast mode* is useful to provide an estimate of the quality of the code base and to pinpoint the files which require further analysis with the *complete mode*.

6. The ROSApplication Design Pattern

In the previous section, we showed how to find suspicious code patterns within a code base with Pangolin. Yet, we can still improve code quality with finer rules to: capitalize on the ones that have proven to be effective; overcome the limitations of the others. To be effective, these finer rules require enforcing a specific design on the code of a package. The requirements on the design take the form of a design pattern called *ROSApplication*. Section 6.1 provides an informal overview of the design pattern and then Sec. 6.2 formalizes it with FO^{++}. Notice that in the design pattern overview, we emphasize the subset of rules we will formalize in the paper but the whole design pattern is available in Pangolin's repository. Finally, we perform a quick review of the design pattern.

6.1. *Design pattern overview*

The *ROSApplication* design pattern constrains the use of ROS C++ API. It is based on the rules R1 and R5 in Sec. 5 and constrains the structure of the code to ensure that finer rules work properly. It also aims for:

- splitting data exchange and processing in order to have ROS agnostic algorithms;
- ensuring consistent behavior with respect to ROS communication (the mapping between publishers or subscribers and topics does not change during an execution);
- being simple to use.

```
struct ROSApplication{
 ROSApplication():rate(10){init();}
 void run(){
  while(ros::ok()){
   ros::spinOnce();
   computation();
   rate.sleep();
 }}
private:
 void init(){
  pub = nh.advertise<Msg>("pub_topic",10);
  sub = nh.subscribe("sub_topic",10,
          &ROSApplication::callback,this);
 }
 void callback(Msg const& m){/*...*/}
 void computation(){
 //...
 Msg m;
 pub.publish(m);
 }
 ros::NodeHandle nh ; ros::Rate rate ;
 ros::Publisher pub ; ros::Subscriber sub;
};

int main(int argc, char *argv[]) {
 ros::init(argc,argv);
 ROSApplication app;
 app.run();
}
```

Listing 2 shows an instance of the design pattern.

It centralizes all ROS-related operations within the ROSApplication class (node handles, publishers or subscribes are forbidden in any other class or functions). Thus, analyzing this class is enough to know publishers and subscribers that are used.

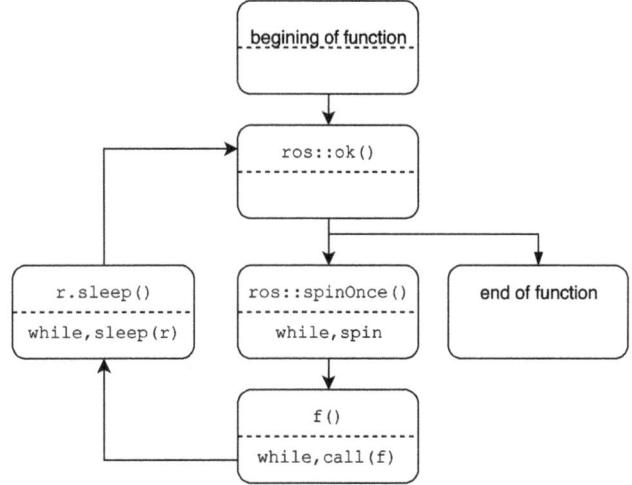

Fig. 2. ROSApplication::run's CFG illustrated. The lower halves of the states are valuations for predicates in PREDCFG.

To ensure a constant mapping between mapping between topics and publishers/subscribers, they should be attribute of the class. To centralize topics-related operation, *there is an* init *method in which each publisher and subscriber is affected*. Also, *all constructors should call* init to ensure the publishers/subscribers are always affected.

To achieve simplicity, global variables are forbidden and callbacks are private functions. More importantly, it also provides a run method which acts as an event loop. In that method, there is a loop which begins with a call to ros::spinOnce and ends with a call to sleep on an attribute whose type is ros::Rate. In between, each function in which there is a call to publish on a variable whose type is Publisher is called in the loop. Figure 2 illustrates ROSApplication::run's CFG. Thus, main is simply reduced to the creation of a ROSAppliction object with a call to run on it.

6.2. *Formal description*

For ease of understanding, the formula is divided into several subformulas.

Looking for a class named ROSApplication is formally expressed as $\exists c$ isClass$(c) \wedge$ name$(c, ROSApplication)$. Looking for a method named init is expressed as $\exists i$ isMemFctOf$(i, c) \wedge$ name$(i, init)$. Ensuring that the init method is call in all constructors is formally expressed as $\forall d$ (isConstructorOf$(d, c) \Rightarrow$ models$_{\text{CTL}}(d, \mathbf{AF}\text{call}(i)))$ where $call(f)$ is an element of PREDCFG such as it will be true on states where there is a call to f. Finally, making sure that each publisher is affected once and only once in init is expressed as

$$\forall p(\text{isAttributeOf}(p, c) \wedge \text{hasType}(p, Publisher)$$
$$\Rightarrow (\exists n(\text{isAttributeOf}(n, c) \wedge \text{hasType}(n, NodeHandle)$$
$$\wedge \text{ models}_{\text{CTL}}(i, \mathbf{AF}(\text{adv}(p, n))$$
$$\wedge \mathbf{AG}(\text{adv}(p, n)) \Rightarrow \mathbf{AX} \mathbf{AG} \neg \text{adv}(p, n))))). \quad (6)$$

The existential quantification for the `NodeHandle` is done after the universal one for publishers because different node handles can refer to different namespaces and there is no obligation to use the same node handle for all publishers and subscribers. $\mathrm{adv}(X, n)$ is another element of PREDCFG such that it will be true on states where there is a statement $X = n$. `advertise(...)`. The constraint that each element is affected at least once is expressed by the first part of the CTL formula **AF** ... (it should happen at least once) whereas at most one constraint is dealt with by the second part of the CTL formula **AG** ... (once it has happened, it should not happen again). The formal property for subscriber attributes is similar.

Equation (7) shows `run`'s formal specification. We first look for a member function named `run` and for an attribute whose type is `Rate`. Then, we look for all functions f such that there is an attribute whose type is `Publisher` on which a message is published within f. For such functions, there is within r the sequence spin $\wedge$ while, while $\wedge$ call(f) and finally while $\wedge$ sleep(s) $\wedge$ **AX**¬while. We forbid all *while loops* before the sequence to ensure that the call to `spin` is the first instruction within the loop. This is why the **seq** operator appears on the right-hand side of $\mathbf{A}[\neg \text{while}\mathbf{U} \ldots]$. The end of the loop is a state where while is true and false all next states. On this state, sleep(s) has to be true:

$$\exists r(\mathrm{isMemFctOf}(r, c)$$
$$\wedge \ \mathrm{name}(r, run) \ \exists s(\mathrm{isAttributeOf}(s, c)$$
$$\wedge \ \mathrm{hasType}(s, Rate) \wedge \forall f(\mathrm{isMemFctOf}(f, c)$$
$$\wedge \ (\exists p(\mathrm{isAttributeOf}(p, c) \wedge \mathrm{hasType}(p, Publisher)$$
$$\wedge \ \mathrm{models}_{\mathrm{CTL}}(f, \mathbf{EF}\mathrm{pub}(p))))$$
$$\Rightarrow \ \mathrm{models}_{\mathrm{CTL}}(r, \mathbf{A}[\neg \text{while U } \mathbf{seq}(\text{spin}$$
$$\wedge \ \text{while, while} \wedge \mathrm{call}(f), \text{while} \wedge \mathrm{sleep}(s)$$
$$\wedge \ \mathbf{AX}\neg \text{while})])))). \quad (7)$$

Equation (8) shows how to specify `main` behavior. The first-order part of the property searches for a free function named for `main` which has a variable whose type is the same as c (i.e., `RosApplication`). callInit is an element of PREDCFG on states with a call to `ros::init`. Hence, `main` must start by a call to `ros::init` and then two steps later there is a call to the previously quantified `run` function on n. The step following the call to `ros::init` is the declaration of the ROSApplication variable and is not specify within the temporal property, hence the use of **AXAX** is to specifically target the third step within `main`:

$$\exists m(\mathrm{isFunction}(m) \wedge \mathrm{name}(m, main)$$
$$\wedge \ \exists n(\mathrm{locallyDeclared}(n, m) \wedge \mathrm{hasType}(n, c)$$
$$\wedge \ \mathrm{models}_{\mathrm{CTL}}(m, \text{callInit} \wedge \mathbf{AXAX}\mathrm{callFct}(n, r))). \quad (8)$$

6.3. *Design pattern review*

This design pattern is partially used in some of the packages we analyzed: 11 of them centralize all ROS communications within a single class. Yet, these classes are most of the time mingled with the algorithms that operate on the data. This design pattern was used to enhance ROS code quality in different packages. For instance, we reimplemented imu_bias_remover.cpp from the package imu_processors. This new implementation maintains the original code behavior while fixing the defects found using Pangolin (10 global variables, three free functions used as callbacks). The new code is available in Pangolin's repository.

7. Conclusion

This paper presents a framework for checking source code compliance with a user-provided specification looking for suspicious code patterns. The framework includes a new logic called FO^{++} which is based on a parametrization of first-order logic with temporal logics. This logic allows us to conjointly specify some properties over the control flow graphs of one or several functions and over the surrounding abstract syntax tree. Once instantiated for C++, it is used as a specification formalism for Pangolin, a verification engine for C++ based on Clang, and nuXmv.

To improve code quality on ROS packages, we showed how five common suspicious patterns could be formalized with FO^{++} and using Pangolin, we checked their absence in 25 ROS packages. We found several occurrences of these patterns in the packages we analyzed, some of which are real design issues. To overcome the limitations of some of the rules, we need to constrain the design of the code and thus we propose a design pattern that addresses all the issues listed before. We show how Pangolin can be used to enforce it.

There are two main directions for improving Pangolin: user interaction and expressive power. The first one would be to provide an input language closer to actual code, and to improve user feedback. The second one would be to add interprocedural and multi-file analysis. This would require extending FO^{++}'s temporal logics to distinguish in which function a statement occurs.

References

[1] M. Quigley, K. Conley, B. Gerkey, J. Faust, T. Foote, J. Leibs, R. Wheeler and A. Y. Ng, ROS: An open-source robot operating system, *Proc. ICRA Workshop on Open Source Software* (2009).

[2] P. Evans and S. Edwards, A disruptive community approach to industrial robotics software, presented at the 2012 RoboBusiness Leadership Summit (2012).

[3] B. W. Boehm, J. R. Brown and M. Lipow, Quantitative evaluation of software quality, *Proc. 2nd Int. Conf. Software Engineering* (1976), pp. 592–605.

[4] V. Rivera, N. Catano, T. Wahls and C. Rueda, Code generation for Event-B, *Int. J. Softw. Tools Technol. Transf.* **19**, 31 (2017).

[5] P. Cuoq, F. Kirchner, N. Kosmatov, V. Prevosto, J. Signoles and B. Yakobowski, Frama-C, *Proc. 10th Int. Conf. Software Engineering and Formal Methods* (2012), pp. 233–247.

[6]MISRA Collab., *MISRA C++: 2008: Guidelines for the Use of the C++ Language in Critical Systems* (MIRA, 2008).

[7]B. Stroustrup and K. Carroll, C++ in safety-critical applications: The jsf++ coding standard Presentation, Lockheed Martin (2006).

[8]Programming Research, Ltd., *High-Integrity C ++ Coding Standard Manual* (Programming Research Group, 2004).

[9]P. Avgustinov, O. de Moor, M. P. Jones and M. Schäfer, QL: Object-oriented queries on relational data, *Proc. 30th European Conf. Object-Oriented Programming* (2016).

[10]P. Pialorsi and M. Russo, *Introducing Microsoft® LINQ* (Microsoft Press, 2007).

[11]J. Brunel, D. Doligez, R. R. Hansen, J. L. Lawall and G. Muller, A foundation for flow-based program matching: Using temporal logic and model checking, *ACM SIGPLAN Not.* **44**, 114 (2009).

[12]N. Palix, G. Thomas, S. Saha, C. Calvès, G. Muller and J. Lawall, Faults in Linux 2.6, *ACM Trans. Comput. Syst.* **32**, 4-1 (2014).

[13]A. Santos, A. Cunha, N. Macedo and C. Lourenço, A framework for quality assessment of ros repositories, *2016 IEEE/RSJ Int. Conf. Intelligent Robots and Systems (IROS)* (2016), pp. 4491–4496.

[14]A. Santos, A. Cunha and N. Macedo, Mining the usage patterns of ROS primitives, *Proc. 2017 IEEE/RSJ Int. Conf. Intelligent Robot and Systems* (2017).

[15]J.-P. Ore, S. Elbaum and C. Detweiler, Dimensional inconsistencies in code and ros messages: A study of 5.9 m lines of code, *Proc. 2017 IEEE/RSJ Int. Conf. Intelligent Robots and Systems (IROS)* (2017), pp. 712–718.

[16]Ros.org, ROS c++ Style Guide (2014), http://wiki.ros.org/CppStyleGuide.

[17]Ros.org, code_quality (2013), http://wiki.ros.org/code_quality.

[18]GitHub, ros/roslint (2013), https://github.com/ros/roslint.

[19]B. C. Lopes and R. Auler, *Getting Started with LLVM Core Libraries* (Packt, 2014).

[20]R. Cavada, A. Cimatti, M. Dorigatti, A. Griggio, A. Mariotti, A. Micheli, S. Mover, M. Roveri and S. Tonetta, The nuXmv symbolic model checker, *Proc. Int. Conf. Computer Aided Verification* (2014), pp. 334–342.

[21]C. Caleiro, C. Sernadas and A. Sernadas, Mechanisms for combining logics, Unpublished paper, IST, Lisbon (1999).

[22]M. Huth and M. Ryan, *Logic in Computer Science: Modelling and Reasoning about Systems* (Cambridge University Press, 2004).

[23]Iso, Iso Standard ISO/IEC 14882:2014 (E): Information Technology: Programming Languages: C++, Geneva (2014).

Integrating planning and reactive behavior by using semantically annotated robot tasks

Andreas Schierl*, Alwin Hoffmann[†], Ludwig Nägele[‡] and Wolfgang Reif[§]

Institute for Software & Systems Engineering, University of Augsburg
Universitätsstr. 6a, 86159 Augsburg, Germany
*schierl@isse.de
[†]hoffmann@isse.de
[‡]naegele@isse.de
[§]reif@isse.de

Tasks that change the physical state of a robot and its environment take a considerable amount of time to execute. However, many robot applications spend the execution time waiting, although the following tasks might require time to prepare. This paper proposes to amend robot tasks with a semantic description of their expected outcomes, which allows planning and preparing successive tasks based on this information. The suggested approach allows sequential and parallel compositions of tasks, as well as reactive behavior modeled as state machines. The paper describes the means of modeling and executing these tasks, details different possibilities of planning in state-machine tasks and evaluates the benefits achievable using the approach on two example scenarios.

Keywords: Planning; reactive behavior; state machine; robot programming; optimization.

1. Introduction

Most classical industrial robot arms are currently used for fixed and recurring preprogrammed tasks. There, planning (at runtime) and reaction to indeterministic events are of minor importance. When it comes to more modern or mobile robots, the situation is different: Due to the greater variability in the environment, these robots must be able to sense external events and react accordingly, and can thus be seen as reactive systems.

In this context, it can be observed that when executing a task, performing its physical actions often requires a considerable amount of time compared to the computations needed to prepare them. Still, it is often possible to tell details about the positive outcome of the task even before the task is fully executed. In contrast, for failures the amount of information that can be given in advance is limited (e.g., where the robot will be when the error occurs is usually unknown). While these details could considerably help to plan ahead, many software frameworks often ignore this potential and spend the execution time waiting for completion (cf. Sec. 2).

To improve this situation, we propose to model tasks including their expected outcome(s) and details about the corresponding situation(s). These expected outcomes include successful execution, but also detectable errors and relevant situations that occur during execution. Using these semantically enriched tasks, it becomes possible to plan the following tasks even before the execution of the current task has completed. These tasks can be combined in a sequential or parallel way, can perform case distinctions and can even be used to model reactive behavior through state machines.[1] Composing these tasks, the programmer can configure details about the relationship between the tasks and the quality of transition: For some events, an immediate reaction may be required, so the corresponding reaction task has to be planned before the event can occur, while other events allow some time, so that more relaxed planning schemes can be used.

To evaluate the approach, the cooperation between two mobile robots was modeled and implemented (cf. Fig. 1). There, two KUKA youBots[2] drive in parallel, and hand over a baton while in motion. To realize the reactive behavior and coordination, state machines are used, while the possibility of preparation during execution reduces the waiting time between the different robot actions. Additionally, a theoretical example with more complex planning is introduced and analyzed, showing the possible advantages of the approach.

The work presented here is based on the conference paper on predictive preparation of state-machine tasks[3] and extends it mainly in three fields:

- Semantic descriptions of task outcomes, describing environment states as well as situations that result from changes therein
- Tasks with multiple possible outcomes and their handling
- Complex task compositions, and the advantages and drawbacks of using state-machine tasks with deferred preparation

---

*Corresponding author.

112   *A. Schierl et al.*

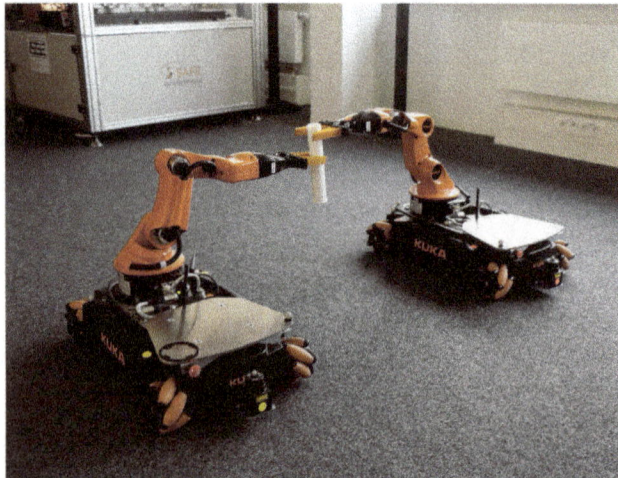

Fig. 1. Robots handing over a baton.

This paper is structured as follows: After an overview of different ways to handle complex robot behavior (Sec. 2), the approach of defining and executing tasks is outlined in Sec. 3. Section 4 describes the experiments conducted to evaluate the approach and points out the corresponding results. Finally, Sec. 5 gives a conclusion and outlook.

2. Related Works

For modeling and composing tasks or device capabilities, different approaches exist in modern robot frameworks such as *ROS* and *OROCOS*. *ROS* actionlib[4] allows to model interruptible tasks through *Actions*. *Actions* represent long-running, preemptible tasks that provide feedback and notify about their result, and that can be canceled when the goal has to be changed. They can be seen as an extensible way to model device capabilities, allowing extensions by introducing new components that provide *Action* servers. To combine multiple *Actions*, a new component can provide an *Action* that invokes the corresponding *Actions* in parallel or sequentially. However, no timing guarantees are given for this type of composition, as it relies on network communication between the different *Action* servers. Furthermore, *Actions* do not share a common interface and thus cannot directly be passed between different components (for example, a planner cannot provide its result as a generic *Action* that can be passed to another component deciding when it should be executed).

For reactive behavior on the task level with *ROS*, Bohren and Cousins introduced *SMACH* state machines.[5] In *SMACH*, *ROS Actions*, *Services* or Python code can be defined as *States*, and *States* can be composed into new *States* in a concurrent, sequential or state-machine form. Therefore, each *State* defines different outcomes that can be used in composite *States* to define reactions or handle error conditions. Furthermore, states in *SMACH* (and thus state machines) can be made available to other components as *actionlib Actions*. This composition mechanism is similar to the one presented in

this paper, however without timing guarantees, so that guaranteed reactions to events cannot be specified this way. Furthermore, *SMACH* (as well as *actionlib*) provides no explicit semantics or metadata for *Action* or *State* post-conditions, making further planning during execution harder to achieve. *SMACH* state machines are executed in a blocking way and thus do not offer the possibility of using the execution time of robot actions for further (motion) planning steps.

For *OROCOS*, restricted finite state machines (*rFSMs*)[6] allow modeling reactive behavior. An *rFSM* describes a hierarchical state machine without parallelism, aimed at the coordination of robot applications. According to the *pure-coordination* pattern,[6] these state machines only process and raise Boolean events which have to be provided by monitor components or handled by configurator components, which in turn manipulate or reconfigure the active components in order to achieve the goal. *rFSM* state machines are implemented on top of the programming language *LUA* with specialized memory allocation and garbage collection, and can thus be executed with real-time guarantees.

As an extension to *rFSM* state machines, Scioni *et al.* describe how to achieve *preview coordination*[7]: In this approach, the execution environment takes hints about execution probabilities based on likelihood labels on some transitions. Using these labels, likely successor states can be prepared (performing some of their work) while the previous state is still active, as long as there is no conflict between the preparation steps and the actions performed in the current state. This allows reducing the execution time, while keeping the action definitions coherent (instead of moving the preparation step into the previous state).

The preview coordination mechanism introduces a form of decoupling between workflow and capability execution and makes use of metadata about conflicts between states, but the *rFSM* mechanism still does not include further semantic descriptions for the results of states. This way, the following tasks cannot analyze and prepare for the expected results of the previous *State*, a powerful and important feature offered by the approach introduced in this paper.

Another related approach has been introduced by Angerer[8] in the form of Robotics API *Activities*. It can be seen as the basis of the approach suggested in this paper, but has a stronger focus on real-time and specific ways of modeling metadata, and does not support reactive behavior in the form of state machines. This approach has been further extended by Schierl[9] and along with further research led to the results presented in this paper.

3. Approach

To facilitate planning during execution, we propose to model individual robot tasks (cf. Sec. 3.1) along with a semantic description of their expected outcome(s) as described in Sec. 3.2. Based on these outcome description(s), it becomes possible to prepare the successor task (or at least start planning) before the current task has been fully executed — maybe even before execution starts. This feature can be exploited during execution, as described in Sec. 3.3. When it comes to more complex tasks with different viable outcomes, an implementation that is based on simple control flow tends to become confusing, so it becomes helpful to be able to compose multiple simple tasks (cf. Sec. 3.4). As an alternative, the desired behavior can be defined in a model-based way, e.g., using the formalism of state machines,[1] as described in Sec. 3.5.

3.1. *Task definition*

Initially, a task to be executed by a specific actuator (e.g., move to a given position, pick up an item or even bring me a beer) is modeled as an *Activity* (cf. Fig. 2). As one task can be performed more than once (if you are with friends), it becomes helpful to keep track of different task executions. So, when a task is to be executed in a given situation, its *Activity* is instantiated through *createHandle*() yielding an *ActivityHandle*.

Being responsible for a single task execution, the *Activity-Handle* offers state tracking that notifies about execution progress, and is also responsible to decide how the desired task is to be executed in a given situation. It receives a semantic description of the situation (see below) the task is to be executed in, and has to provide instructions on how to act, and what outcome to expect after the action has been performed. Thus, the *ActivityHandle* takes a situation description called *ActivityResult*, and provides an *ActivitySchedule* that describes the task to execute, as well as a set of possible *ActivityResults* that can occur while/after executing the task. This way, it provides possible post-conditions for the execution of the task, while the preconditions are derived from the results of the previous task. If the previous task has multiple possible outcomes, the *ActivityHandle* is provided with the different acceptable outcomes, and can plan for each of the envisioned situations. This differs from typical planning mechanisms that are based on pre/post-conditions and try to combine fitting actions, but instead allows applying a given task in many different situations.

To describe the expected situation, the *ActivityResult* contains status information about each individual actuator contributing to the task, as well as a description about the envisioned geometric and logical situation of the environment, i.e., the world state. The actuator-specific information may contain details about a gripper, e.g., whether it is open or closed, as well as joint positions, velocities and accelerations for robot arms.

3.2. *Task outcomes*

To talk about the world state, a model as described by Schierl *et al.*[10,11] is used. It is based on physical objects and geometric features (modeled as frames), along with relations that are defined between the frames and objects and form an undirected multigraph. Each relation describes an aspect about the relationship between two objects (such as physical objects or coordinate frames), and can give quantitative information. Between two physical objects, a containment relation tells that a screw is part of a workpiece, or a link is part of a robot. Other relations bring together geometric features (modeled as coordinate frames) and their physical object, while relations between coordinate frames describe (quantitative or qualitative) geometric aspects. Between two frames, a logical relation describes the type of connection (e.g., persistent for the connection between two links of a robot or transient for the connection between a gripper and the grasped workpiece), while a geometric relation describes the exact position where the frames are relative to each other (or gives a sensor or computation that provides that information). Together with containment and geometric feature relations, this semantic modeling allows to derive that a gripper is connected to the robot, that the robot has grasped a workpiece and that the workpiece is at a specific position in space.

While these relations are known by their corresponding objects, they are not static during runtime (an object grasped by a gripper can be placed on the ground, removing one relation and establishing up another). Thus, situations with different (geometric) relations have to be modeled. One

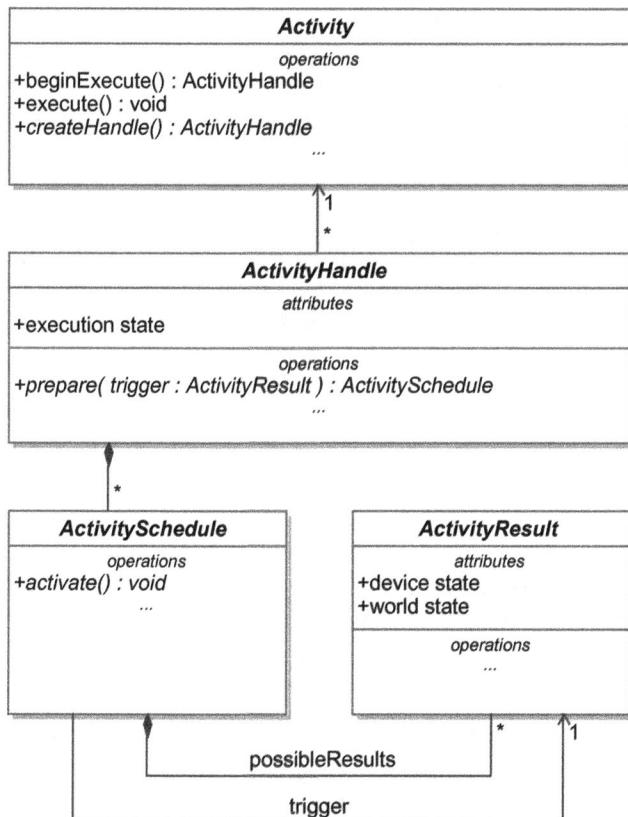

Fig. 2. Relevant classes for task definition and execution.

situation of established relations is modeled as a *FrameTopology*. A *FrameTopology* stores the existing relations for each of the objects modeled in the environment, and allows to find all relations for a given object, as well as relation sequences between given objects.

To give a description about the expected world state after the execution of an *Activity*, the view of the world model is expressed as a *FrameTopology*. When a robot picks up an object from the ground, the relation between ground and object is removed, and a new relation between gripper and object is established. These topology changes are relevant for further planning, because a grasped object changes the shape of a robot: some motions that were previously possible now result in collisions. Thus, the world view modeled in the *FrameTopology* provided by the task includes sets of removed and added relations.

Apart from the established relations, a *FrameTopology* can also give hints about the expected sensor values for geometric aspects to express the positions. This allows to model not only that the robot has grasped an object, but also the position where the robot is expected to be. For tasks that include robot motions, the resulting *FrameTopology* thus contains a position change that gives the (expected) new pose of the robot or all of its components. This geometric information is used whenever performing geometric calculations based on the world view, e.g., when planning the next motion.

To model different situations, the *FrameTopology* does not contain a full snapshot of all relations, but rather works by storing the difference compared to another topology. This is especially important when trying to apply the topology changes after successful task execution: If each *FrameTopology* contained the complete set of relations, it would be impossible to execute two tasks at the same time, because the second task would overwrite the changes applied by the first task, because these changes would not be present in the snapshot taken before the execution of the first task. Instead, modeling *FrameTopologies* as a difference to a previous world state allows to keep all unrelated parts of the situation unchanged. In contrast, the position changes stored in the *FrameTopology* do not have to be applied, because they only describe expected sensor values for the corresponding time. Thus, after successfully executing the task the position of the robot has changed as defined, so that the real sensor values for the robot position are similar to the values predicted by the task.

3.3. *Task execution*

Once a task has been defined, i.e., created as an *Activity*, it can be executed in a blocking or nonblocking way. For a specific execution instance, the *Activity* creates an *ActivityHandle* that takes the *ActivityResults* of the previous *Activity* and prepares *ActivitySchedules* for each of the situations. The *ActivitySchedules* are activated so that they trigger the execution of the respective task once their *ActivityResult* is reached.

Additionally, the *ActivitySchedules* provide their possible *ActivityResults* for use with following *Activities*.

The first option is to execute the task in a blocking way. Figure 3 shows the tasks (A) and (B), which are executed sequentially in a blocking way. In this case, the method *execute*() of task (A) blocks until a completion *ActivityResult* of a corresponding *ActivitySchedule* has been reached.

Figure 4 shows the timing of this behavior. The *Activities* are shown as horizontal lifelines with time running from left to right, while the different boxes denote the phases preparation (P) and execution (E). However, in this case the situation description contained in the *ActivityResult* is of little use, because once the result has been reached, the described situation is already reached, and the execution time has already been wasted waiting.

To use the execution time, nonblocking execution can be used. Figure 5 shows task (A) followed by task (B) in a nonblocking way. Executing tasks with *beginExecute*() allows the control flow to continue once any of the *ActivitySchedules* has been triggered and the execution of the task has thus started. The following task can then be prepared for different possible start situations (based on the *ActivityResults* of the currently running task), and immediately execute one of the planned solutions if the corresponding situation occurs.

This situation is shown in Fig. 6 — here, the dotted lines denote times where (B) is already planned and waiting to be executed. The preparation of (B) happens while (A) is running, and (B) starts running once (A) is completed.

In contrast, if task (B) is independent of (A), e.g., if it controls different devices, it starts immediately when prepared, as shown in Fig. 7. However, this execution mode does not guarantee parallelism, because preparation of (B) starts

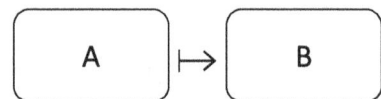

Fig. 3. Blocking execution of task (A), followed by task (B).

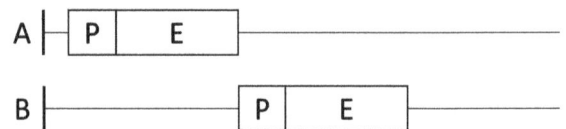

Fig. 4. Blocking execution of task (A), followed by task (B), with preparation (P) and execution (E).

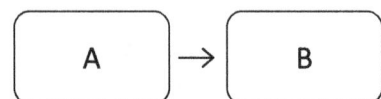

Fig. 5. Nonblocking execution of task (A), followed by task (B).

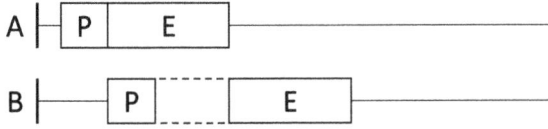

Fig. 6. Nonblocking execution of task (A), followed by task (B), with preparation (P) and execution (E).

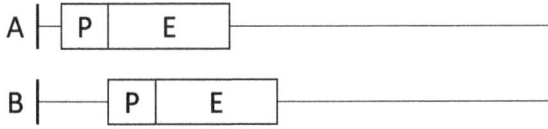

Fig. 7. Timing of nonblocking execution of independent tasks, with preparation (P) and execution (E).

after execution of (A) has started, resulting in a certain delay between the two execution starts.

While called nonblocking, this execution scheme is not fully asynchronous, because preparation for the following task only starts after the current task has started. This behavior becomes obvious when having a look at a sequence of three tasks (as shown in Fig. 8), each of which has two possible results. In this situation, preparation of (B) starts once (A) has started, while (C) is prepared once (B) has started.

Using the proposed nonblocking semantics (as shown in Fig. 9), (B) has to be prepared for both results of (A), yielding two *ActivitySchedules*. However, since preparation of (C) is postponed to the time when (one concrete schedule for) (B) is started, (C) only has to be planned for the chosen (B) schedule and its two results.

In a fully asynchronous execution, in contrast, preparation had to be performed for both *ActivityResults* of each *ActivitySchedules* of (B), resulting in four *ActivityResults* to plan for and an exponential growth in preparation work for longer sequences.

3.4. *Task composition*

In addition to different execution types, multiple tasks can be combined to provide better guarantees about execution (e.g., reliable parallel or sequential execution).

When sequentially combining two tasks (as shown in Fig. 10), the created sequential task provides the second task with all the *ActivityResults* of the first task's *ActivitySchedules*, so that the second task can react to all expected outcomes of the first task, while the results of the second task are provided as results of the sequence. In cases where determinism is required, this kind of preparation can even allow real-time guarantees (given a corresponding software framework that supports scheduling of real-time tasks, e.g., the Robot Control Core introduced by Vistein *et al.*[12]): preparing the second task for all possible outcomes of the first task before the first task's execution starts guarantees that the second task can immediately take over no matter how long or short the first task takes.

In Fig. 11, the sequence *S* is executed after (A), and the preparation of *S*1 and *S*2 completely happens before *S*1 is started, guaranteeing that *S*2 can be started immediately when *S*1 ends. However, this determinism is only possible if the outcomes are fully known in advance: if details of an outcome description depend on sensor data, these *ActivityResults* can only be provided when the sensor data becomes available, and preparation of the next task is delayed to this moment (no longer guaranteeing determinism).

However, these sequences do not delay preparation, so the second task has to be planned for all results of the first task, which has to be planned for all results of the previous task, leading to an exponential amount of preparation required for longer sequences.

As a second combination option, parallel execution can be used (as shown in Fig. 12). The created parallel task then provides both subtasks with the same initial situation (as *ActivityResults*), and combines resulting *ActivitySchedules*

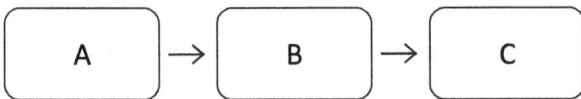

Fig. 8. Nonblocking execution of multiple tasks.

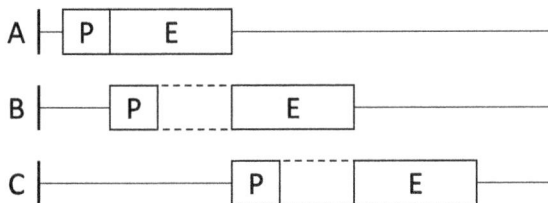

Fig. 10. Simple task, followed by a sequential task.

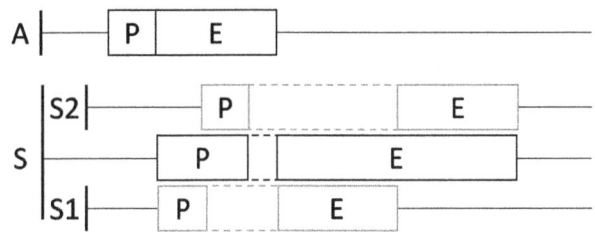

Fig. 9. Timing of nonblocking execution of multiple tasks, with preparation (P) and execution (E).

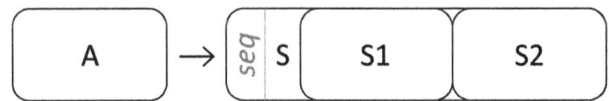

Fig. 11. Timing of sequential task with preparation (P) and execution (E).

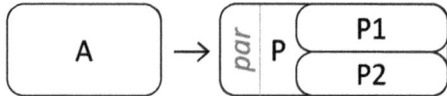

Fig. 12. Simple task, followed by a parallel task.

(and optionally *ActivityResults*) to create the *ActivitySchedule* and *ActivityResults* of the parallel task.

This situation is shown in Fig. 13, where the parallel tasks *P*1 and *P*2 are planned and thus allow *P* to start them at the same time after (A) is completed. Of course, the parallel task has to make sure that the subtasks do not conflict, e.g., by checking that they control different devices.

Additionally, simple case distinctions can be made based on the possible *ActivityResults*, as shown in Fig. 14. A task with case distinction classifies the *ActivityResults* of the previous task and forwards them to the corresponding subtask depending on the decision condition.

Figure 15 shows the execution of this situation, with (C) as a case distinction for the result of (A), deciding to execute (C2) and unload (C1).

Combined tasks can further be composed to handle more complex scenarios. Figure 16 shows a task *S* that first executes (A1) and continues with (C1) in case of success, or with (B1) followed by (C1) if (A1) fails.

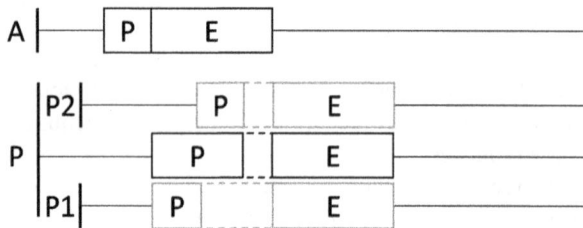

Fig. 13. Timing of parallel task with preparation (P) and execution (E).

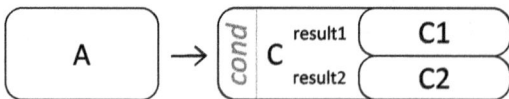

Fig. 14. Simple task, followed by a conditional task.

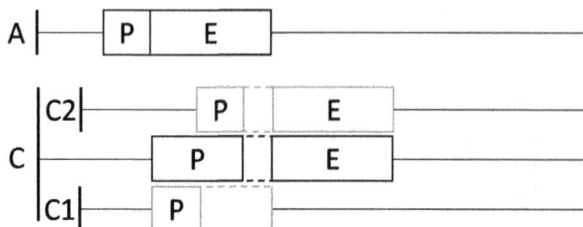

Fig. 15. Timing of parallel task with preparation (P) and execution (E).

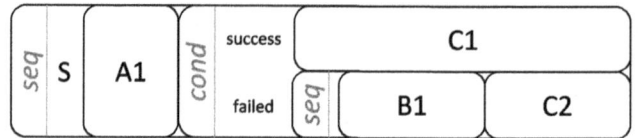

Fig. 16. Complex task with sequential and conditional compositions.

For execution, first (A1) is prepared. For its error result, (B1) has to decide how to react, while (C1) handles all results of (B1). Additionally, (C2) has to be prepared for the success case of (A1). Once all these preparations are completed, execution of *S* can begin, first executing (A1), and then switching to (B1) [because (A1) failed] and (C1). This behavior is shown in Fig. 17.

This entire composition can provide real-time guarantees for execution, because every possible case has been prepared beforehand; however as a downside all possible cases have to be planned, which requires a lot of preparation time before execution can start. Additionally, this type of composition does not allow for unbounded loops, because preparation requires to unroll the loop to plan for every possible amount of repetitions, and quickly limits bounded loops by the exponential amount of results to plan for.

Instead, these repetitions should be handled through nonblocking execution of separate tasks that are repeated through control flow mechanisms wherever timing guarantees are less important, or through state machines that allow to better handle reactive behavior as described in the following subsection.

3.5. *State machines for reactive behavior*

For longer or repeating task sequences, these composition mechanisms that fully plan ahead may reach their limit, because the number of tasks that has to be prepared grows exponentially with the number of successive tasks (given that each has more than one possible result). Thus, it becomes helpful to limit the amount of preparation. One way to achieve this while supporting complex compositions of tasks is the use of state machines — a step often taken when specifying reactive behavior.

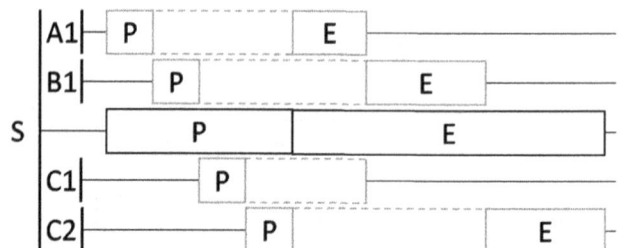

Fig. 17. Timing of complex task execution with preparation (P) and execution (E).

The proposed approach uses *Activities* to define states of the state machine. For each state (or *Activity*), transitions can be given that specify a switch to another *Activity* if certain *ActivityResults* occur. Additionally, one distinct *Activity* is chosen as start state, and *ActivityResults* of some *Activities* can be defined as transitions to a final state.

As each state can be entered more than once during execution, for its *Activity* multiple *ActivityHandles*, *ActivitySchedules* and also *ActivityResults* are prepared. The resulting set of *ActivityResults* for an *Activity* is dynamic, so the transition cannot give a complete list. Instead, it works as a predicate to *ActivityResults* that chooses whether a given *ActivityResult* qualifies for the transition, which can be as generic as *any error* or as specific as *any successful execution of the given grasp*.

As an example, Fig. 18 shows a state machine of the three tasks (A), (B) and (C), switching from (A) to (C) if (A) succeeds, and to (B) if it fails. State (B) is always followed by (C), leading to two different paths reaching (C).

Regarding execution, state-machine tasks differ from the composed tasks mentioned previously: While for sequential, parallel or conditional tasks preparation can completely be performed upfront, state-machine tasks require some preparation while the task is running. The *Activity* for the initial state is prepared traditionally, based on the *ActivityResults* of the previous task. Then, the *ActivityResults* of the created *ActivitySchedule* are compared against the transitions originating from the start state to decide whether they have to be handled by switching to another state (i.e., *Activity*). However, for these transitions and *Activities*, different options exist when to prepare them for the corresponding results, as a trade-off between preparation of unnecessary traces and not having prepared a required task. For the scope of this paper, the preparation is usually delayed until their originating state is entered (i.e., its *Activity* started running), preparing the target states for outgoing transitions using the reported *ActivityResults* of the current state, so that they can react to the occurrence of the corresponding result. Additionally, the outgoing transitions of the target state are analyzed, so that they can be handled once the target state is entered. This way, the states are prepared with a look-ahead of one transition, which works fine as long as the typical execution time of a state is longer than the preparation times for all following states.

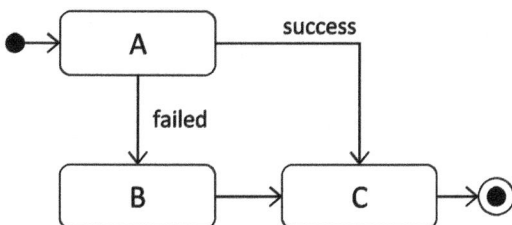

Fig. 19. Execution of state machine from Fig. 18 with preparation (P) and execution (E).

Figure 19 shows the preparation timing resulting for the state machine, numbering the different traces leading to state (C) as (C1) and (C2). When executing the state machine, (B1) and (C1) are prepared once (A1) is started, and when the failure occurs, (B1) is started and (C1) is unloaded. A new instance (C2) has to be prepared for the new situation resulting of (B1), which is then executed once (B1) finishes. Using this method yields a behavior that is similar to the one of nonblocking execution (cf. Fig. 6), where the following task is prepared once the previous is started.

However, this mechanism of delaying preparation for outgoing transitions once the state is entered is not sufficient to guarantee that a certain transition can be taken without further delay, especially if the result happens quickly after entering the state. Thus, for critical transitions, preparation may not be delayed until this time.

To allow this, transitions can also be annotated with qualifiers that direct the order or timing of preparation. For transitions that may not be missed (such as error recovery strategies that bring the robot into a safe state), a transition can be marked as reliable, requesting it to be prepared before the originating *Activity* starts. These reliable transitions are handled similar to sequential tasks described before, making sure that the following state can be executed no matter how long or short the first state takes. While having the same advantages of sequential tasks, reliable transitions also inherit their drawbacks, especially the amount of preparation necessary if more than one *ActivityResult* is handled by a reliable transition, as well as the fact that cycles in the reliable transition graph are forbidden.

Figure 20 amends the transition from (A) to (B) with a stereotype ⟨⟨*reliable*⟩⟩, defining that the execution environment has to guarantee that the transition will be taken immediately if the failure occurs.

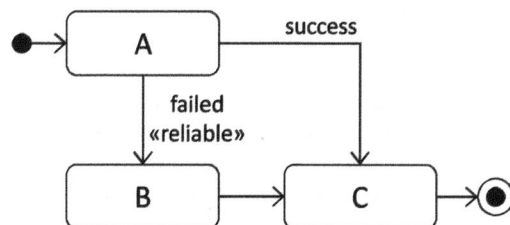

Fig. 18. Definition of a state-machine task consisting of three tasks.

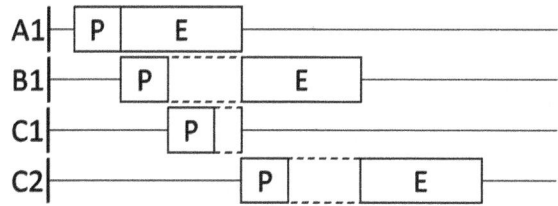

Fig. 20. Definition of a state-machine task with a reliable transition.

118 A. Schierl et al.

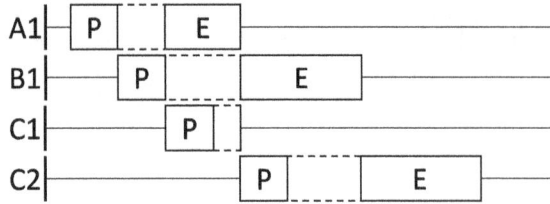

Fig. 21. Execution of reliable transitions from Fig. 20 with preparation (P) and execution (E).

For execution, reliable transitions and their target state have to be prepared along with their originating state, before the originating state is entered. This behavior is shown in Fig. 21. Here, (A1) and (B1) are planned before (A1) is started, making sure that the occurrence of the failure cannot happen while (B1) is still being planned.

When all transitions in the state machine are marked with the stereotype ⟨⟨reliable⟩⟩, the defined task has the same semantics as the complex composition shown in Fig. 16, guaranteeing that every transition can be taken immediately. For execution, all traces through this state machine have to be prepared beforehand, so preparation looks like the one shown for the complex composition in Fig. 17. This leads to long preparation time before execution can start, and inherits the same restrictions that long reliable sequences (with multiple results per task) lead to exponential growth in preparation time, and reliable cycles are forbidden.

4. Experimental Results

The proposed approach has been evaluated in two examples. First as a real-world example, the interaction between two mobile robots has been modeled. Two mobile robots drive parallel to each other and hand over a baton while in motion. The implementation uses two KUKA youBots and is based on the Robotics API.[8] There, both robots are controlled using their on-board computers using a C++ control core,[12] while the high-level coordination and task execution is implemented in Java and performed from a laptop computer connected to the youBots through a wireless network.

Because the handover example (apart from error handling) is purely sequential, the first implementation was based on separate tasks for moving the arm and gripper and the execution model described in Sec. 3.3. For all tasks that do not depend on the second robot, nonblocking execution was used, while the correct order of the gripping and releasing tasks was performed through blocking execution of the corresponding tasks in a common control flow. While working fine in simulation, this implementation led to unintended delays between the grasp and release operations when executed on real robots, and thus required more space for parallel driving than necessary. This was mainly caused by the unreliable network connection and inefficient network code that required multiple communication roundtrips to transmit and start the tasks.

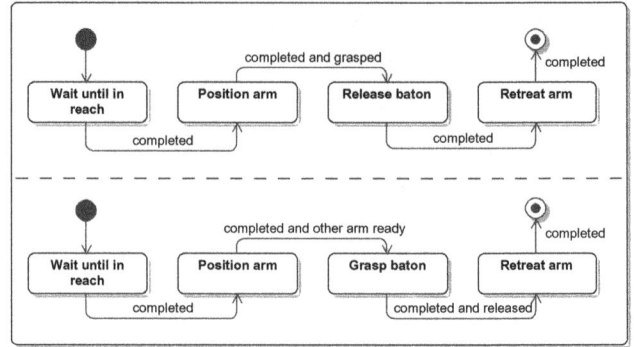

Fig. 22. State machine: robots interaction.

A second implementation modeled the expected behavior through two linked state machines (cf. Fig. 22). Here, the following tasks could already be prepared while the second youBot was waiting for the first youBot, so the delay between the grasp and release operation as well as the required space for the interaction were significantly reduced.

The second, theoretical, scenario was performed without connection to a robot framework, but based on a prototypical implementation with support for reliable transitions. Here, the scenario was to drive a mobile robot to a pick-up area, take a large object and return to its start position. In addition, during all steps a sensor had to be observed, and the robot had to stop once the sensor detected a dangerous situation. To complicate things, in the environment the shortest path to the pick-up zone could not be taken while carrying the large object, but instead a detour was required.

Figure 23 shows a state-machine model of the task. The top row of states gives the main success flow, while reliable transitions handle danger occurring during this task. Additionally, a *Back off* state has been added, which is executed after the robot has stopped in case of emergency. As after stopping the robot is already in a safe state, this transition is not time-critical and thus does not need to be reliable.

The motion planning tasks, *Go to pick-up zone* and *Return home*, were implemented as tasks that take considerable time to plan (3 s for collision-free motion planning) and to execute (5 s to move the robot along the planned path), while *pick up*

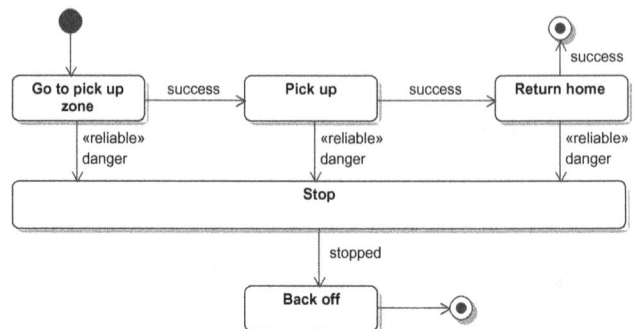

Fig. 23. State machine: second scenario.

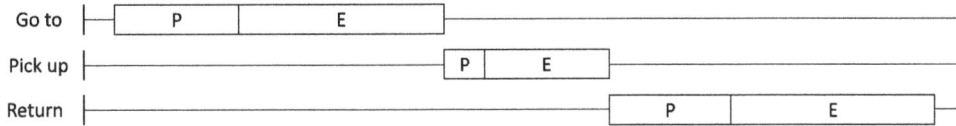

Fig. 24. Sequential execution of the second scenario, with preparation (P) and execution (E).

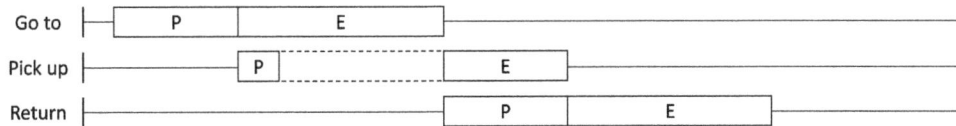

Fig. 25. State-machine execution of the second scenario, with preparation (P) and execution (E).

only takes time to execute (3 s to move the arm and gripper), but plans quickly (1 s). As a comparative reference simulating a system that does not support planning ahead based on result metadata, the three main success tasks were executed in a blocking way as a sequence.

When executed, the state machine of the second scenario implementation completed after about an average of 16.05 s, while the sequential execution took an average of 20.08 s. Figure 25 shows the resulting lifelines for those different execution models, clearly showing where time can be saved by preparation during execution.

5. Conclusion

Working with robots, the performance of applications is not only limited by the available processing power of the computer, but also by the physical limitations of controlled devices. Usually, the time needed to execute a task is significantly longer than the mere computation time, but still some planning also takes considerable time. In this paper, we proposed to amend robot tasks with descriptions of their expected outcomes. This allows preparing the following task while the current task is still running, thus reducing delays between the execution of successive robot tasks and avoiding the situation that execution time goes unused although preparation steps may be pending.

These amended tasks can be executed in a blocking or nonblocking way, or combined into complex tasks. Besides sequential and parallel compositions, state machines promise to allow the specification of more complex reactive behavior including recurring subtasks, while still behaving like regular tasks, so that further composition or preparation remains possible.

A prototypical implementation of this mechanism has been created based on the Robotics API[8] and shown to work for cooperating robots. Additionally, the approach promises to accelerate the execution of tasks where considerable time is spent preparing the next steps, such as collision-free motion planning, and also allows specifying events for which a timely reaction has to be guaranteed (given a capable

execution environment, such as a Realtime Robot Control Core[12] used with the Robotics API).

Still, this paper is limited to simple preparation strategies of state machines (transitions can be guaranteed, or planned when the state is entered). As a part of further research, more complex strategies might benefit from likelihood annotations[7] or estimated preparation times to decide which transitions in the state machine will likely happen and should be prepared first (to reduce the risk of missed transitions). Additionally, longer sequences of transitions could be prepared in corresponding situations, e.g., to skip over states that take a very short time to plan and execute. In the given theoretical example, this could help if the grasping task took shorter, because then the execution time of the *Go to* task could be used to plan the *Return* task.

References

[1] D. Harel, Statecharts: A visual formalism for complex systems, *Sci. Comput. Program.* **8**, 231 (1987).

[2] R. Bischoff, U. Huggenberger and E. Prassler, KUKA youBot: A mobile manipulator for research and education, *Proc. 2011 IEEE Int. Conf. Robotics and Automation* (IEEE, 2011), pp. 1–4.

[3] A. Schierl, A. Hoffmann, L. Nägele and W. Reif, Integrating reactive behavior and planning: Optimizing execution time through predictive preparation of state machine tasks, *Proc. 2018 Second IEEE Int. Conf. Robotic Computing (IRC)* (2018).

[4] E. Fernandez, E. Marder-Eppstein and V. Pradeep, actionlib (2016), http://wiki.ros.org/actionlib.

[5] J. Bohren and S. Cousins, The SMACH high-level executive, *IEEE Robot. Autom. Mag.* **17**, 18 (2010).

[6] M. Klotzbücher and H. Bruyninckx, Coordinating robotic tasks and systems with rFSM statecharts, *J. Softw. Eng. Robot.* **3**, 28 (2012).

[7] E. Scioni, M. Klotzbuecher, T. De Laet, H. Bruyninckx and M. Bonfe, Preview coordination: An enhanced execution model for online scheduling of mobile manipulation tasks, *Proc. 2013 IEEE/RSJ Int. Conf. Intelligent Robots and Systems (IROS 2013)* (2013), pp. 5779–5786.

[8] A. Angerer, A. Hoffmann, A. Schierl, M. Vistein and W. Reif, Robotics API: Object-oriented software development for industrial robots, *J. Softw. Eng. Robot.* **4**, 1 (2013).

[9]A. Schierl, Object-oriented modeling and coordination of mobile robots, Ph.D. thesis, Universität Augsburg (2017).

[10]A. Schierl, A. Angerer, A. Hoffmann and W. Reif, Consistent world models for cooperating robots: Separating logical relationships, sensor interpretation and estimation, *Proc. 2017 First IEEE Int. Conf. Robotic Computing (IRC)* (2017), pp. 101–108.

[11]A. Schierl, A. Hoffman and W. Reif, Consistent geometric estimation based on a world model describing logical relationships and sensor interpretation, *J. Softw. Eng. Robot.* **8**, 104 (2017).

[12]M. Vistein, A. Angerer, A. Hoffmann, A. Schierl and W. Reif, Flexible and continuous execution of real-time critical robotic tasks, *Int. J. Mechatronics Autom.* **4**, 27 (2014).

Portmanteau word-play for vocabulary enhancement with humanoid robot support

Daniele Schicchi[*,‡] and Giovanni Pilato[†,§]

Dipartimento di Matematica e Informatica
University of Palermo, Via Archirafi 34, 90123 Palermo, Italy

†*ICAR-CNR, Italian National Research Council*
Via Ugo La Malfa 153, 90146 Palermom, Italy
‡daniele.schicchi@unipa.it
§giovanni.pilato@cnr.it

Word-play is as powerful learning and motivation tool often used by educators for teaching the ability of reading, which is a complex activity. In this paper, we introduce a system that exploits a Pepper humanoid robot acting as a playfellow in a word-play game based on portmanteau words. The robot shows the ability to play with children using a conversation engine, a portmanteau creation engine, and a definition engine. In this manner, Pepper can integrate itself within a group of kids, and it can support a teacher in her activities. The humanoid can be involved in a word-based round-game in which it can play the role of either answerer or generator of new words.

Keywords: Social robotics; educational robotics; portmanteau creation.

1. Introduction

The ability of reading is a complex activity that takes into account core components such as phonemic awareness, fluency, comprehension, and vocabulary.[1] In particular, building a broad vocabulary is an important task that allows a human being to access his background knowledge, express ideas, and communicate and learn new concepts.

An efficient and motivational learning method used by teachers for vocabulary enhancement is *word-play*. Word-play (or play-on-words) is a literary technique and a form of cleverness in which the words that are chosen become the main subject of the work, primarily for the purpose of intended effect or amusement.[2] Blachowicz and Fisher[3] show the effectiveness of the use of word-play in a classroom, underlining that word-play motivates and enriches the word-patrimony of the classroom and encourages the students to make a cognitive effort in order to understand the meaning of new words, of their parts, and the context in which they are used, engaging the students by making them active learners.

The last few years have marked great progress in the direction of technological education. The interest towards the possible benefits coming from the use of humanoid robots in the educational context has grown mainly because robots with their physical embodiment provide an unforgettable experience to the learner, their cost in the last decade has dropped exponentially and, last but not the least, of the possibility of setting up a robotic system in a plug-and-play manner. Moreover, the use of a robot whose behavior is similar to that of human beings makes the human–robot interaction more natural and amusing.[5]

The development of educational robots is still in the initial stage: in 2009 RUBI, a social robot able to teach a foreign language to children was presented presented[16]; "Robovie" is a humanoid robot used to teach English in a Japanese elementary school.[17]

More recently, Westlund and Breazeal[18] have developed a system based on robotic learning/teaching companion that supports children's language development by playing a storytelling game. The results have shown an improvement of the desire to play again with the robot at the same game helping the vocabulary instruction. Alemi *et al.*[19] in 2014 used an NAO robot in an Iranian classroom as teacher support for the learning of English.

Hameed *et al.*[20] in 2018 proposed a robotic system able to teach a lesson by presenting slides while generating movements mimicking those made by a human body. The system can measure the students' attention level using its camera, being able to switch from teaching mode to entertainment mode to attract the attention of students.

In this paper, we introduce a system that facilitates the vocabulary instruction through the use of the Pepper humanoid robot[15] and word-play game based on portmanteau words (PWs). The use of a robot allows for a full immersion experience by kids, maintaining the classical structure of

*Corresponding author.

word-play that provides the interaction among more than two physical participants and it demonstrates the capabilities of Pepper to integrate itself with a group of kids, to support the teacher in her activities, and to face the difficult problem of working with portmanteau words.

We have built a system which uses such a robot as an interface to play, in a catchy and amusing manner, with children by using the fundamental rules of word-play and interacting with them like a social component of the group.

The robot is capable of playing a round-game inspired by the one presented by Morin,[6] which involves at least two players and it is played on real or invented portmanteau words.

To tackle this task, we have developed three "engines": a *conversation* engine, a *creation* engine, and a *definition* engine.

The first results are encouraging, and they are illustrated in Sec. 4. To the best of our knowledge, this is the first time that a robotic playfellow based on word-play is tested for vocabulary instruction.

The remainder of the paper is structured as follows: Section 2 deals with the illustration of the portmanteau words game. Section 3 illustrates the proposed system, and Sec. 4 shows an example of interaction as well as the experimental results obtained by the two main engines that make the robot playing possible. Conclusion are given in Sec. 5.

2. Playing with Portmanteau Words

An essential component of vocabulary learning is the ability to manipulate, combine and recombine the parts of words.[4] Indeed, if a student encounters a new word whose meaning he does not know, he could examine the elements of the word such as *prefixes*, *suffixes*, *root*, and so on to make an inference for understanding the meaning of the new word. In this manner, a student has the possibility of enriching his vocabulary starting from the knowledge of only chunks of words.

A common lexical form that is interesting to vocabulary instruction is the *portmanteau*. "Portmanteau words" are words based on blending devices in which a new lexeme is created by merging parts of at least two other source words, e.g., *brunch* is the composition of *breakfast* and *lunch*.

Word-play based on PWs can activate the aforementioned cognitive processes. Analyzing this kind of words allows the learner to improve his cognitive flexibility, to find out new words, and to understand a manner to create words, which are all useful instruments for vocabulary enhancement.

The word-play that we use in our system is a round-game, it involves at least two players, and it is divided into three game modes[6]:

(1) A player chooses two words that form a *real* PW and he communicates them to the other players. The goal is to guess the *real* PW, giving its definition.
(2) A player chooses a *real* PW and he communicates it to the other players. The goal is to guess the source words and give their definitions.
(3) A player *creates* his own PW and he communicates it together with its "invented" definition to the other players. The goal is retrieving the source words of the newly created PW.

With the three modes defined above, the game facilitates vocabulary building and word recognition, improving both cognitive flexibility and word manipulation abilities.

3. System Description

The system exploits a Pepper robot. Pepper is a SoftBank human-shaped robot built to play the role of a humanoid companion so that it can naturally and intuitively interact with human users, making use of its body movements and its voice.

In the approach illustrated in this paper, Pepper exploits its Internet connection capability to interface itself with a family of "ad hoc" built modules that help it to accomplish a series of tasks required by the game, making the robot a useful tool at supporting teachers.

In this section, we describe the system operation, its structure, that is shown in Fig. 1, and every single module that constitutes the whole system.

3.1. *System operation*

The system is thought to be an instructive playfellow that can play with word-play games by directly interacting with children that play with him. The hearing and speaking abilities of Pepper, together with its humanoid looking, allow the integration with children and the participation at the

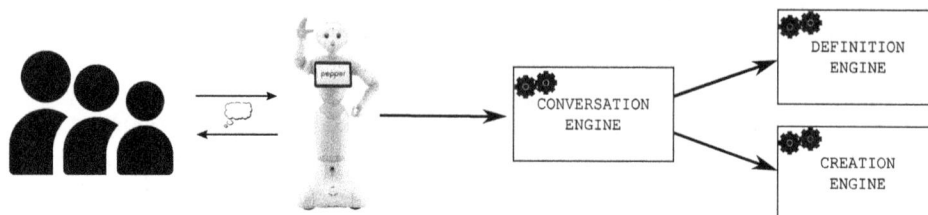

Fig. 1. System infrastructure: Lines between rectangles identify the interaction between Pepper and engines. The arrows with a bubble-cloud identify the interaction between Pepper and the players.

round-based PW-game described in the previous section, playing all the roles expected in the game.

The teacher could focus the game on a specific set of words (e.g., words describing an animal), which have to be thoroughly investigated. For these reasons, she can set up the game session either through a dialog interface or a web-based interface that advises her about the candidate words to be used. For example, the teacher could focus on details about *animals*. To do this, she can choose *animal* category from the Pepper's tablet among all the available categories. Subsequently, it shows a list of words representative of animals. Otherwise, the teacher is free to choose a mix of concepts contained in the knowledge base of Pepper. All the terms are ordered by their frequencies in common English. Indeed, a higher frequency identifies the term as more common and so simpler to use or identify. This way, the system guides the student on the exploration of only specific words. After the choice of terms and categories, the teacher has to set the level of difficulty about the game. There are three levels of difficulty (low, medium, and high) and they regard mainly the level of correctness of the definition that will give the children about the portmanteau words. The same choices can be done through a dialog-based interface.

Figure 2 shows a structure of the graphical user interface.

Before the beginning of the game, it is required to set the game mode so that Pepper knows how to play the game. This is done by saying *Pepper, let us play in mode X. You play the role of Y* where *X* can be "one," "two," or "three" and *Y* can be "answerer" or "generator."

For each game mode, Pepper changes its behavior by selecting an appropriate set of question–answer modules, which are bound to a set of functions that perform the required task. Moreover, even if the robot, in some cases, is capable of calculating the answer in less than a few seconds, it waits for a random time, between 10 s and 1 min, before giving the solution to keep high the interest of the children on the game. The possible roles played by the robot within the three modes of the game are described below:

- Mode 1
  - *Pepper as answerer*: Pepper is waiting for two source words, and, once it hears them, it creates all possible PWs by using the creation engine and checking in its

Fig. 2. Graphical User Interface on Pepper's tablet. On the left, the possibility of the choice of category of words. On the right the possibility of looking for terms from Pepper's knowledge base.

knowledge base to answer with the retrieved PW and its definition.
  - *Pepper as generator*: Pepper chooses a PW from its knowledge base and it communicates the source words to the other players. The list of PWs that Pepper use can be set up by the teacher before the game starts using a specific graphical user interface.

- Mode 2
  - *Pepper as answerer*: Pepper looks up for the PW from its knowledge base, and it provides the source words. If it does not know the portmanteau word, it will not answer; however, it will store the word in its knowledge base for further retrieval. Before adding the unknown portmanteau word to its knowledge, the new word must be checked and approved by the teacher.
  - *Pepper as generator*: Pepper chooses a real PW from a list of words selected from its knowledge base that satisfies the teacher requests and communicates it to the other players, listening for answers. Once it detects the correct source words, it gives the confirmation that the right answer has been given.

- Mode 3
  - *Pepper as answerer*: Pepper uses a definition engine to retrieve the main words from the given definition. Then, it creates all possible PWs by invoking the creation engine and sorting them by exploiting a properly defined score function, trying to guess the right source words.
  - *Pepper as generator*: Pepper creates a catchy and creative PW from a list of concepts, set by the teacher, using the creation engine and then uses the definition engine to define a brief description to communicate to other players. Once it listens to the correct source words, it gives the confirmation that the right answer has been given.

3.2. *Conversational engine*

The conversational engine exploits a speech-recognition module which makes use of the Google speech recognition APIs; after the speech-to-text task is performed, the recognized string is sent to a dialog manager. A set of question–answer rules are set up into the conversation engine to guide the dialog and to invoke adequate procedures of the creation engine or the definition engine, according to the task to be executed.

The conversational agent engine allows for natural human–robot interaction. The conversational engine that we have developed makes use of RiveScript,[10] which is a simple scripting language for realizing chatbots and other conversational entities. Such engine, instead of the standard conversational module embedded in Pepper, allows us to quickly connect it to other kinds of robots or different kind of services. A RiveScript knowledge base is made up of

Triggers/Replies pairs. *Triggers* are identified by a "+" sign, while *Replies* are denoted by a "−" sign. At the beginning of the game, a specific RiveScript *Topic* is activated. Topics are logical groupings of triggers. When the conversation is bound in a topic, what the user says can only match triggers that belong to the activated topic.[10]

```
+ let us play in mode one you play the role of
  answerer
- Ok. We will play mode one and I will answer.
  {topic=m1ans}

> topic m1ans
 // set of Trigger/Reply pairs
 // within the mode one, answerer function
 + ...
 - ....
< topic
```

A set of specific commands have been added to perform operations on the two other engines; for example:

```
+ pepper tell me the generated portmanteau word
- RETRIEVE-PW
```

In this case, for example, when the user says "Pepper, tell me the generated portmanteau word," the "RETRIEVE-PW" command is invoked, which calls a procedure in the creation engine.

```
+ pepper tell me the definition of *
- RETRIEVE-DEFINITION <star>
```

In this case, for example, when the user says "Pepper, tell me the definition of motel," the "RETRIEVE-DEFINITION" command with the "motel" parameter is invoked, calling the associated procedure of the definition engine.

3.3. *Definition engine*

The definition engine has three main functionalities: the *first* one is to build a definition about a created PW; the *second* one is to find the most important words that are related with the given definition; the *third* one is to check if the definition given by a player is correct or not.

The first task is addressed by defining both words masquerading the source words if they are present in their definitions. For example, if the system has to describe the PW "brunch" (breakfast + lunch) it will provide a definition like this: "The word meaning is the mix between that thing that is the first meal of the day and that thing that is a meal eaten in the middle of the day." To the best of our knowledge, this is the first time that the problem of automatically describing a new PW is addressed and it needs a deeper study that is out of the scope of this paper.

The second task has been tackled through the use of WordNet[11] and a semantic network. The semantic network consists of 4000 words that represent concepts and 15,000 relations among them such as *is part of*, *is located to*, *is a*, and so on. In our case, each node of the network is a word that is connected with other words through edges representing semantic relations. Such concepts and relations have been taken from ConceptNet5[9] deleting the multi-words nodes. ConceptNet is a semantic network containing common sense, cultural, and scientific knowledge. The nodes of ConceptNet are words or a short phrase connected with some other node through relations such as *is a, desires, created by*, and so on.

The basic idea of our solution is to retrieve a set of concepts from the given definition to allow for the creation of all possible PWs between the retrieved concepts. Indeed, it is logical that the PW is the representation of its definition and the definition is based on concepts that have retrieved following logical path driven by the definition's words. For example, let us examine the definition of *motel* (motor + hotel): "a hotel providing travelers with lodging and free parking facilities." The motor word is not explained but following the logical paths (*traveler → vehicle → motor*), it results clear the meaning of motor concept in PW.

To find the set of candidate concept words, we have created an algorithm that combines both cognitive and lexical processes by performing the following steps:

(1) Delete the stop words from the definition.
(2) For each word *w*, find the synset of *w* from WordNet and for each element of the synset, let us collect its lemma.
(3) For each word *w* and lemma *l*, let us collect the words that are directly connected with *w* in the semantic network used by the system.

The core of our approach is in the second step because the algorithm uses the aforementioned semantic network to retrieve words that are semantically correlated to the bag of words that have been collected. Using the semantic network is a powerful manner of exploring the semantic neighborhood of the main concept, and in our problem, this approach gives a high probability to retrieve concepts useful to create the PW. For example, a definition of the PW *brunch* (breakfast + lunch) could be: "A meal usually taken late in the morning." Although the definition does not contain either the words *breakfast* or *lunch*, our system can retrieve both of them using the semantic network, since *breakfast* and *lunch* are both linked to the concept "meal" through the "is a" relation, adding them to the bag of candidate words. At the end of the process, the engine gives as output a set of words that will be processed by the creation engine to create PW.

In this approach, the main problem could be the huge number of words retrieved from the definition engine that will be transferred to the creation engine. Indeed, the creation engine computes all the possible PWs relying on all the pairs of words. To limit the number of pairs of words, the definition

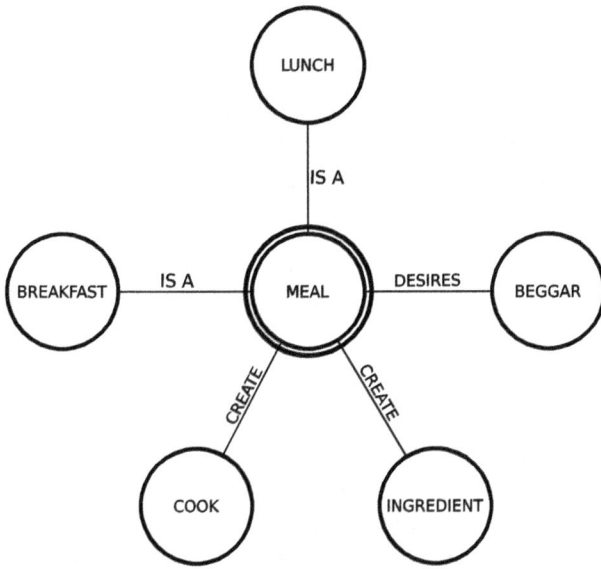

Fig. 3. Extract of semantic network centered on meal concept. The relation has to be read from the external nodes towards the central node. For example, "lunch is a meal."

engine selects the first 25 most common words in the English corpus.

The third task is addressed using the well-known TF–IDF model together with the cosine rule.[21] The definition engine holds a huge vocabulary of words associated with their TF–IDF score calculated on Wikipedia corpus[22] assigning a weight $w_{i,j}$ to each word using the following formula:

$$w_{i,j} = \begin{cases} (1 + \log f_{i,j}) \times \log \dfrac{N}{n_i} & \text{if } f_{i,j} > 0, \\ 0 & \text{otherwise,} \end{cases}$$

where $f_{i,j}$ is the frequency of the term t_i in the document d_j, N is the total number of documents (the Wikipedia pages), and n_i is how many times the term t_i is contained in the documents collection.

The system maps the PW definition and the definition given by a player into two n-dimensional vectors, respectively, d_{own} and d_p in which each component is weighted following the TF–IDF model. Then it uses the cosine rule

$$\cos(\text{def}_p, \text{def}_{\text{own}}) = \frac{\langle \text{def}_p, \text{def}_{\text{own}} \rangle}{\|\text{def}_p\| * \|\text{def}_{\text{own}}\|},$$

where $\langle \text{def}_p, \text{def}_{\text{own}} \rangle$ is the dot product and $\|\text{def}_p\|$ and $\|\text{def}_{\text{own}}\|$ are the Euclidean norms, to understand the similarity between the definition given by the player and the right definition expanded with synonyms taken from WordNet.

Pepper judges the definition as being correct if the $\cos(d_p, d_{\text{own}})$ is above a certain threshold set up by the teacher. The possible thresholds are *low, medium,* and *high* that correspond to similarity values of 70%, 80%, and 90%. The teacher can set the threshold by the user interface that proposes three kinds of circles colored respectively green,

yellow, and red. The same choice can be done through a dialog guided by the robot. The use of a low level allows the student to make more mistakes: this could be useful at the initial stage of a formative path. Other levels could be used as an intermediate step and the end of the formative path.

3.4. *Creation engine*

The main purpose of the creation engine is to create a list of PWs that can be used during the game. Bakaradze[12] has studied the most common blending devices that are used for the creation of PWs[12] and the authors of this paper have explained their implementation aspects, realizing them in their system, named WORDY.[7] WORDY is a methodology and a system aimed at inventing catchy, meaningful, and creative portmanteau words. WORDY uses both semantic networks to represent cognitive processes and according to Bakaradze, the most common "blending" linguistic devices that allow, respectively, to retrieve words that are related to each other and to create the final artifacts.

Furthermore, a set of ranking methodologies have been defined in WORDY to measure the effectiveness of the obtained results. Hence, we use as creation engine the part of the system developed in Ref. 7 that deals with the creation of PWs.

Our creation engine recalls the three blending devices[7] that combine two source words differently. For the sake of simplicity, we call these linguistic devices as *Blending* 1, *Blending* 2, and *Blending* 3.

Their core functions are explained below:

- *Blending* 1: It combines two words to create a portmanteau word that is the concatenation of the prefixes of both words. For example, words *biography* and *picture* can be concatenated by using prefixes for the creation of the *biopic* word.
- *Blending* 2: It combines two words to create a portmanteau word that is the concatenation of the prefix of the first word and the suffix of the second word. For example, the words *lion* and *tiger* can be concatenated by using prefixes and suffixes for the creation of *liger* PW.
- *Blending* 3: It uses the overlapping of the two words to be blended. The words that it creates have the following structure: the first part of the first word is concatenated, by an overlap of the two words, with the last part of the second word. An example is the word *dramedy* that is a portmanteau word created by the combination of *comedy* and *drama* words using *m* as overlap.

The difficulties with the implementation of Blending 1 and Blending 2 are focused on the right choice of prefixes and suffixes that the system should concatenate. Indeed, for each word, there are some prefixes/suffixes equal to the length of the word itself. To face the problem, Blending 1 and Blending 2 use the recognition point concept defined as *a point of a word W that identifies a part T of W for which a majority of speakers* (e.g., 80%) *can recognize W from T*. For example, the recognition point of *mexico* word could be *mex* because a high percentage of people when listening to *mex* think of *mexico* word. Using the recognition point assures the choice of the most representative and meaningful prefixes/suffixes of the source words that allow maintaining the footprint of them.

Blending 3 is applicable only if the source words have some parts in common. If this condition is met, the system uses the well-known Longest Common Substring (LCS) algorithm based on dynamic programming to find the maximum overlap to create a portmanteau word based on Blending 3. Once the maximum overlap **o** is found, the system extracts the prefix **p** before the overlap from the first source word and the suffix **s** after overlap from the second word and then it creates a portmanteau word concatenating **p**, **o**, and **s**.

In mode 3, acting as an answerer, Pepper has to retrieve the PW's source words given PW's definition. The creation engine, before starting the blending procedures, carries out a filtering operation cutting off the words, retrieved by the definition engine, in which the prefixes or suffixes do not appear as prefix or suffix of the given PW. For example, let *pw* be *cyborg* and we suppose that the definition engine has retrieved the following words w_i: *cybernetic, fictional, organism, person,* and *physical*. The creation engine calculates the prefixes *c, cy, cyb, cybo, cybor, cyborg* and suffixes *g, rg, org, borg, yborg, cyborg* of *pw* and then checks for each w_i whether the word w_i starts with a prefix of *pw* or it ends with a suffix (or prefix) of *pw*. If w_i neither starts with any prefix of *pw* nor ends with any suffix (or prefix) of *pw* then w_i will be cut off. In our example, the words taken into account are *cybernetic* and *organism*.

Taking into account the limitations of the definition engine, and adopting proper constraints on the creation engine, it is possible to lower the computational load to get acceptable results in a reasonable time.

4. Testing of the System

The evaluation of the system is complex. Below we report a sample of a possible interaction among the Pepper robot, the Teacher, and the Kids. Subsequently, we report examples of the system running and our first results about the performances of the creation and the definition engines.

```
Teacher: Let us play in mode one you, play the role
         of answerer.
 Pepper: Ok. We will play mode one, and I will
         answer.
    Kid: I thought my portmanteau word.
 Pepper: Great! Could you please tell me the words
         contained in it?
```

```
    Kid: They are lion and tiger.
 Pepper: I know your portmanteau word: It is liger.
```

The most interesting case is Pepper that plays in mode 3. In this example, we report the interaction of the system after running over the two aspects: generator and answerer.
Generator (*mode*-3):
In generator mode, Pepper creates a catchy and creative PW from a list of concepts set by the teacher through the creation engine and then it uses the definition engine to define a brief description to communicate to other players. Once it listens to the correct source words, it gives the confirmation that the right answer has been given.

```
Human > Pepper, please tell me the list of concepts
Pepper> The concepts that I have used are biology,
        electronic, robot.
Human > Pepper tell me the generated portmanteau
        word?
Pepper> The generated portmanteau word is elebot.
Human > Pepper tell me the source words?
Pepper> The source words were electronic, robot.
Human > Pepper tell me the definition of the word
        that you have generated?
Pepper> The word meaning is the mix between (of a
        device) having or operating with compo-
        nents such as microchips and transistors
        that control and direct electric current
        AND (especially in science fiction) a
        machine resembling a human being and able
        to replicate certain human movements and
        functions automatically.
```

Answerer (mode 3):

Pepper uses a definition engine to retrieve the neighborhood of the definition's words; then it invokes the creation engine to calculate PWs.

```
Human > Pepper tell me the generated portmanteau
        word.
Pepper> The generated portmanteau word is motel.
Human > Pepper tell me the definition.
Pepper> The definition is a hotel providing tra-
        velers with lodging and free parking
        facilities.
Human > Pepper tell me the semantic neighborhood
Pepper> The semantic neighborhood is leave, room,
        hit, rest, sleep, city, free, hot, beat,
        offer, road, hotel, charge, stick, calm,
        space, park, sad, rid, innocent, allow,
        file, dry, sunday, restaurant, desk,
        loose, release, ghost, supply, confer-
        ence, spare, bull, vacation, expensive,
        provide, desert, parking, elevator, fuel,
```

```
facility, motor, deposit, reception,
facilities, lodge, liberal, folder, dis-
charge, housing, providing, justify, re-
lieve, driveway, squirrel, resign,
freeway, justified, playground, render,
hilton, traveler, landscape, installa-
tion, wedge, dynamic, cactus, accommo-
date, barren, furnish, travelers,
readiness, detached, caveman, complimen-
tary, frisbee, exempt, lodging, travel-
ler, monotonous, cater, liberate,
relinquish, disengage, wasteland, desti-
tute, devoid, absolve, unfreeze, gratis,
ply, loiter, quickness, gratuitous, dis-
lodge, unblock, unloosen, gpl, unloose,
deftness, costless, adroitness, adept-
ness, lodgment.
Human > Pepper tell me the source words.
Pepper> The source words were motor, hotel.
```

4.1. *Creation engine test*

The creation engine has the main objective of creating PWs to allow Pepper participating in *mode* 1 and *mode* 2 of the portmanteau game.

Measuring how much a created PW is catchy, is a hard task. The authors have tackled this problem by using two methods, among those available, which have been empirically tested in WORDY,[7] showing that their use can help to identify catchy and creative PW. In particular:

(1) The system sorts the list of created PWs by the length of overlaps. This method can be used only for Blending 3.
(2) The system calculates the Levenshtein distance between the PW and a set of common English words, sorting the resulting words by distance in an ascending manner. This method can be used for all the considered types of blending.

The methods have been empirically tested[7] and it has been shown that the use of the overlap as quality index is a good idea because long overlaps represent well both source words. The ranking based on the Levenshtein distance seems to be useful mainly for identifying catchy portmanteau words; indeed, a portmanteau word with a low distance from a common English word usually has a nice sound.

In order to test the effectiveness of blending devices, we have executed the test using a dataset containing 1624 PWs with their source words inspired by Gangal *et al.*'s work.[13] We have run the blending device procedures by using the information in the dataset to investigate the ability of these algorithms to recreate the same PWs in the dataset. The results are very encouraging, as shown in Fig. 4, we have obtained a recall (i.e., words that the device has been able to recreate out of the total number of PWs in the dataset) of 2.6% for Blending 1, 50.2% for Blending 2, and 41.4% for

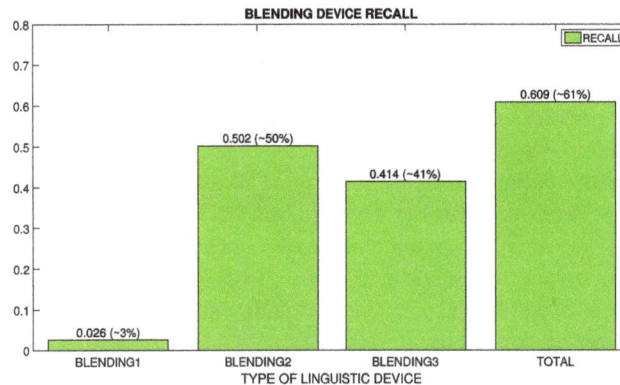

Fig. 4. (a) Recall is intended as the percentage of recreated PWs out the total PWs contained in the dataset. (b) Each bar represents the percentage of recreated words by the corresponding language device. (c) The last bar "total" represents the percentage of recreated words by all linguistic devices out the number of PWs in the dataset.

Blending 3, reaching a 60.9% as the percentage of the total words recreated using the three linguistic devices.

4.2. *Definition engine test*

The definition engine allows Pepper to play the third game mode. Due to the difficulty of the problem and the absence of state of the art about it, through the following test, we would like to understand if our algorithm at the base of definition engine is worthy to be through and if it can give a contribution to the improvement of the response capability of Pepper.

The experiment aims at testing the ability of the robot to recreate a PW reading its definition. For the test, we have used a set of 382 PWs associated with their definition. Because of the unavailability of datasets that contain PWs adequately defined, we have extracted the definition from Oxford Dictionary.[14]

From the result set, we have run the system formed by the union of creation and definition engines to understand how many words it has been able to recreate. The first results, with 18% of total words recreated, have not been very encouraging, but investigating about this percentage it appeared that the main problem was related to the semantic network that we have used. Trying to test the system with more number of concepts, we have manually added to the semantic network about 20 new concepts such as *agriculture*, *radical*, *luxurious*, *click*, and so on, we have verified an improvement of the recreated words percentage reaching about 21%.

5. Conclusions

In this paper, we have presented a preliminary prototype of a social humanoid robot that can play the role of a playfellow for vocabulary enhancement of children by using portmanteau word-play. The main objective of the paper was to understand whether a robot can address the problems related to such kind of word-play and whether it can be a valid support to teacher's activity. Nevertheless, there are many interesting points that are outside the scope of this paper which offer room for the research such us the children–robot interaction or methods for the comprehension of the definition.

The tests regarding the definition engine of the presented system have shown that the use of a semantic network allows the system to know the meaning of PW using its definition. The exploitation of a big semantic network can further improve the results.

The system is still under development, and while the creation engine and the dialog engine are significantly developed, more work has to be done on the definition engine. Experimentation in a real environment will be done to understand the true advantages and limitations of the approach, as well as the effectiveness concerning the traditional game approach.

References

[1] National Reading Panel, Teaching children to read: An evidence-based assessment of scientific research literature on reading and its implications for reading instruction, Report of the Subgroups, National Institutes of Health, Bethesda, MD (2000).

[2] Wikipedia, Word play (2017), https://en.wikipedia.org/wiki/Word_play.

[3] C. L. Blachowicz and P. Fisher, Keep the fun in fundamental, *Vocabulary Instruction: Research to Practice*, Chapter 13 (Guilford, 2004), pp. 218–237.

[4] W. Tunmer, M. Herriman and A. Nesdale, Metalinguistic Abilities and Beginning Reading, *Read. Res. Q.* **23**(2) 134 (1988).

[5] I. Infantino, G. Pilato, R. Rizzo and F. Vella, I feel blue: Robots and humans sharing color representation for emotional cognitive interaction, *Biologically Inspired Cognitive Architectures*, Chapter 30 (Springer, Berlin, 2013), pp. 161–166.

[6] A. Morin, Portmanteau words games for kids (2017), www.thespruce.com/portmanteau-words-games-for-kids-2086509.

[7] D. Schicchi and G. Pilato, WORDY: A semi-automatic methodology aimed at the creation of neologisms based on a semantic network and blending devices, *Complex, Intelligent, and Software Intensive Systems: CISIS 2017*, Advances in Intelligent Systems and Computing, Vol. 611 (Springer, Cham, 2018), pp. 236–248.

[8] J. B. Weinberg and X. Yu, Robotics in education: Low cost platforms for teaching integrated systems, *IEEE Robot. Autom. Mag.* **10**, 4 (2003).

[9] H. Liu and P. Singh, Conceptnet: A practical commonsense reasoning tool-kit, *BT Technol. J.* **22**, 211 (2004).

[10] N. Petherbridge, *Artifical Intelligence Scripting Language* (2018), https://www.rivescript.com.

[11] G. Miller, WordNet: A lexical database for English, *Commun. ACM* **38**, 39 (1995).

[12] E. K. Bakaradze, Principle of the least eort: Telescopic wordformation, *Int. J. Arts Sci.* **3**, 86 (2010).

[13] V. Gangal, H. Jhamtani, G. Neubig, E. Hovy and E. Nyberg, Charmanteau: Character embedding models for portmanteau creation, arXiv:1707.01176 [es.CL].

[14] Oxford University Press, Oxford Dictionaries (2017), https://www.oxforddictionaries.com/our-story/our-products.

[15] SoftBank Robotics, Robots (2017), https://www.ald.softbankrobotics.com/en/robots/pepper.

[16] J. Movellan, M. Eckhardt, M. Virnes and A. Rodriguez, Sociable robot improves toddler vocabulary skills, *Proc. 4th ACM/IEEE Int. Conf. Human Robot Interaction* (ACM, 2009), pp. 307–308.

[17] Z. J. You, C. Y. Shen, C. W. Chang, B. J. Liu and G. D. Chen, A robot as a teaching assistant in an English class, *Proc. Sixth IEEE Int. Conf. Advanced Learning Technologies* (2006), pp. 87–91.

[18] J. K. Westlund and C. Breazeal, The interplay of robot language level with children's language learning during storytelling, *Proc. Tenth Annu. ACM/IEEE Int. Conf. Human-Robot Interaction Extended Abstracts* (ACM, 2015), pp. 65–66.

[19] M. Alemi, A. Meghdari and M. Ghazisaedy, Employing humanoid robots for teaching English language in Iranian junior high-schools, *Int. J. Humanoid Robot.* **11**, 1450022 (2014).

[20] I. A. Hameed, G. Strazdins, H. A. M. Hatlemark, I. S. Jakobsen and J. O. Damdam, Robots that can mix serious with fun, *Proc. Int. Conf. Advanced Machine Learning Technologies and Applications*, eds. A. Hassanien, M. Tolba, M. Elhoseny and M. Mustafa, Advances in Intelligent Systems and Computing, Vol. 723 (Springer, Cham, 2018), pp. 595–604.

[21] G. Salton and M. McGill (eds.), *Introduction to Modern Information Retrieval.* (McGraw-Hill, 1983).

[22] https://dumps.wikimedia.org/enwiki/20171120/.

[23] Icons made by http://www.freepik.com and http://www.onlinewebfonts.com/icon.

Enhanced navigation using computer vision-based steering angle calculation for autonomous vehicles

M. Ranjith Rochan[*], K. Aarthi Alagammai[†] and J. Sujatha[‡]

Wipro Technologies, Ltd.
No. 72, Keonics Electronic City, Hosur Road
Bengaluru 560100, Karnataka, India
[*]ranjithrochan@gmail.com
[†]aarthialagammai21@gmail.com
[‡]sujatha.gkm@gmail.com

A key requirement in the development of intelligent and driverless vehicles is steering angle computation for efficient navigation. This paper presents a novel method for computing steering angle for driverless vehicles using computer vision-based techniques of relatively lower computing cost. The proposed system consists of four major stages. The first stage includes dynamic road region extraction using Gaussian mixture model and expectation maximization algorithm. The second stage is to compute the steering angle based on the extracted road region. Subsequently, Kalman filtering technique is used to cancel spurious angle transition noises. In addition, future steering angle is estimated which in turn gives informative feedback for smooth navigation of the vehicle. The proposed algorithm was tested both on a simulator and real-time images and was found to give a good estimation of actual steering angle required for navigation. Further, it was also observed that this works in different lighting conditions as well as for both structured and unstructured road scenarios.

Keywords: Steering angle; drivable road region; spurious angle transition cancellation; navigation; speed analysis.

1. Introduction

The number of self-driving cars that will hit the road in 2020 will be 10 million, with one in four cars being a self-driving one by 2030 according to a research by Garret.[1] A crucial requirement for intelligent, driverless vehicles is to maneuver without moving out of its drivable region of the road. Steering angle calculation plays an important role in maintaining the vehicle in the center of the road or within the boundary lanes to meet safety-critical requirements. Usage of sensors like LIDAR and RADAR for steering angle calculation imposes a high cost on the system which would practically diminish the usability of such autonomous vehicles, especially in developing countries. This paper presents our effort in creating a novel approach for steering angle calculation for autonomous vehicles using computer vision techniques.

2. Literature Survey

Generally, steering angle calculation for autonomous vehicles consists of two major approaches: (1) neural network-based approach and (2) computer vision-based approach. The former includes high-cost computing requirements and sensors like LIDAR, RADAR, etc. The latter approach follows two-step traditional method which includes road boundary extraction (road region extraction) followed by steering angle computation with only few robust approaches for the same.

Steering angle calculation using only camera, independent of sensors like LIDAR and RADAR, has been proposed in Ref. 2. They have used Euclidean method to calculate the steering angle. However, it also seems to be limited to slow and smooth turns rather than sharp turns. The method proposed by Ref. 2 is limited to the case where both the boundaries or either one of the boundaries is visible. Also, most of the works for steering angle calculation use road boundary extraction before computation of steering angle. In Ref. 2 they use Robert cross-edge detection algorithm for boundary detection.

An earlier method of road boundary detection[3] uses simple road boundary detection algorithms such as the canny edge detection and the Hough transform, where Ref. 3 has control over all the parameters. Reference 3 explains extraction of visible edges which hampers the usage in unstructured and shadowed regions.

In Ref. 6, drivable road region is found using classification and regression tree (CART) method and it uses epipole to calculate the steering angle using voting method. The negative and positive samples to calculate the road region are set manually here which poses a problem in terrain change and initial start of the vehicle every time. Moreover, the epipole calculation depends upon the position of the camera, Ref. 6 sets the intrinsic and extrinsic parameters of the camera manually. The horizontal position of the epipole is then used for calculating the steering angle.

---

[*]Corresponding author.

Deep-learning neural network approach for determining the steering angle has gained popularity in recent times. In Ref. 8 steering angle calculation is done using two different neural network approaches: (1) 3D CNN with RNN and (2) transfer learning. By feeding large amount of training data to the network, a model is made to learn the possible estimation of steering angle. The model requires different kinds and a wide variety of input data like curved roads, intersection, highways, etc., which requires collection of large data in developing countries where there are different kinds of road structures with different terrains and harsh road conditions. In contrast, the proposed method helps in efficient steering angle calculation with minimum hardware requirements and low-cost computation. This paper provides a novel method for calculating the steering angle using a monocular camera and aligning the vehicle accordingly with respect to the road's orientation.

3. Proposed Method

The proposed system involves four major steps which include dynamic road region extraction, steering angle computation, spurious angle transition cancellation and estimation of future steering angle. The entire system architecture is shown in Fig. 1.

3.1. *Dynamic road region extraction*

The boundaries of the road are obtained using the Gaussian mixture model using expectation maximization (GMM-EM) method. From the input image, a patch is chosen, for which Gaussian mixture model is created in RGB color space using individual pixels. The parameters of GMM like mean and covariance matrix are improved using EM algorithm iteratively.

To find the free road region (drivable), maximum likelihood method is used. Pixels whose log of maximum likelihood estimate (MLE) values of RGB color space is within the threshold are chosen to be the best candidate pixels for road region.

From the above drivable road region, canny edge detection is applied to find the road boundaries. GMM-EM is a dynamic learning algorithm which is found to be robust in all weather conditions, shadowed road regions as shown in Figs. 2(a) and 2(b). Moreover, GMM-EM method works for both structured and unstructured road region as shown in Figs. 2(a) and 2(b). This reduces the usage of shadow processing.

3.2. *Steering angle calculation*

The required steering angle will be the deviation of the point of intersection of the boundaries from the orientation of the vehicle (midline of the image) as shown in Fig. 3.

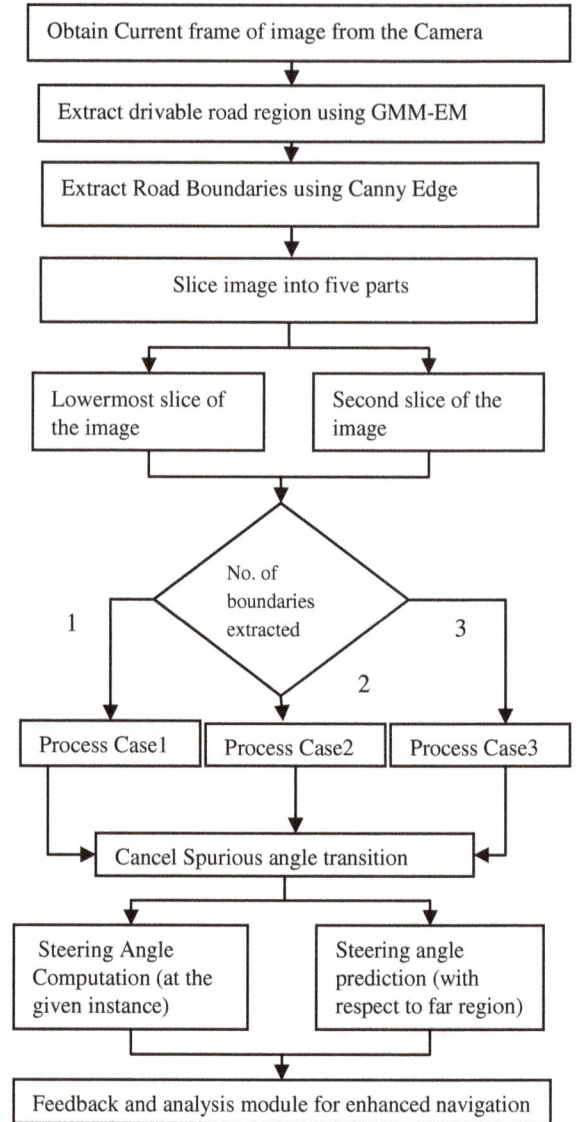

Fig. 1. Proposed system architecture.

Fig. 2. Road region extraction: (a) input images and (b) GMM-EM.

Fig. 3. Steering angle calculation.

(a) Input image.　　(b) GMM-EM output.　　(c) Output.

Fig. 4. Steering angle calculation for Case 1.

In GMM-EM method the required boundaries are obtained using canny edge detection algorithm. After detecting the boundary, the image is sliced into five equal parts along the rows. It is divided into five parts based on the resolution of the image given in Udacity dataset and it is also found that the width of the road is small, where the lowermost part is considered for steering angle calculation since it gives the current orientation of the vehicle according to which the vehicle is to be aligned. Based on the field of view of the camera and road conditions there will be three cases of boundary detection: (1) both the boundaries are detected; (2) one of the boundaries is detected; and (3) neither of the boundaries are detected. The following procedure explains how to calculate the steering angle for all the three cases.

3.2.1. Case 1

Both the boundaries are visible and have been extracted using the mentioned road region extraction process. From the extracted boundaries, the slope of the lines (boundaries) are calculated. The calculation of slope and position of the extracted lines is necessary to decide whether extracted boundary is towards the left or the right of the vehicle. The line with a negative slope and in the left half of the image will be the left boundary and the line with a positive slope and in the right half of the image will be the right boundary. The common point of intersection is calculated from the slope of both boundaries using (5) and (6).

The next step is to calculate the angle of deviation corresponding to the vehicle heading direction which is calculated with the help of point of intersection of the boundaries. This angle of deviation represents the required steering angle as shown in (8). Equations (1)–(8) are used to calculate the orientation of the vehicle with respect to the road boundaries and the minimum angle of turn required to keep the vehicle in the center of the lane which is shown in Fig. 4.

In Fig. 3, markings 1–4 represent the coordinate points obtained from the road region extraction. In Fig. 3, marking 5 represents the point of intersection of both the boundaries as shown in (5) and (6). In Fig. 3, marking 6 represents the required steering angle as calculated in (8). White dotted lines in Fig. 3 represent the midline of the image:

$$\text{Slope}_{\text{left}} = \frac{Y_2 - Y_1}{X_2 - X_1}, \tag{1}$$

$$\text{Slope}_{\text{right}} = \frac{Y_4 - Y_3}{X_4 - X_3}, \tag{2}$$

$$C_1 = Y_{1\text{left}} - (\text{Slope}_{\text{left}} * X_{1\text{left}}), \tag{3}$$

$$C_2 = Y_{1\text{right}} - (\text{Slope}_{\text{right}} * X_{1\text{right}}), \tag{4}$$

$$\text{Intersection } X = (C_2 - C_1)/(\text{Slope}_{\text{left}} - \text{Slope}_{\text{right}}), \tag{5}$$

$$\text{Intersection } Y = ((\text{Slope}_{\text{left}} * \text{Intersection } X) + C_1), \tag{6}$$

$$S = ((\text{Intersection } X - (w/2)) - (\text{Intersection } Y - h), \tag{7}$$

$$\text{SteeringAngle} = 90 - (\tan^{-1}(1/S) * (180/\pi)), \tag{8}$$

where Y_1, Y_2, Y_3, Y_4 and X_1, X_2, X_3, X_4 are the four coordinate points of markings 1–4 in Fig. 3. C_1, C_2 are the constants of the obtained boundaries. Intersection X, Intersection Y are the coordinates of the point of intersection of both the boundaries. S is the slope between the vehicle heading direction and the point of intersection of the boundaries, whereas w and h are the width and height of the image in pixels.

3.2.2. Case 2

Only one of the boundaries is visible within the field of view of the camera. Two approaches can be used to solve this problem:

(a) by maintaining the vehicle in the center of the road;
(b) by maintaining the vehicle towards the visible boundary.

3.2.2.1. Case 2A

In this case the vehicle is maintained at the center of the road. If only one of the boundaries is visible within the field of view of the camera, then the other boundary is usually taken to be the last column of pixels on the other side of the image. For example, if only left boundary is visible and right boundary is not visible, then the last column of pixels in the right-hand side of the frame is considered to be the right boundary.

(a) Input image.   (b) GMM-EM output.   (c) Output.

Fig. 5. Steering angle calculation for Case 2A.

Since slope will be too large for the right-hand side boundary, to make it computationally easy we move the last column value by one pixel so that getting the slope value becomes convenient for calculation purposes. After calculating the slope, the angle of deviation is computed using the same method as mentioned for case 1 and maintained in the center of the obtained boundaries which is shown in Fig. 5.

3.2.2.2. Case 2B

In this case the vehicle is maintained at a certain distance from the visible boundary. In case of unstructured road or if the width of the road is too large it will be difficult to maintain it in the center of the road since the vehicle will move in a zig-zag manner if it follows case 2A approach.

In order to avoid such random movement of the vehicle, it is safe to maneuver along the visible boundary. Since a monocular camera is used, distance to the boundary cannot be computed accurately. The angle computed from the distance obtained from the camera is not reliable. To maintain the vehicle at a certain distance from the boundary a base offset is necessary.

The base offset is calculated as follows. Let a distance to be maintained be $2m$ from the required boundary. An imaginary straight line boundary is assumed whose two edges will be $[x_1, y_1, z_1] = [x_1, 2, \text{height at which the camera is placed}]$, $[x_2, y_2, z_2] = [x_1 + \text{displacement}, 2, \text{height at which the camera is placed}]$. The points are converted into camera coordinates using (9)–(16) to find the base offset angle using (17):

$$\theta_1 = (180/\pi) * \left(\tan^{-1} \left(\frac{y_1}{x_1} \right) \right), \tag{9}$$

$$\phi_1 = 90 - \left\{ (180/\pi) * \tan^{-1} \left(\frac{(\sqrt{x_1^2 + y_1^2})}{z_1} \right) \right\}, \tag{10}$$

$$l_x = \frac{\text{rows}}{\text{horizontal field of view}}, \tag{11}$$

$$l_y = \frac{\text{cols}}{\text{vertical field of view}}, \tag{12}$$

(x_{im1}, y_{im1})
$$= [\text{rows} - (c_x + (\theta_1 * l_x)), \text{cols} - (c_y + (\phi_1 * l_y))], \tag{13}$$

$$\theta_2 = (180/\pi) * \left(\tan^{-1} \left(\frac{y_2}{x_2} \right) \right), \tag{14}$$

$$\phi_2 = 90 - \left\{ (180/\pi) * \tan^{-1} \left(\frac{(\sqrt{x_2^2 + y_2^2})}{z_2} \right) \right\}, \tag{15}$$

(x_{im2}, y_{im2})
$$= [\text{rows} - (c_x + (\theta_2 * l_x)), \text{cols} - (c_y + (\phi_2 * l_y))], \tag{16}$$

$$\theta_b = \tan^{-1} \left\{ \frac{(y_{im2} - y_{im1})}{(x_{im2} - x_{im1})} \right\} * (180/3.14), \tag{17}$$

where $x_1, y_1, z_1, x_2, y_2, z_2$ are the assumed edges of the boundary, $x_{im1}, y_{im1}, x_{im2}, y_{im2}$ are the corresponding image coordinates, rows are the width of the image, cols are the height of the image, c_x is the center x-coordinate of the image (rows/2), c_y is the center y-coordinate of the image (cols/2), l_x is the horizontal pixel per degree information, l_y is the horizontal pixel per degree information and θ_b is the base offset angle in degrees.

This base offset angle should be maintained until it is required to maneuver the vehicle with one boundary. The angle between the boundary which is visible and the frame of reference (or a line parallel to the frame of reference) is calculated. For every successive frame, the base offset angle is to be maintained. The difference between the base offset angle and the angle between the boundary which is visible and the frame of reference will be the required steering angle.

For example, in Fig. 6(e) the base offset angle is calculated to be $29°$, to maintain a distance of 2 m from the vehicle. The angle between the left boundary and the frame of reference is $25°$. The vehicle should be steered $4°$ towards left to maintain the calculated base offset angle such that the vehicle is oriented towards left boundary.

3.2.3. *Case 3*

When both the boundaries are not present, the vehicle is made to move straight along its path till any one of the boundaries

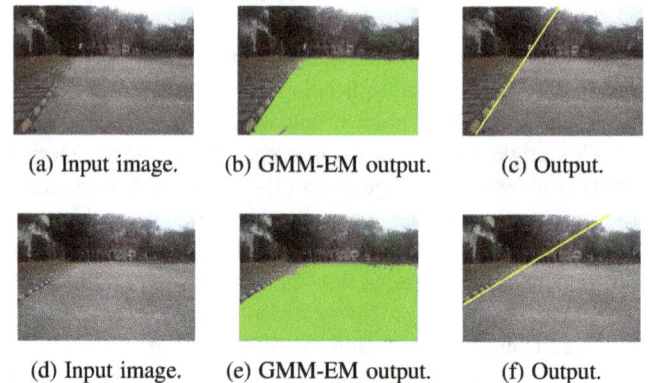

(a) Input image.   (b) GMM-EM output.   (c) Output.

(d) Input image.   (e) GMM-EM output.   (f) Output.

Fig. 6. Steering angle calculation for Case 2B.

(a) Input image.　　(b) GMM-EM output.

Fig. 7. Steering angle calculation for Case 3.

is identified which is shown in Fig. 7. Since no significant boundaries are found, the steering angle calculation algorithm is not used and the vehicle is made to move forward with zero steering angle.

3.3. *Spurious angle transition cancellation*

While the vehicle is moving, it is subjected to various random noises due to environmental conditions like road structures. Also, the presence of humps, potholes, sharp turns, etc. in the road might cause sudden spurious changes in steering angle which lead to rugged movement of the vehicle. To avoid these spurious transitions, we use Kalman filter. Active noise cancellation plays an important role in smooth movement of the vehicle. Kalman filter is used to remove these active noises. It is also used to continuously track the steering angle with appropriate measurement noise parameter. Any drastic changes occurring due to above-mentioned environmental issues are taken care by Kalman filter such that it does not impact the steering angle and smooth navigation of the vehicle is ensured.

3.4. *Estimation of future steering angle*

After detecting the boundary through GMM-EM method and canny edge detection algorithm, the image is sliced into five parts along the rows. It is divided into five parts based on the resolution of the image given in Udacity dataset and it is also found that the width of the road is small. The lowermost part is considered for steering angle calculation with respect to the current frame of image as it gives the existing orientation of the vehicle at that instant (near region). The second slice of the image can be used to estimate the steering angle (far region) with which the vehicle needs to turn. The estimated steering angle will be the angle at which the vehicle has to steer from the current orientation in order to maintain its navigation in the center of the road, this will be used as a feedback for navigation.

As mentioned earlier in Sec. 3.2, based on the field of view of the camera and road conditions, steering angle estimation is in accordance with these three cases.

Sometimes, the boundary (road) ahead will not be visible due to the position and field of view of the camera, during sharp turns, heavy traffic conditions, etc. In such cases, the estimation of the immediate steering angle is not

possible and the vehicle is made to move at minimum possible speed.

4. Experimental Results

The experimental results are tested on both Gazebo simulator and real-time images using Udacity dataset which are mentioned below.

4.1. *Simulator*

The steering angle calculation algorithm initially has been tested in Gazebo[9] simulator (Robot simulator). A world has been built with structures having both curved and straight road regions. A car model has been developed and made to navigate through the path. Visually the car was able to maneuver through the path successfully along the center of the path. Figure 8 shows the Gazebo model which was tested.

Similarly, for Case 2B a separate world was built in which only one of the boundaries was visible in the camera. It has been observed that the car was able to maneuver along the visible boundary at a safe distance of 3.5 m as shown in Fig. 9.

However, there was no standard method and possible ground truth values to evaluate the Gazebo model, so the following method was used to evaluate the proposed algorithm.

4.2. *Udacity dataset for Cases 1–3*

The steering angle calculation algorithm has been tested over a sample dataset collected from Udacity.[10] The steering angle calculation challenge comprises the images taken during both day and night. Since the scores for Udacity challenge were reported in root mean square error (RMSE), which is the square root mean of sum of the squared differences between actual and predicted results, the same was used to find accuracy. The RMSE value calculated using (19) is found to

Fig. 8. The steering angle experiment with Gazebo.

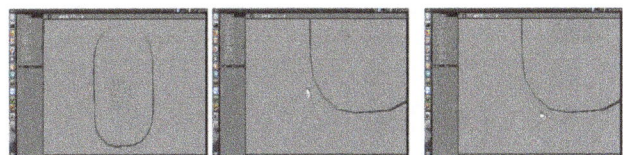

Fig. 9. The steering angle experiment for Case 2B.

be 2.598:

$$MSE = (1/N) * \Sigma(y_l - \hat{y}_l)^2, \qquad (18)$$

$$RMSE = \sqrt{(1/N) * \Sigma(y_l - \hat{y}_l)^2}. \qquad (19)$$

In Ref. 4 a neural network-based approach using 3D-CNN with LSTM and Resnet deep-learning techniques was used to compute steering angle on Udacity dataset. The RMSE value on test set for 3D-CNN with LSTM is 0.1123 and for Resnet transfer model it is 0.0709.

Figure 10(a) shows the sample inputs obtained from Udacity which include all three cases. Figures 10(b) and 10(c) show the corresponding GMM-EM output and the steering angle algorithm output, respectively. Table 1 shows the comparison between the actual and estimated steering angles in Fig. 10.

In addition, Fig. 11 shows the comparison graph between the actual and the estimated steering angle values over 500 frames of a video chosen from the Udacity dataset. A maximum of $\pm 5°$ of deviation is observed. This maximum deviation is mostly observed during sharp turns.

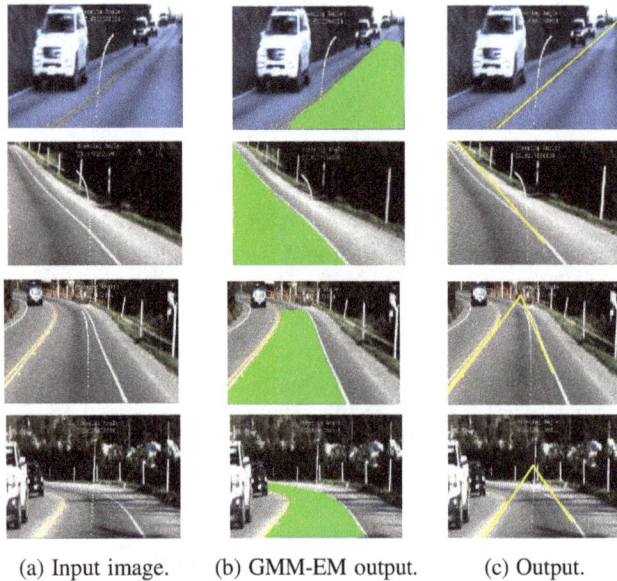

Fig. 11. Comparison of actual and estimated steering angles.

4.3. *Speed control analysis with prediction of steering angle*

It is important to ensure that speed of the vehicle is in control for safe navigation. One of the major parameters that controls the speed of the vehicle is the steering angle. The predicted steering angle as discussed can be used as one of the input parameters to control the speed of the vehicle. In addition to this, other information like presence of obstacles, terrain of the road, traffic conditions, etc. also play a role in finial decision about the speed of the vehicle.

In Fig. 12(1a), it is seen that the road is structured and has a sharp turn, the steering angle at that instant is $-5.984°$ ($-$ve sign indicating turn towards right). As the vehicle approaches towards the turn as shown in Fig. 12(1d) it is estimated that the angle changes to $-13.38°$. So, a trigger is sent to

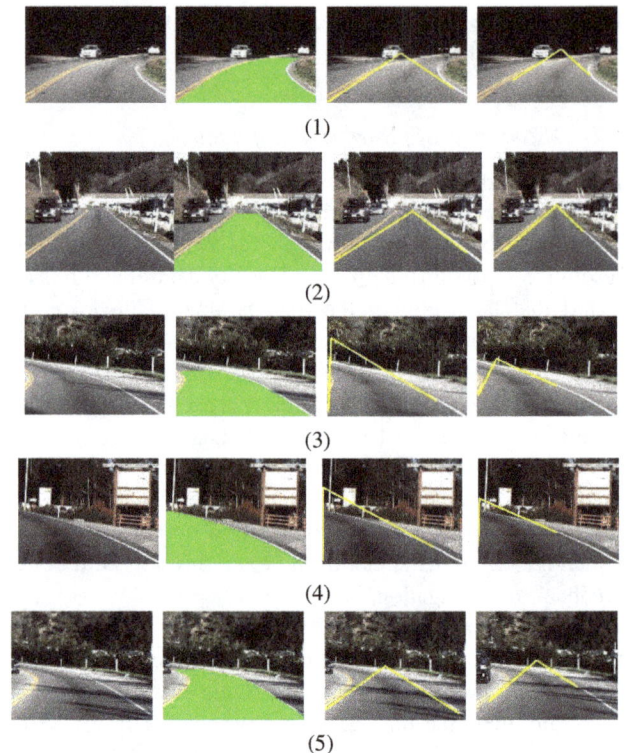

(a) Input image.   (b) GMM-EM output.   (c) Output.

Fig. 10. Sample dataset inputs and outputs.

Table 1. Comparison of actual and estimated angles in Fig. 10.

Fig. 10	Actual angle (deg)	Estimated angle (deg)
C1	−17.63	−23.87
C2	20.91	25.0
C3	12.197	10.1
C4	0.201	0.3

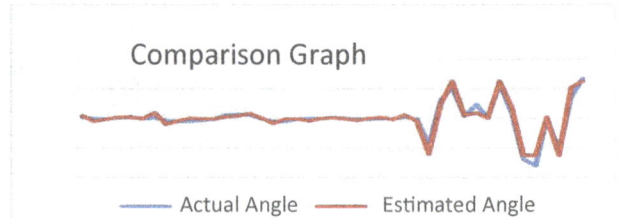

(1)

(2)

(3)

(4)

(5)

Fig. 12. Sample dataset inputs and outputs. (a) Input image. (b) GMM-EM. (c) Output. (d) Future steering angle.

Table 2. Comparison of actual and estimated angles [see Figs. 12(1)–12(5)].

Fig. 12	Actual steering angle with respect to current frame of image (near region) in deg	Computed steering angle with respect to current frame of image (near region) in deg	Estimated steering angle with respect to next frame of image (far region) in deg
1	−7.963	−5.984	−13.38
2	0.346	−4.41	−1.072
3	21.138	25.0	25
4	22.58	25.0	25.0
5	14.114	10.70	19.598

navigation stack along with the steering value to reduce the speed of the vehicle.

In Fig. 12(2a), it is seen that the road is structured and it is a straight road with no turns, the steering angle at that instant is −4.41°. The future angle is predicted to be −1.072° as shown in Fig. 12(2d) with a minimal change (±4° is considered as minimal change which does not require change in speed). Consequently, the steering angle information is sent to the navigation stack of the vehicle to maintain the existing speed.

Figures 12(3a) and 12(4a) depict similar situations to that of Fig. 12(2a). Even though the road is curved, as the change in turn is gradual the current and the estimated steering angles difference is small resulting in no change of speed.

In Fig. 12(5a) it is seen that the road has an immediate sharp turn towards left, the steering angle at that instant is 14.114°. The future angle is predicted to be 19.598° in Fig. 12(5d). So, a trigger to reduce the speed of the vehicle is sent to navigation stack.

Figure 12(a) shows the input dataset from Udacity, Figs. 12(b) and 12(c) show the corresponding GMM-EM output and the steering angle algorithm output at that instant and Fig. 12(d) shows the future steering angle algorithm estimation.

Table 2 show the estimated and actual angles for current and future instants. It is observed from the table that a maximum of ±5° deviation is obtained.

5. Conclusion

The proposed system provides a novel method for the calculation of steering angle for an autonomous car. This method makes use of computer vision-based techniques and requires minimum hardware. The design also takes care of the situation where only one or no boundaries of the road appear in the captured image of the vehicle. Thus, this is independent of the structure of the road.

Extending the steering angle computation to prediction of steering angle with respect to the next instant of time, this method is able to provide crucial information required for smooth and safe navigation of the vehicle. Experimental results both on simulator and Udacity dataset have been encouraging.

This presented algorithm can be extended by enhancing the road region extraction to be more robust by making the system to extract the patch and Gaussian mixtures dynamically. Also, the approach can be extended for very sharp and steep turns. More importantly, estimation of future steering angle can be used as an information for analysis of road information/vehicle behavior in addition to the speed control.

The system proposed above can be used in the development of self-driving cars allowing it to move independently and safely.

References

[1]O. Garret, 10 million self-driving cars will hit the road by 2020: Here's how to profit (2017), https://www.forbes.com/sites/oliviergarret/2017/03/03/10million-self-driving-cars-will-hit-the-road-by-2020-hereshow-to-profit.

[2]V. Umamaheswari, S. Amarjyoti, T. Bakshi and A. Singh, Steering angle estimation for autonomous vehicle navigation using Hough and Euclidean transform, *Proc. IEEE Int. Conf. Signal Processing Informatics Communication and Energy Systems* (2015), pp. 1–5.

[3]B. Yu and A. K. Jain, Lane boundary detection using a multiresolution hough transform, *Proc. Int. Conf. Image Processing* (1997).

[4]S. Du, H. Guo and A. Simpson, Self-driving car steering angle prediction based on image recognition (2017), cs231n.stanford.edu/reports/2017/pdfs/626.pdf.

[5]O. Ramstrom and H. Christensen, A method for following unmarked roads, *Proc. IEEE Intelligent Vehicles Symp.* (2005), pp. 650–655.

[6]K. W. Her, J. E. Ha and W. H. Lee, Steering angle determination using cart and voting, *Proc. 11th Int. Conf. Control, Automation and Systems* (2011), pp. 26–29.

[7]J. Lee and C. Crane, Road following in an unstructured desert environment based on the EM (expectation-maximization) algorithm, *Proc. SICE-ICASE Int. Joint Conf.* (2006), pp. 2969–2974.

[8]M. Bojarski, D. Del Testa, D. Dworakowski, B. Firner, B. Flepp, P. Goyal, L. D. Jackel, M. Monfort, U. Muller, J. Zhang, X. Zhang, J. Zhao and K. Zieba, End to end learning for self-driving cars, arXiv:1604.07316[cs.CV].

[9]Gazebo, Gazebo: Robot simulation made easy (2017), www.gazebosim.org.

[10]GitHub, Udacity/self-driving-car (2016), https://github.com/udacity/self-drivingcar/tree/master/datas.

Efficient resource allocation for decentralized heterogeneous system using density estimation approach

J. Senthilnath[*,‡], K. Harikumar[*] and S. Suresh[*,†]

ST Engineering-NTU Corporate Laboratory
School of Electrical and Electronic Engineering
Nanyang Technological University
Singapore 639798, Singapore

[†]*School of Computer Science and Engineering*
Nanyang Technological University
Singapore 639798, Singapore
[‡]senthil.iiscb@gmail.com

This paper focuses on enhancing the mission duration by deploying secondary agents to coordinate with the primary agents to accomplish the mission with a minimal interruption. The interruption considered here is due to limited fuel carrying capability of primary agents. In this study, primary and secondary agents refer to unmanned aerial vehicles (UAVs) and unmanned ground vehicles (UGVs), respectively. Conventionally, UAVs are refueled with the fixed main charging stations which lead to interruption during the ongoing mission. In this work, we propose two-stage density estimation approach for efficiently distributing the swarm of UGVs to act as mobile refueling stations for UAVs. In the first stage, the optimal number of UGVs and their initial placement are computed. In the final stage, the UGVs minimize the average distance for the nearest UAVs to refuel. The performance of the proposed method is compared with the state of the art. The numerical simulation shows a better performance with the distributed UGVs than the state of the art.

Keywords: Heterogeneous system; resource allocation; density estimation.

1. Introduction

Recent advances in the unmanned aerial vehicle (UAV) technology lead to a plethora of applications mainly but not limited to defense community.[1] Many researches focused on building lightweight UAVs, with high endurance and low maintenance.[2,3] UAVs are used in a variety of applications like remote sensing,[4] surveillance,[5] patrol[6] and data collection over areas that are dangerous for human intervention.[7] They are capable of flying autonomously in different environmental conditions and are usually equipped with sensor payload like a camera for environmental monitoring.[5] Dedicated communication hardware devices are used to exchange data with other UAVs and also with the central control station (CCS).[8]

The major work contribution in the field of multi-UAV is to increase the efficiency under communication constraints and cooperate with one another to accomplish the assigned mission.[1,6,9] Communications between the UAVs and between individual UAV and CCS are other matters of contention when the area of operation is considerably large. The power required for data transmission is proportional to the square of the distance between the transmitter and receiver. To overcome the long distance data transmission problem, sensor nodes can be configured as intermediate links for establishing communication between UAVs and also between individual UAV and CCS. The usage of intermediate communication links abbreviates the power requirements for communication of UAVs, thereby increasing the endurance. It also minimizes information loss due to multipath reflections and maximizes RF energy received at CCS by intermediate boosting operations.[10]

The cooperation among the agents performing a mission is a well-studied problem in the literature. However, most of the previous works focus on the effectiveness of cooperative missions under various communication network topologies and distance constraints.[11,12] The problem of enhancing the overall mission duration by coordinating unmanned ground vehicles (UGVs) with nearby UAVs is seldom addressed in the literature. As per authors' knowledge, there are only few research works addressing the possibility of mobile recharging platforms for UAVs.[13,14] For continuity of such mission, each of these UAVs needs to be periodically refueled. So automated maintenance and refueling are of prime importance for UAV systems to perform smooth operations with minimal interruption and also without human intervention.

Synchronized coordination of heterogeneous systems is gaining popularity.[15] Tokekar *et al.*[16] developed sensor planning for a symbiotic UAV and UGV for the application in

---

[‡]Corresponding author.

precision agriculture. A swarm of UAVs perform shared tasks by coordinating the required inter-vehicle actions using UGVs.

This paper aims to coordinate UGVs with the UAVs undergoing a mission. A preliminary version of this paper was presented in Ref. 17. The proposed algorithm efficiently distributes the swarm of UGVs to act as mobile refueling stations for UAVs. This is achieved with the two-stage density estimation approach. The first stage generates the optimal number and distributed placement of UGVs. In the second stage, velocity commands are generated for the motion of each UGV within the Voronoi region to coordinate with the nearest UAVs for resource allocation. The performance of the algorithm is compared using a statistical significance test for the number of UGVs distributed and the advantage over the placement of charging stations.

The remaining sections of this paper are organized as follows: Section 2 explains the proposed methodology for UGVs distribution and subsequent motion based on UAVs density. The simulation results are discussed in Sec. 3. The paper is concluded in Sec. 4.

2. Methodology

This section mainly focuses on various challenges involved in the distributed placement of UGVs and subsequent motion of UGVs to coordinate with UAVs for resource allocation.

2.1. *Problem formulation*

Multiple UAVs are performing some missions within some time limit. The main necessity is to maximize the area coverage, due to limited endurance to accomplish the complete task without delay.

These problems can be overcome by placing distributed UGVs which act as resource allocators. Further, the issues involved in the placement of UGVs at a given instant can be of two extremes, such as (i) limited resource allocation using single UGV to coordinate with all UAVs and (ii) individual UGV placement for UAVs is more expensive. Hence there is a need to find an optimal number of UGVs and their position to provide maximum coverage for an efficient resource allocation for UAVs so that the mission can be continued without reducing UAVs.

Our proposed density estimation approach uses two stages to solve this problem: (1) optimal placement of UGVs during launch phase and (2) coordinated motion of UGVs with the movement of UAVs during the mission.

2.2. *Optimal placement of UGVs*

UAVs have different endurances based on their fuel and payload capabilities. Hence there is a need to know where to place the UGVs. This depends on the number of UAVs and their launch positions, with this the estimation of the optimal number of UGVs and their positions can be achieved. The optimal placement of UGVs is performed considering the initial launch positions of UAVs.

Given n UAVs $x_i, i = 1, 2, \ldots, n$, on an IR^2-dimensional space, the stationary allocations are used to estimate considering the UAV density for the placement of UGVs. Convergence occurs by following the initial launching positions of UAVs which lie within their sensor range (r_s) and have the optimal positions of UGVs. The two phases of the density estimation algorithm are:

(i) *Selection phase*: Considering the distance of the UAVs with the sensor range (r_s) neighboring UAVs are selected as given in following equations:

$$d_{ij} = \|(x_i - x_j, y_i - y_j)\|_2 \quad \text{for } j = 1, 2, \ldots, m, \quad (1)$$

$$V_{ij} = \begin{cases} 1, & \text{if } d_{ij} \le r_s, \\ 0, & \text{if } d_{ij} > r_s, \end{cases} \quad (2)$$

where d_{ij} contains distance of each UAV with other UAVs, $\|\cdot\|_2$ is the Euclidean norm operator and vicinity range (V_{ij}) is obtained using the distance d_{ij} to retrieve the number of UAVs within a sensor range (r_s).

(ii) *Placement phase*: During this phase, placement of each UGV depending on UAVs launching positions is considered. Each UAV depending on sensor range finds its neighbors. Keeping this into consideration, the position and number of UGVs need to be computed as follows:

$$U_j(it + 1) = U_j(it) + s * \|\mu_j(it) - U_j(it)\|_2, \quad (3)$$

$$\mu_j = \frac{\sum_{i=1}^{n} \sum_{j=1}^{n} V_{ij} U_j}{\sum_{i=1}^{n} \sum_{j=1}^{n} V_{ij}}, \quad (4)$$

where $U_j(it)$ is the location of the UAV j for itth iteration, s is the step size and $\|\mu_j(it) - U_j(it)\|_2$ gives the distance of movement.

With these UAVs, density is estimated to place UGVs (Voronoi points) by splitting into regions covered by each UGV (Voronoi partition).

2.3. *Coordinated motion of UGVs towards nearest UAVs*

Let the pair $(x_i(t), y_i(t))$ represent the position coordinates of ith UAV at a given time t, where $i \in \{1, 2, 3, \ldots, n\}$. The velocity coordinates are represented by the pair $(\dot{x}_i(t), \dot{y}_i(t))$. Similarly, the position and velocity coordinates of ith UGV are represented by $(x_{gi}(t), y_{gi}(t))$ and $(\dot{x}_{gi}(t), \dot{y}_{gi}(t))$, respectively. Here $j \in \{1, 2, 3, \ldots, m\}$. The distance between ith UAV and jth UGV is given by

$$d_{ij}(t) = \|(x_i(t) - x_{gj}(t), y_i(t) - y_{gj}(t))\|_2. \quad (5)$$

An ith UAV belongs to the neighborhood of jth UGV if

$$\min(d_{i1}(t), d_{i2}(t), \ldots, d_{ij}(t), \ldots, d_{im}(t)) = d_{ij}(t). \quad (6)$$

An optimal trajectory for UGV is planned based on the minimization of predicted miss distance at a prediction

Algorithm 1. Placement of UGVs and coordination with nearest UAVs

---

Compute UGVs initial positions using Eqs. (1)–(4).
while *time*< t_{end} **do**

> Identify all the UAVs in its neighborhood using Eq. (6);
> Minimize the performance index with its velocity as decision variable for a fixed prediction horizon T_p using Eq. (7);
> The value of the decision variable $(\dot{x}_{gj}^*, \dot{y}_{gj}^*)$ is updated and given as reference input to inner loop velocity control, where
> $P=(x_i, y_i, \dot{x}_i, \dot{y}_i)...(x_{n_j}, y_{n_j}, \dot{x}_{n_j}, \dot{y}_{n_j})$;

end

---

horizon time T_p. The performance index to be minimized for *j*th UGV is given by

$$J_j = \sum_{k=1}^{n_j} \frac{1}{2}(d_{kj}(t + T_p))^2, \qquad (7)$$

where n_j is the number of UAVs in the neighborhood of *j*th UGV. Let the change in velocity of UAV be governed by first-order dynamics with a time constant τ_u. If the prediction horizon $T_p \ll \tau_u$, then

$$d_{kj}(t + T_p) = \|(\Delta X(t + T_p), \Delta Y(t + T_p))\|_2. \qquad (8)$$

The expressions for $\Delta X(t + T_p)$ and $\Delta Y(t + T_p)$ appearing in Eq. (8) are given in Eqs. (9) and (10), respectively:

$$\Delta X(t + T_p) = x_k(t) - x_{gj}(t) + (\dot{x}_k(t) - \dot{x}_{gj}(t))T_p, \qquad (9)$$

$$\Delta Y(t + T_p) = y_k(t) - y_{gj}(t) + (\dot{y}_k(t) - \dot{y}_{gj}(t))T_p. \qquad (10)$$

The optimization problem is formulated as min J_j with decision variable as $(\dot{x}_{gi}(t), \dot{y}_{gi}(t))$.

Algorithm 1 executes till the end of the mission denoted by t_{end}.

The parameter n_j varies as the search operation of UAV continues. The whole area can be treated as Voronoi partitioned with each UGV at its centroid. The controller architecture for individual UGV is shown in Fig. 1. The inertial positions and velocities of the neighborhood UAVs are obtained through a wireless communication channel. The inertial positions and velocities of UGVs are obtained through global positioning system (GPS) and inertial measurement unit (IMU) sensor mounted on each UGV. The optimization block takes the positions and velocities of all the neighborhood UAVs and UGVs positions and velocities to compute the optimal velocity command. The governing equations of motion for UGV with closed-loop velocity control are given as Eqs. (11) and (12). The time constant of the closed-loop velocity control system is represented by τ_j:

$$\ddot{x}_{gj}(t) = -\frac{1}{(\tau_j)\dot{x}_{gj}(t) + \left(\frac{1}{\tau_j}\right)\dot{x}_{gj}^*}, \qquad (11)$$

$$\ddot{y}_{gj}(t) = \left(-\frac{1}{\tau_j}\right)\dot{y}_{gj}(t) + \left(\frac{1}{\tau_j}\right)\dot{y}_{gj}^*. \qquad (12)$$

3. Numerical Simulation

This section compares the evaluations of the optimal placement and movement of UGVs based on density estimation of UAVs initial launch and subsequent movement being simulated.

3.1. *Optimal placement of UGVs*

Figure 2 shows initial launching positions of UAVs. With these initial UAVs positions, the optimal number of UGVs is analyzed using cost function.[18] The cost function is the sum of the distance from UAVs to UGVs and that between UGVs. The minimum value of the cost function corresponds to the optimum number of UGVs. From Fig. 3, it is observed that the required number of UGVs for this scenario can be chosen to be two or three UGVs.

The number of UGVs distribution is statistically validated using cluster index.[19,20] In the literature, the Calinski–Harabasz index[21] has been widely used for the number validation considering the data distribution. In this study, the Calinski–Harabasz index is used for the validation of the UGV number along with a cost function. The Calinski–Harabasz index is used as a measure for comparison and verification along with the cost function. The Calinski–Harabasz index function uses the ratio of the intra-UAVs to

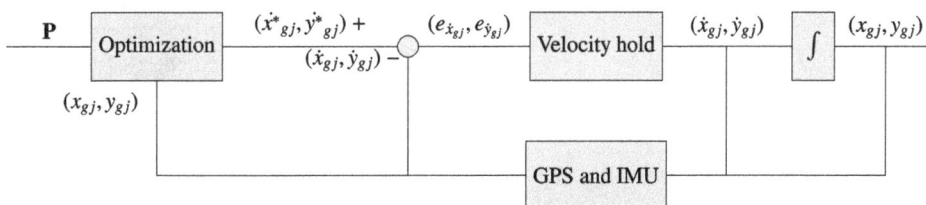

Fig. 1. Controller architecture for the UGV.

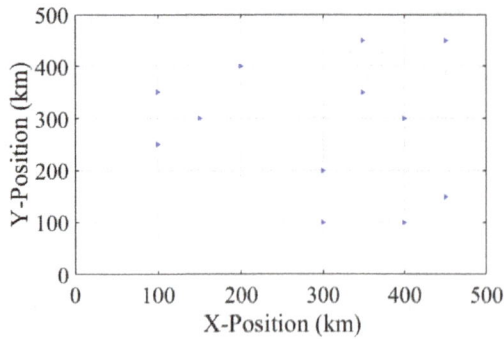

Fig. 2. Initial launching positions of UAVs.

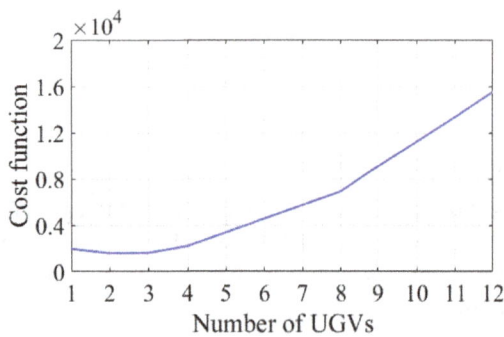

Fig. 5. Distributed placement of UGVs covering UAVs vicinity radius.

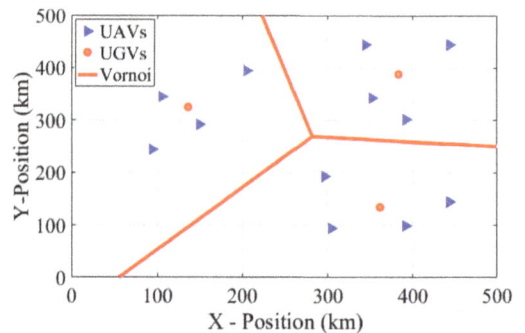

Fig. 3. Cost for UGVs placement.

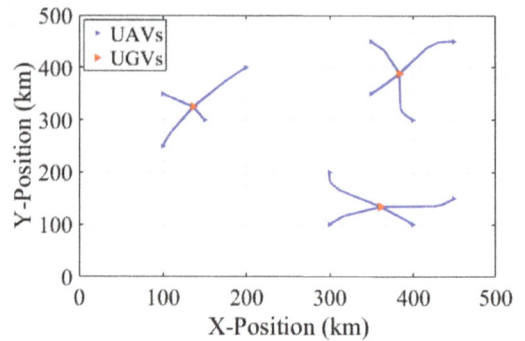

Fig. 6. Voronoi partition for each UGV covering UAVs.

inter-UGVs separation. For the given launching positions of UGVs, the main objective is to maximize the Calinski–Harabasz index resulting in proper placement of UGVs.

From Fig. 4, we can observe that the maximum Calinski–Harabasz value is obtained for UAV position with the number of UGVs as 3. This value can also be observed using the cost function. This clearly validates the cost function and statistical validity method (Calinski–Harabasz index) for UAVs positioning. Overall, the proposed density estimation method using cost function has been validated using Calinski–Harabasz index for the random distribution of the UAVs positions.

The placement of UGVs is obtained by considering the density estimation of UAVs. The optimal number and

distributed placement of UGVs is estimated based on the convergence points of UAVs within the vicinity radius as shown in Fig. 5.

Figure 6 shows the positions of UAVs and UGVs at a particular time instant. In the figure, red line indicates the Voronoi partition, i.e., a UGV covering that area.

3.2. Coordinated motion of UGVs

The UAVs search an area of size 500 km × 500 km randomly. The UAVs are assumed to be flying at different altitudes to avoid collision among themselves. The UAVs are flying with a velocity of 10 km/h. The maximum speed of a UGV is 20 km/h. A first-order dynamics is considered for velocity change in the case of both UAVs and UGVs. Total mission duration inclusive of the time taken by UAVs for multiple refueling operations is taken as 300 h. The positions of UAVs and UGVs at different time instants of the mission are shown in Fig. 7. The blue triangle indicates the UAV and red circle indicates the UGV. The Voronoi partition is shown using red lines. We can see that the intersection point of the Voronoi lines lies almost at the center of the total area for all the six time instants. This indicates that the area covered by each UGV is almost equal during the mission with the Voronoi centroids shifting with time. The numerical simulation is performed in MATLAB (R2016a).[a]

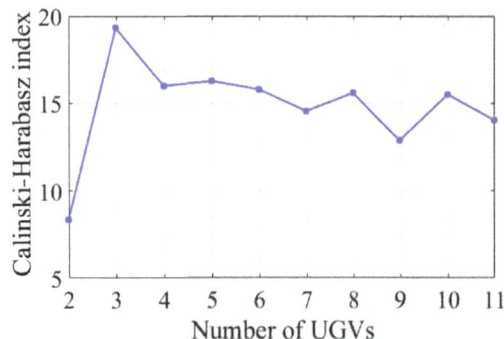

Fig. 4. Calinski–Harabasz index analysis.

[a]https://youtu.be/3-pQVd91Ss4

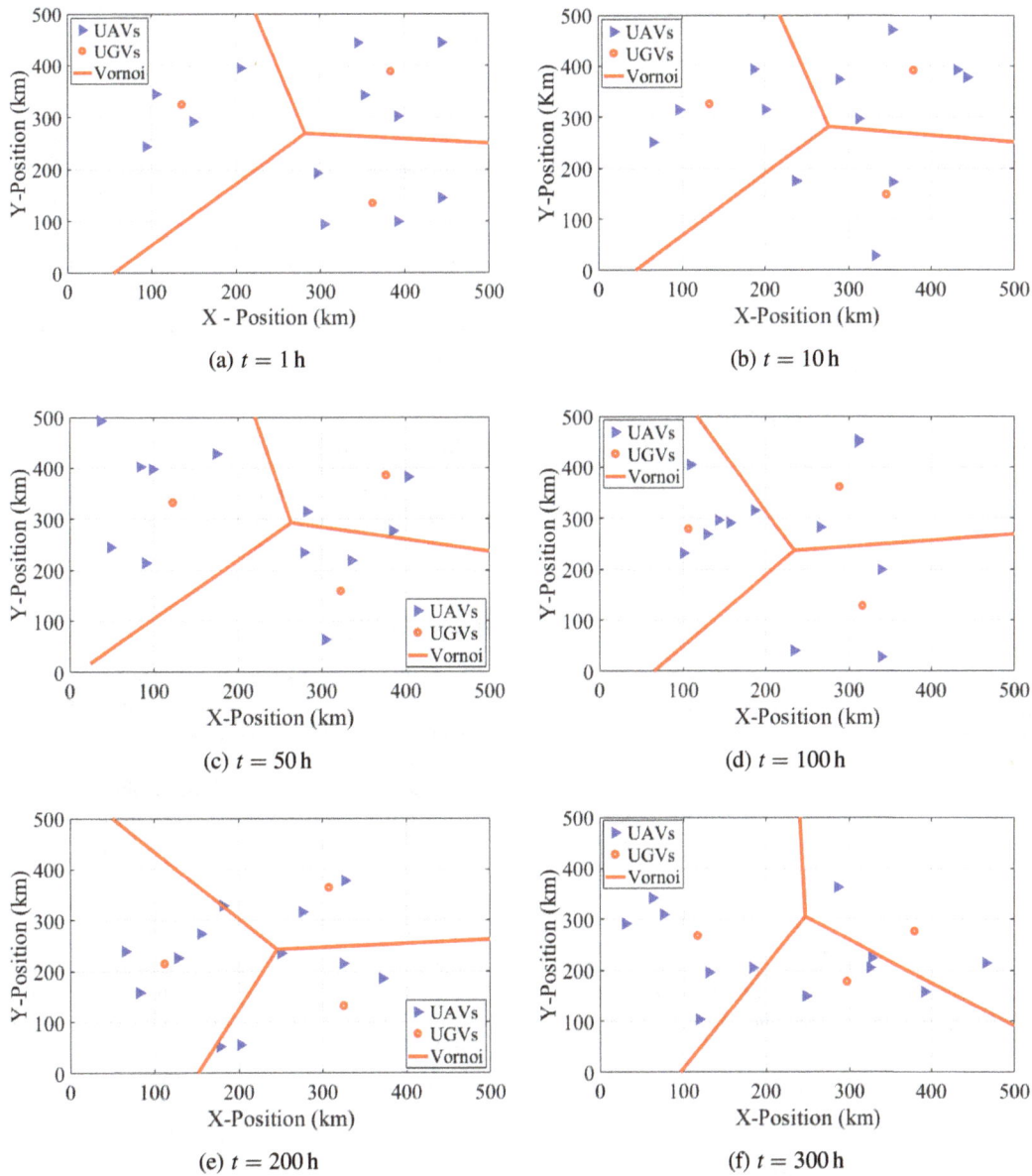

(a) $t = 1$ h

(b) $t = 10$ h

(c) $t = 50$ h

(d) $t = 100$ h

(e) $t = 200$ h

(f) $t = 300$ h

Fig. 7. Simulation results for three UGVs and 12 UAVs at different time instants between $t = 1$ h and $t = 300$ h.

For comparing the performances of mobile and stationary maintenance and refueling stations, three stationary refueling stations are placed at the locations (125 km, 250 km), (250 km, 125 km) and (375 km, 250 km), respectively. These three points are along the center line that constitutes the whole 500 km × 500 km area. A plot showing the average distance to be traveled by a UAV to reach the nearest refueling station is given in Fig. 8. The curve shown in red indicates the case for stationary refueling station and the curve in blue color indicates the case for mobile UGV stations. The maximum value of the average distance to be traveled for stationary refueling station is close to 130 km and the minimum value is close to 70 km, whereas, for mobile

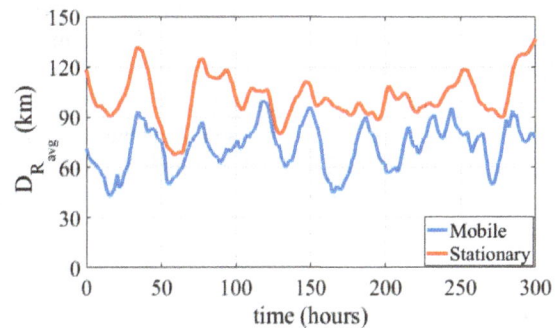

Fig. 8. Average distance to be traveled by a UAV to the nearest refueling station.

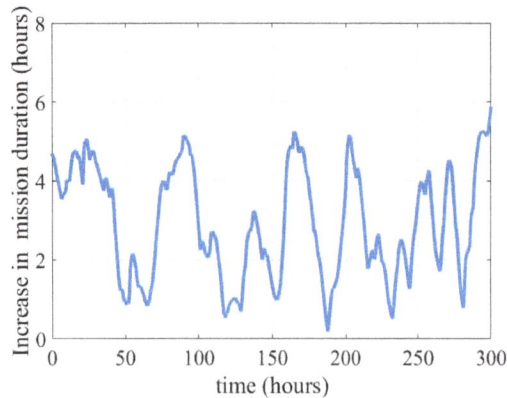

Fig. 9. Increase in mission duration with UGVs in comparison with the stationary refueling station.

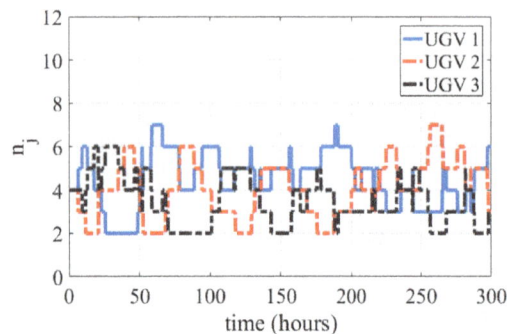

Fig. 10. Number of UAVs assigned to each UGV during the mission.

refueling station the maximum and minimum distances to be traveled by UAVs are close to 90 km and 50 km, respectively.

Figure 9 shows the increase in mission duration when UGVs are deployed instead of stationary refueling stations. The increase in mission duration is due to the reduction in the distance to be traveled by UAVs for refueling. A maximum increase of about 5 h in mission duration is achievable with an average increase of about 2.5 h for a single UAV. For the case of 12 UAVs with n_r times for refueling operations, an average of $30n_r$ h increase in overall mission duration is achieved.

The numbers of UAVs in the neighborhood of each UGV during the entire mission is shown in Fig. 10. The number of UAVs in the neighborhood goes as low as 2 and also as high as 7. The statistics of neighborhood UAVs indicates the resource utilization pattern and is used to decide the capacity constraint for each UGV. Overall, the mobile UGVs performance is better in comparison to the same number of stationary stations.

4. Conclusion

In this paper, we developed a density estimation approach for efficient resource allocation by coordinating UGVs with their nearest UAVs. The proposed density estimation approach

consists of two stages. In the first stage, the optimal number and placement of UGVs required to support the UAVs mission are obtained. In the second stage, each UGV within the Voronoi region moves to coordinate with the nearest UAVs for providing resources. Simulation results show that the proposed method is able to find the optimal number of UGVs and this has been evaluated using statistical significance test. Further, the importance of UGVs coordinating with their nearest UAVs is compared with the stationary resource allocation stations. Overall, the proposed method performs better than the state of the art in terms of the increase in UAVs useful mission time.

Acknowledgments

The research was fully supported by the ST Engineering-NTU Corporate Lab, Singapore, under the Project CRP3-P2P.

References

[1] T. Shima, S. J. Rasmussen, A. G. Sparks and K. M. Passino, Multiple task assignments for cooperating uninhabited aerial vehicles using genetic algorithms, *Comput. Oper. Res.* **33**, 3252 (2006).
[2] K. Harikumar, S. Dhall and M. S. Bhat, Nonlinear modeling and control of coupled dynamics of a fixed wing micro air vehicle, *Proc. Indian Control Conf. (ICC)* (2016), pp. 318–323.
[3] K. Harikumar, J. V. Pushpangathan, T. Bera, S. Dhall and M. S. Bhat, Modeling and closed loop flight testing of a fixed wing micro air vehicle, *Micromachines* **9**, 1 (2018).
[4] J. Senthilnath, A. Dokania, M. Kandukuri, K. N. Ramesh, G. Anand and S. N. Omkar, Detection of tomatoes using spectral-spatial methods in remotely sensed RGB images captured by UAV, *Biosyst. Eng.* **146**, 16 (2016).
[5] J. Senthilnath, M. Kandukuri, A. Dokania and K. N. Ramesh, Application of UAV imaging platform for vegetation analysis based on spectral-spatial methods, *Comput. Electron. Agric.* **140**, 8 (2017).
[6] J. Senthilnath, S. N. Omkar, V. Mani and A. R. Katti, Cooperative communication of UAV to perform multi-task using nature inspired techniques, *Proc. IEEE Symp. Computational Intelligence for Security and Defense Applications (CISDA)* (2013), pp. 45–50.
[7] R. Bhola, N. H. Krishna, K. N. Ramesh, J. Senthilnath and G. Anand, Detection of the power lines in UAV remote sensed images using spectral-spatial methods, *J. Environ. Manage.* **206**, 1233 (2018).
[8] Y. Zeng, R. Zhang and T. J. Lim, Wireless communications with unmanned aerial vehicles: Opportunities and challenges, *IEEE Commun. Mag.* **54**, 36 (2016).
[9] T. K. Venugopalan, K. Subramanian and S. Sundaram, Multi-UAV task allocation: A team-based approach, *Proc. IEEE Symp. Ser. Computational Intelligence* (2015), pp. 45–50.
[10] M. Dong, K. Ota, M. Lin, Z. Tang, S. Du and H. Zhu, UAV-assisted data gathering in wireless sensor networks, *J. Supercomput.* **70**, 1142 (2014).
[11] B. Rajarshi, B. Titas and S. Suresh, A decentralized game theoretic approach for team formation and task assignment by autonomous

unmanned aerial vehicles, *Proc. Int. Conf. Unmanned Aircraft Systems* (2017).

[12]P. B. Sujit, J. G. Manathara, D. Ghose and J. B. de Sousa, Decentralized multi-UAV coalition formation with limited communication ranges, *Handbook of Unmanned Aerial Vehicles*, Section XV, Chapter 3 (Springer, 2014), pp. 2021–2048.

[13]J. Leonard, S. Aldhaher, A. Savvaris and A. Tsourdos, Automated recharging station for swarm of unmanned aerial vehicles, *Proc. ASME Int. Mechanical Engineering Congr. and Exposition* (2012), pp. 607–614.

[14]C. H. Choi, H. J. Jang, S. G. Lim, H. C. Lim, S. H. Cho and I. Gaponov, Automatic wireless drone charging station creating essential environment for continuous drone operation, *Proc. Int. Conf. Control, Automation and Information Sciences (ICCAIS)* (2016), pp. 132–136.

[15]K. Harikumar, T. Bera, R. Bardhan and S. Sundaram, Autonomous navigation and sensorless obstacle avoidance for UGV with environment information from UAV, *Proc. Second IEEE Int. Conf. Robotic Computing* (2018), pp. 266–269.

[16]P. Tokekar, J. V. Hook, D. Mulla and V. Isler, Sensor planning for a symbiotic UAV and UGV system for precision agriculture, *IEEE Trans. Robot.* **32**, 1498 (2016).

[17]J. Senthilnath, K. Harikumar and S. Suresh, Dynamic area coverage for multi-UAV using distributed UGVs: A two-stage density estimation approach, *Proc. Second IEEE Int. Conf. Robotic Computing* (2018), pp. 165–166.

[18]J. Senthilnath, S. Kulkarni, J. A. Benediktsson and X. S. Yang, A novel approach for multi-spectral satellite image classification based on the bat algorithm, *IEEE Geosci. Remote Sens. Lett.* **13**, 599 (2016).

[19]J. Senthilnath, D. Kumar, J. A. Benediktsson and X. Zhang, A novel hierarchical clustering technique based on splitting and merging, *Int. J. Image Data Fusion* **7**, 19 (2016).

[20]J. Senthilnath, S. C. Sumanth, G. Nagaraj, M. Thapa and M. Indiramma, BELMKN: Bayesian extreme learning machines Kohonen network, *Algorithms* **11**, 1 (2018).

[21]T. Calinski and J. Harabasz, A dendrite method for cluster analysis, *Commun. Stat.-Theory Methods* **3**, 1 (1974).

Robotic control for cognitive UWB radar

Stefan Brüggenwirth* and Fernando Rial†

Fraunhofer Institute for High Frequency Physics and Radar Techniques (FHR)
Fraunhoferstr. 20, 53343 Wachtberg, Germany
*stefan.brueggenwirth@fhr.fraunhofer.de
†fernando.rial@fhr.fraunhofer.de

In the paper, we describe a trajectory planning problem for a six-DoF robotic manipulator arm that carries an ultra-wideband (UWB) radar sensor with synthetic aperture (SAR). The resolution depends on the trajectory and velocity profile of the sensor head. The constraints can be modeled as an optimization problem to obtain a feasible, collision-free target trajectory of the end-effector of the manipulator arm in Cartesian coordinates that minimizes observation time. For 3D reconstruction, the target is observed in multiple height slices. For through-the-wall radar the sensor can be operated in sliding mode for scanning larger areas. For IED inspection the spotlight mode is preferred, constantly pointing the antennas towards the target to obtain maximum azimuth resolution. UWB sensors typically use a wide spectrum shared by other RF communication systems. This may become a limiting factor on system sensitivity and severely degrade the image quality. Cognitive radars can adapt dynamically their bandwidth, frequency and other transmit parameters to the radio frequency environment to avoid interference with primary users.

Keywords: Cognitive radar; robot control; signal processing.

1. Introduction

The concept of a cognitive radar[1,2] architecture at Fraunhofer FHR[3] is based on the three-layer model by Rasmussen[4] as shown in Fig. 1.

The *skill-based* behavior represents the basic signal generation and processing capabilities of the system. This provides the subsymbolic, continuous stream of input data to the architecture. The *rule-based* abstraction layer applies machine-learning methods to recognize certain prestored cues in the perceived scene. Upon match, the architecture will reactively execute a prestored procedure on its actuators or modify its sensor parameter settings. The *knowledge-based* abstraction layer provides long-term deliberation and goal-based planning.

In this paper, we will focus on the generation of constraints for trajectory planning for a robotic arm carrying a ultra-wideband (UWB) radar sensor. All three layers need to interact to compute the trajectory according to the scene and execute it by commanding the sensor and the manipulator joints and finally form an image.

2. Robotic UWB Sensor Setup

In this section, we introduce the necessary computations related with the problem of trajectory planning for a robotic manipulator arm that carries a UWB sensor able to work in synthetic aperture radar (SAR) mode. We used a stock ST Robotics R17 six-degree of freedom (DoF) industrial-type manipulator arm shown in Fig. 2.

High-resolution imaging can be only achieved with an even higher precision positioning. The 3D trajectory of the sensor needs to be measured and synchronized with the sensor data. For that purpose, accelerometers and gyroscopes from an attached IMU are used. The IMU drift is additionally stabilized using the hardware readout of optical encoders of the robot arm joints controlled by step motors.

The sensor has one transmitter and one receiver in a typical common-offset arrangement as shown in Fig. 3. The horn-type antennas can be rotated to exploit polarization diversity.

The spatial resolution and processing gain that the system can achieve ultimately depend on the trajectory and velocity profile of the sensor head. The constraints can be modeled as an optimization problem to obtain a feasible, collision-free trajectory of the end-effector of the manipulator arm in Cartesian coordinates that minimizes observation time.

3. Sensor Characteristics and Trajectory Constraints

The trajectory constraints are linked to the sensor characteristics. The chirped radar sensor uses a selectable center frequency from 3 GHz to 8 GHz and 4 GHz of bandwidth (B),

*Corresponding author.

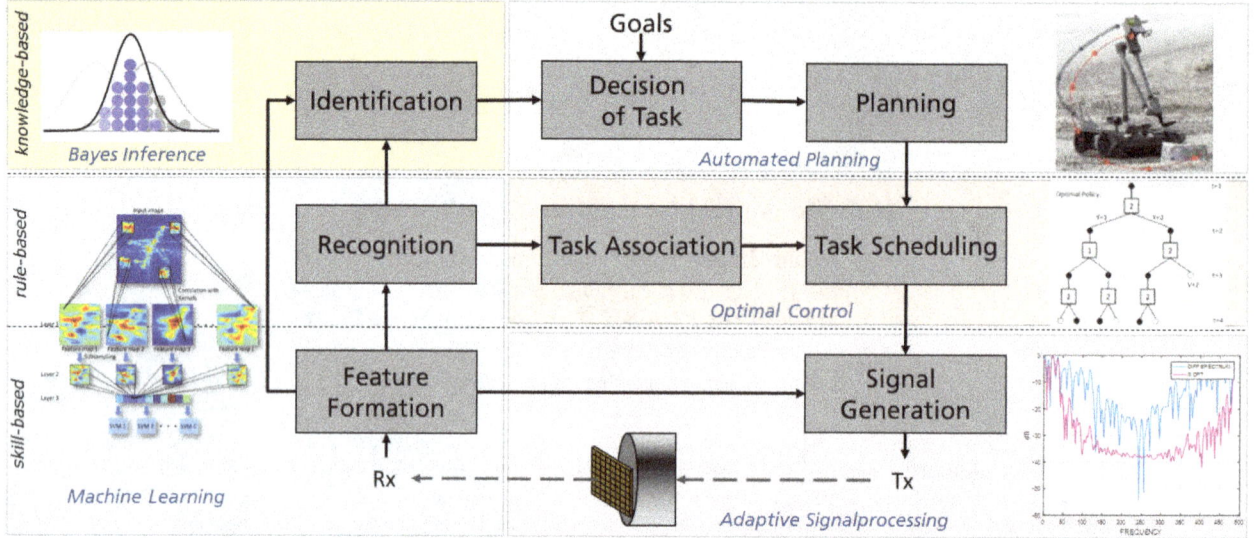

Fig. 1. Three-layer model of a cognitive radar architecture.

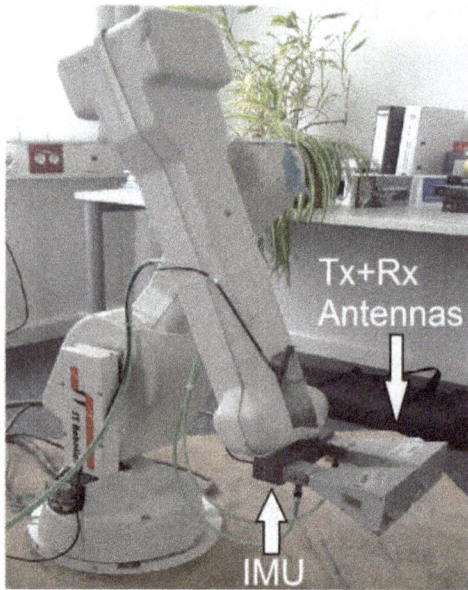

Fig. 2. Transmit (Tx) and receive (Rx) antennas of UWB radar sensor mounted to robotic manipulator arm. Photo also shows the inertial measurement unit (IMU) position.

resulting in $\delta_r = 3.75$ cm of range resolution given by

$$\delta_r \simeq \frac{c}{2B}, \qquad (1)$$

where c denotes the speed of light in vacuum. The center frequency can be tuned according to a particular target or propagation environment (ground penetration, through-the-wall imaging, IED inspection, etc.).

The sensor is able to operate in stripmap or spotlight SAR modes using linear trajectories. The differences between stripmap and spotlight modes can be observed in Fig. 4. In spotlight mode the antenna is always pointing to a target.

This mode generates images at higher resolution than stripmap but the area covered is only determined by the antenna beam θ. In stripmap mode the antenna is fixed to cover a wider area but at lower resolution. The mobility of the arm can be further exploited to generate circular trajectories around a target in spotlight mode to obtain a more accurate object reconstruction. Several parallel trajectories can be combined to define a 2D aperture suitable for 3D imaging of objects.

Considering the case of 2D acquisition geometry working in SAR mode (as in Fig. 5), then the cross-range resolutions $\delta_{x,y}$ can be obtained as

$$\delta_x \simeq R\frac{\lambda_c}{2L_x}, \qquad (2)$$

$$\delta_y \simeq R\frac{\lambda_c}{2L_y}, \qquad (3)$$

where R is the distance to the target, λ_c is the wavelength at the center frequency and L_x and L_y are the lengths of the 2D synthetic aperture, respectively.

A perhaps more convenient way to express the cross-range resolution refers to the actual antenna footprint:

$$\delta_x \simeq \frac{\lambda_c}{2\Theta_x}, \qquad (4)$$

$$\delta_y \simeq \frac{\lambda_c}{2\Theta_y}, \qquad (5)$$

where Θ_x, Θ_y can represent either the antenna beam angles in radians in the x- and y-directions (stripmap) or the total viewing angle ($\Delta\varphi$) (spotlight).

In order to obtain a similar resolution in range and in cross-range, the trajectory planning must aim to create at least an aperture of 0.5–1.3 times the distance R to the target in

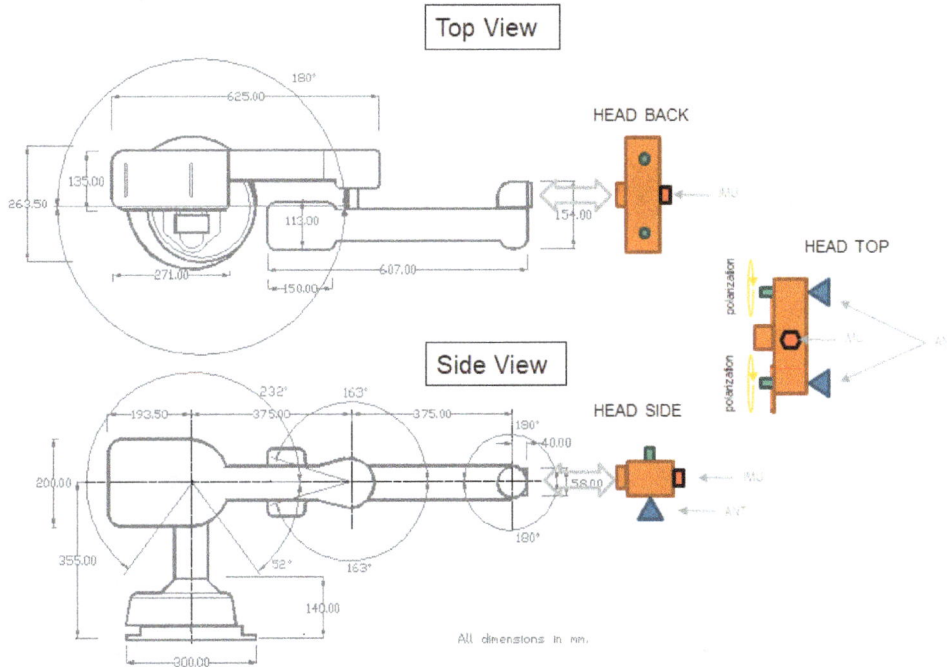

Fig. 3. Schematic of the robot arm and polarimetric antenna mounting.

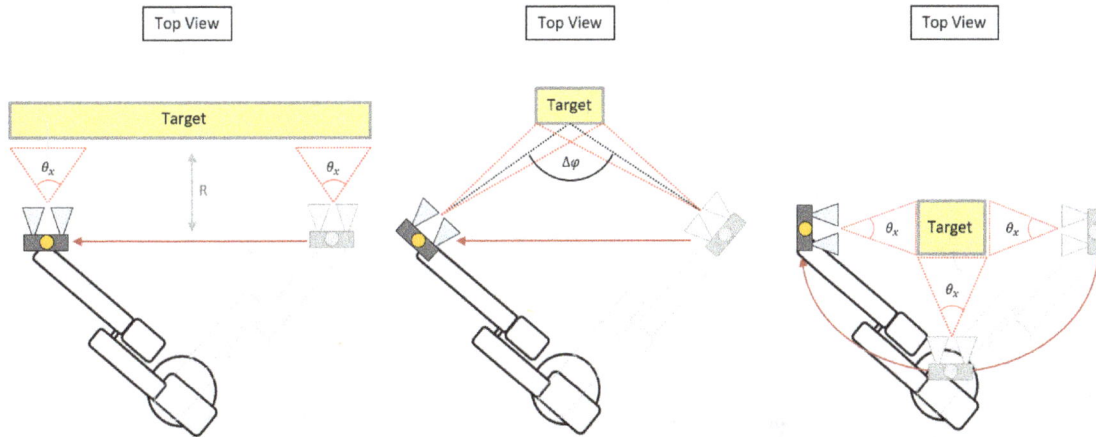

Fig. 4. Left: Linear trajectory (stripmap), middle: linear trajectory (spotlight) and right: circular trajectory (spotlight).

both directions, depending on the center frequency used by the system (8–3 GHz, respectively). In the particular case of a sensor setup with the central frequency of 8 GHz and antenna beamwidth of both $\Theta_x, \Theta_y = 20°$, the expected resolution will be about 5.4 cm. In practice, data is windowed to reduce sidelobes in the imaging and as a result the actual resolutions are usually poorer than those given by the formulas.

Two other important parameters to be taken into account are the optimal size of the scanning area and the sampling requirements. For planar acquisition geometries working in stripmap mode, to obtain full-resolution imaging of the total area of interest an additional half-beam aperture must be added in both dimensions as shown in Fig. 6.

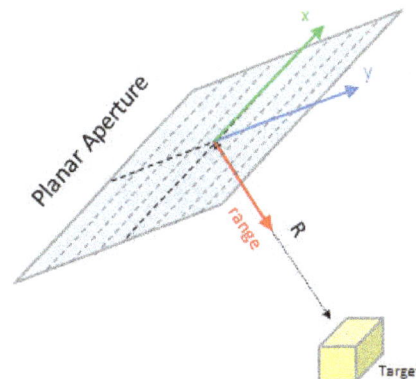

Fig. 5. Planar aperture using linear trajectories.

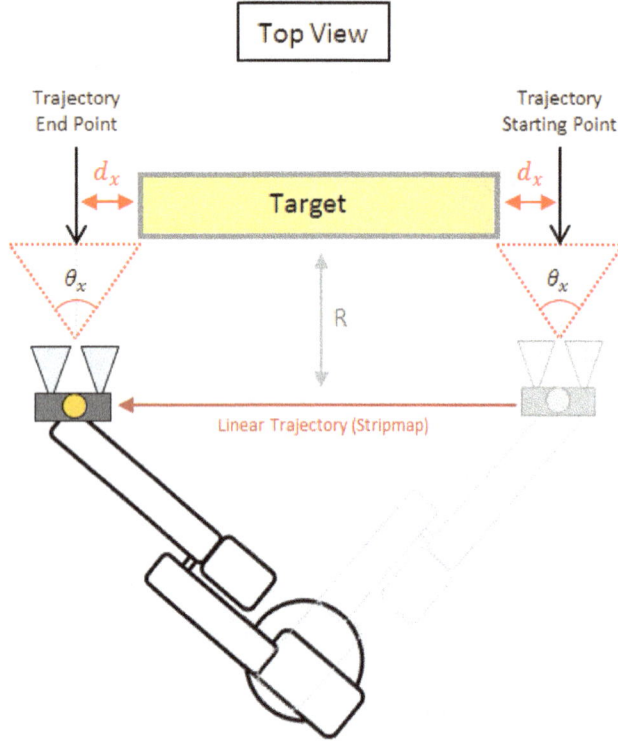

Fig. 6. The x- and z-directions of the scanning area.

This additional distance can be calculated with the expressions

$$d_x = R \tan \frac{\Theta_x}{2}, \tag{6}$$

$$d_y = R \tan \frac{\Theta_y}{2}, \tag{7}$$

that clearly depend on the distance between the sensor and the object of interest.

In the particular case of an object of width $D_x = 0.5$ m located at $R = 0.5$ m from the sensor, with a central frequency of 8 GHz and antenna beamwidth of both $\Theta_x, \Theta_y = 20°$, the scanning trajectory should be extended by about 9 cm in both dimensions. In the following, we will use $L_x = 0.8$ m as linear displacement in the x-direction per height slice.

Another important parameter is related with the sampling requirements of a particular acquisition. The measurement positions in the synthetic radar aperture require a minimum spacing in order to sample adequately the phase history associated with all the scatterers. If the distance between measurements is too large, the Nyquist criterion is not fulfilled and artifacts may appear in the reconstructed image.

Assuming that the targeted area is confined within a rectangular box of dimensions D_x, D_y and that L_x and L_y are the lengths of the 2D synthetic aperture, the required sampling spacings Δ_x, Δ_y in the measurement to satisfy the

Nyquist criterion are given by[5]

$$\Delta_x \leq \frac{\lambda_{\min}}{2} \cdot \frac{\sqrt{\frac{(L_x + D_x)^2}{4} + R^2}}{L_x + D_x}, \tag{8}$$

$$\Delta_y \leq \frac{\lambda_{\min}}{2} \cdot \frac{\sqrt{\frac{(L_y + D_y)^2}{4} + R^2}}{L_y + D_y}, \tag{9}$$

where $\lambda_{\min}$ is the wavelength at the maximum working frequency. The sampling spacing can be also related with the antenna beam using the equations

$$\Delta_x \leq \frac{\lambda_{\min}}{2 \cdot \sin \Theta_x}, \tag{10}$$

$$\Delta_y \leq \frac{\lambda_{\min}}{2 \cdot \sin \Theta_y}, \tag{11}$$

where Θ_x, Θ_y can represent either the antenna beam angles in radians in the x- and y-directions (stripmap) or the total viewing angle ($\Delta\varphi$) (spotlight).

In our example, the sensor head must not move more than $\Delta_x = 0.0095$ m per pulse. Since the pulse repetition frequency (PRF) of our system is fixed to 30 Hz, the resulting maximum velocity $v_{\max} = \Delta_x \cdot \text{PRF} = 0.285$ m/s.

Here it must be considered also that signal propagation in dielectric materials (ground, wall) will shrink the wavelengths and then sampling requirements become even more stringent.[6] A previous estimation of the dielectric permittivity of the propagation media may further optimize the acquisition geometry.

4. UWB Spectrum Sensing

UWB sensors typically use low operating frequencies to penetrate into the materials but at the same time using a large bandwidth needed for high-resolution imaging. Due to the fact that wireless communication technologies are of significant importance in modern times, the available radio frequency spectrum has become a valuable resource for radar. As example, in the US, parts of the S-band (1695–1710 MHz and 3550–3650 MHz) are allocated to wireless communication whereas parts of the C-band (5150–5350 MHz and 5470–5725 MHz) are being used by weather radars and 5 GHz-WiFi. These bands, although allocated, are underutilized providing opportunities for secondary (unlicensed) users to share the bands without harming the primary users, becoming a limiting factor on system sensitivity and severely degrading the image quality.

Spectrum sensing techniques provide algorithms to identify spectrum opportunities, i.e., to decide whether a frequency band is occupied or not. When provided with this information, cognitive radars can adapt dynamically their bandwidth, frequency and other transmit parameters to the radio frequency environment. The simplest spectrum sensing method is the energy detector.[7]

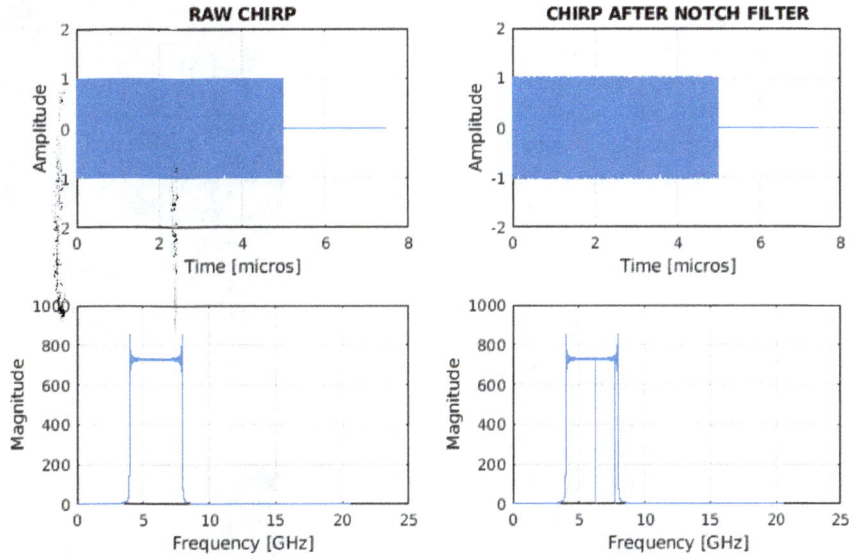

Fig. 7. Example of a chirp using notch filters after spectrum sensing where two RF interferers have been detected at 6.34 GHz and 7.79 GHz, respectively.

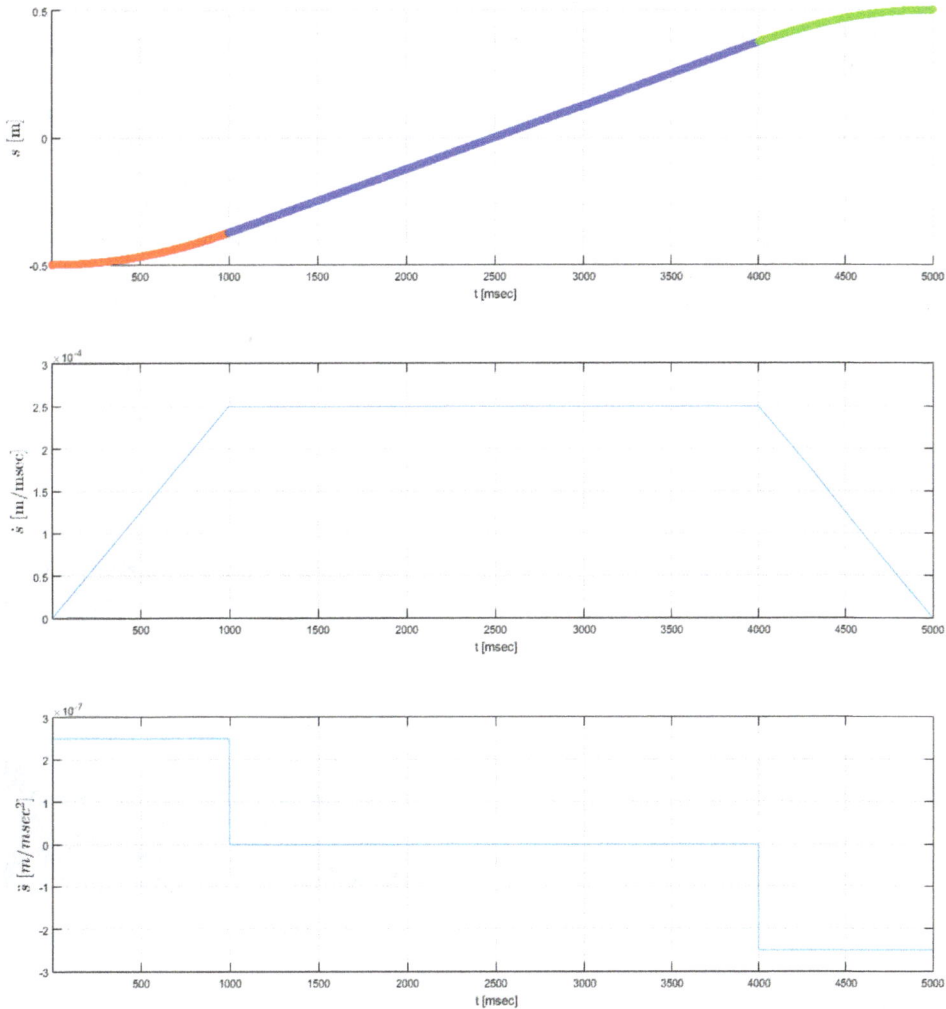

Fig. 8. Trapezoidal velocity profile for linear trajectory of the sensor head.

As a part of the cognitive architecture, before every transmission, an adaptable chirp (notched) is generated with an arbitrary waveform generator (AWG) taking the information from the previous sensing step into account. An example of a notched chirp is shown in Fig. 7 where two notches are added at two specific narrow-frequency bands to avoid interference.

5. Trajectory Generation

In order to maximize the acquisition speed, we selected a trapezoidal velocity[8] profile shown in Fig. 8, with a constant target velocity of $v = 0.25\,\text{m/s} < v_{\text{max}}$. The trajectory accelerates for the first 1000 ms, then reaches constant acquisition speed v for $L_x = 0.8\,\text{m}$ and then decelerates again.

The linear trajectory can be used to form an SAR image in stripmap mode for each height slice. As shown in Fig. 9, the pose of the robot end-effector follows a meander-shaped pattern composed of linear segments for the different height slices. For a 6-DoF robot, the inverse kinematics can easily be computed as illustrated in the figure for joints 3 and 6. The ST Robotics controller already contains routines to command the pose of the end-effector in Cartesian coordinates with the desired trapezoidal velocity profile.

6. Experimental Results

Figure 10 shows an example of an image obtained with the robot arm using some reference objects inside a plastic suitcase. The trajectory followed by the sensor has been planned considering the constraints previously mentioned to obtain unaliased high-resolution images of the total area of interest. The suitcase defines the scanning area with a width of

Fig. 10. Photo of the objects inside the suitcase.

$D_x = 0.5\,\text{m}$ located at $R = 0.5\,\text{m}$ from the sensor. Three reflecting spheres and two metal links are placed on top of absorbing material, creating a reflecting target area of 40 cm in width and 25 cm in range. The image is well focused and the range and cross-range resolutions are sufficient to resolve the individual objects separated by about 5 cm (Fig. 11).

The measurements were taken in a chamber free of RFI interference. Figure 12 shows the results from the 3D reconstruction using the meander trajectory. More accurate 3D object representations can be obtained with nonlinear trajectories using wider range of aspect angles.

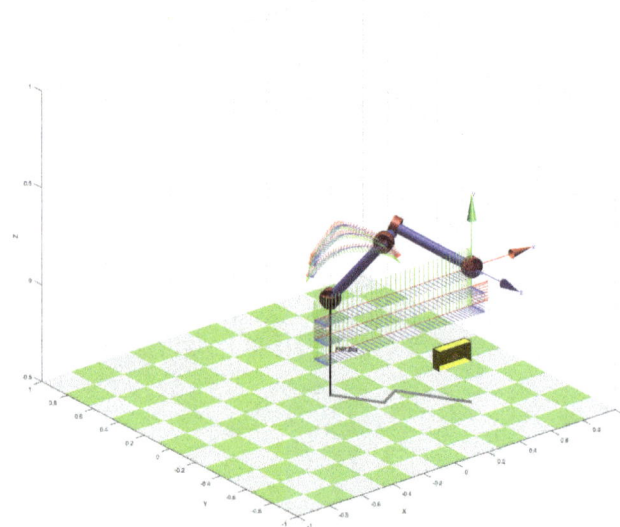

Fig. 9. Resulting inverse kinematics of six-DoF robot joint for linear SAR trajectory.

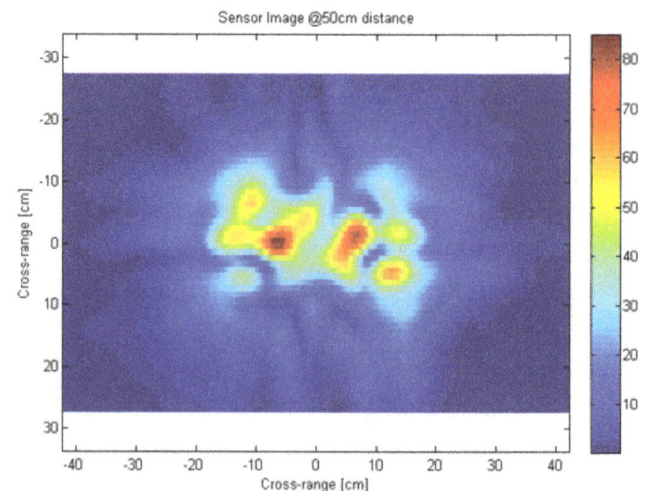

Fig. 11. SAR image of the objects using the robot arm (one height cut).

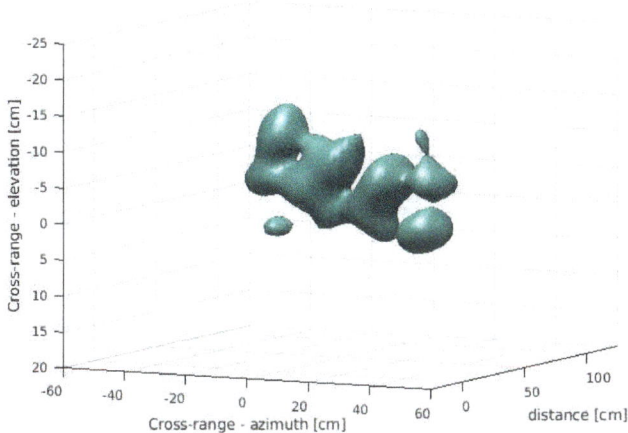

Fig. 12. The 3D reconstruction of the objects inside the suitcase.

7. Conclusions and Future Work

We presented a concept for sensor-controlled trajectory planning for cognitive radar. In this application a UWB radar sensor was used in stripmap SAR mode for 3D imaging. We presented first results from an experiment with a UWB sensor at 8-GHz center frequency controlled by a six-DoF manipulator arm. The test objects inside an open suitcase could be focused and resolved for one height level. Next, we investigate nonlinear trajectories with additional constraints on the workspace of the manipulator arm. Spotlight SAR mode can be used for IED detection inside the backpacks,[9] using higher center frequencies. A detailed comparison of the results subjected to RF interference will be presented in a following paper.

References

[1]S. Haykin, *Cognitive Dynamic Systems: Perception–Action Cycle, Radar and Radio* (Cambridge University Press, 2012).

[2]J. Guerci, *Cognitive Radar: The Knowledge-Aided Fully Adaptive Approach* (Artech House, 2010).

[3]J. Ender and S. Brüggenwirth, Cognitive radar-enabling techniques for next generation radar systems, *Proc. 2015 16th Int. Radar Symp. (IRS)* (IEEE, 2015), pp. 3–12.

[4]J. Rasmussen, Skills, rules, and knowledge; signals, signs, and symbols, and other distinctions in human performance models, *IEEE Trans. Syst. Man Cybern.* **SMC-13** 257 (1983).

[5]J. Fortuny-Guasch, A novel 3-D subsurface radar imaging technique, *IEEE Trans. Geosci. Remote Sens.* **40** 443 (2002).

[6]M. Grasmueck, R. Weger and H. Horstmeyer, Full-resolution 3D GPR imaging, *Geophysics* **70** K12 (2005).

[7]R. Tandra and A. Sahai, SNR walls for signal detection, *IEEE J. Sel. Top. Signal Proc.* **2** 4 (2008).

[8]P. Corke, *Robotics, Vision and Control*, Springer Tracts in Advanced Robotics (Springer-Verlag, Berlin, 2011).

[9]S. Lang and R. Herschel, Schnelle aufklärung mit dem usbv inspektor (2017), http://www.fhr.fraunhofer.de/de/geschaefts-felder/sicherheit/schnelle-aufklaerung-mit-/-dem-USBV-Inspektor.html.

A mixed real-time robot hardware abstraction layer (R-HAL)

Giuseppe F. Rigano*,‡, Luca Muratore*,†, Arturo Laurenzi*,
Enrico M. Hoffman* and Nikos G. Tsagarakis*

*Advanced Robotics Department (ADVR)
Istituto Italiano di Tecnologia, Via Morego 30
16163 Genova, Italy
†School of Electrical and Electronic Engineering
The University of Manchester, Manchester M13 9PL, UK
‡giuseppe.rigano@iit.it

The rapid advances in robotics have recently led to the developments of a wide range of robotic platforms that exhibit significant differences at the hardware components level. Consequently, this poses a significant challenge to robot software developers since they have to know how every hardware device in the robot works to ensure their software's compatibility when transferring/reusing their code on different robots. In this paper we present a new Robot Hardware Abstraction Layer (R-HAL) that permits to seamlessly program and control any robotic platform powered by the *XBot* control software framework. The implementation details of the R-HAL are introduced. The R-HAL is extensively validated through simulation trials and experiments with a wide range of dissimilar robotic platforms, among them the COMAN and WALK-MAN humanoids, the KUKA LWR and the CENTAURO upper body. The results attained demonstrate in practice the gained benefits in terms of code compatibility, reuse and portability, and finally unified application programming even for robots with significantly diverse hardware.

Keywords: HAL; software architecture; real-time; robot control; robot programming.

1. Introduction

A robotic platform can be considered as a complex distributed system composed of a set of hardware devices communicating through different fieldbus systems and a set of interfaces and protocols of diverse instruction and data fields. The aforementioned complexity is mainly due to hardware-specific operation protocols, hardware distribution and synchronization requirements.[3] As a result, frequently during software development in robotics it is very challenging to develop code, e.g., control software modules, that can be reused with no changes on different platforms. However, efficiently developing control and application software, that can be shared, ported and reused in these diverse range of robotic hardware with minimum effort, is a fundamental requisite for advancing fast the capabilities of these machines.

An important component, needed to achieve this, is the Hardware Abstraction Layer (HAL), which can be incorporated to mask the physical robot hardware differences and limitations (e.g., kinematics model, sensor, update frequency, etc.) varying from one robotic hardware to another. Thus, the robotic software architectural decisions should seriously consider the key role played by HAL for the interaction and the coordination of robot hardware and software control modules. HAL can provide a relatively uniform abstraction that could hide the specifics of the underlying hardware such that the underlying robotic hardware is transparent to the robot control software.[2] Furthermore, HAL assures portability and code-reuse: it efficiently permits robot control software developed for one robot to be ported to another. In one example shown in the Ref. 2, the HAL permits the same navigation algorithm to be ported from a wheeled robot to a humanoid legged robot.

In a robotic platform different fieldbus systems can be employed to permit communication among the robot hardware devices. *CAN* bus is one example. Among the others, the iCub humanoid from Ref. 4 and the HyQ quadruped from Ref. 7 are exploiting this protocol to communicate with the actuators' control boards. Nevertheless, the *CAN* bus limits the data throughput and has higher roundtrip (reaction) time with respect to Ethernet-based solutions.[18]

Even though *Ethernet* (defined in IEEE 802.3) is nondeterministic and thus is unsuitable for hard real-time (RT) applications due to its random delays and potential transmission failures, it is used in several robotic platforms, e.g., on COMAN humanoid shown in Ref. 6 or the recent upgrade of the iCub, called iCub2.[5] An example of a fieldbus system with RT communication capabilities is *EtherCAT* (Ethernet Control Automation Technology), an industrial protocol built on the *Ethernet* specifications. It assures high transmission rate, minimum roundtrip time with respect to other industrial protocol and precise synchronization ($\ll 1\,\mu s$), and demonstrates

‡Corresponding author.

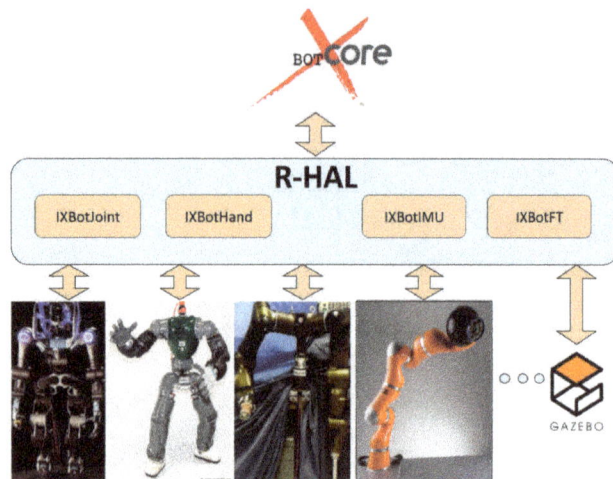

Fig. 1. The R-HAL assures high flexibility towards any type of robotic platform or simulation environment.

flexible topologies, easy configuration, implementation and cost effectiveness.[16] WALK-MAN humanoid robot described in Ref. 17 is using the *EtherCAT* technology.

On the basis of the above-mentioned considerations, the *XBot* RT software platform in Ref. 1 was designed only for *EtherCAT*-based robots. However to support a wider variety of robotic systems it is essential to incorporate a Robot Hardware Abstraction Layer (R-HAL) interface inside the framework.

In the work presented herein we will show the design choices taken for the implementation of R-HAL and the capabilities of this software component, which enable us to efficiently port and run the same control software modules on different robots, both on simulation and on the real hardware platforms. The rest of the paper is organized as follows: Section 2 presents the related works, while Sec. 3 gives a detailed description of the proposed HAL. Section 4 presents the experimental results. Finally Sec. 5 addresses the conclusions.

2. Related Works

In computer science, the *separation of concern* principle followed by the modern software design patterns has been a relevant topic even before the object-oriented programming (OOP) languages. The main goal is to enhance the reuse of the code using abstractions of real objects and hiding implementation details. Although programming by interface is a good practice to use for any type of software development, the need of abstraction layer is strong when interfacing with hardware devices because of the several communication protocols.

For the above reasons, the *Object Management Group* in Ref. 8, not-for-profit technology standards consortium, last year announced the adoption of the *Hardware Abstraction Layer for Robotic Technology* (*HAL* 4 *RT*) from Ref. 9 as an open standard for the implementation of robotics and control software systems. By standardizing on HAL4RT it will be possible to port or reuse drivers on different robotics hardware.

The approach used by some of the RT robotic frameworks follows the use of making abstraction of the low-level analog and digital I/O devices and exploiting the polymorphism to change the specific implementation at runtime.

The *OROCOS* framework shown in Ref. 19 relies on the Device Interfaces[a] as a means to achieve hardware abstraction modeling the concept of low-level analog/digital I/O as well as a bit upper level axis interface. However it requires more effort from the user side in order to interface an entire robot.

EEROS from Ref. 10, an industry-ready open-source RT robotics software, uses a similar methodology trying to wrap each specific hardware libraries using a configuration file to choose the low-level I/O parameters. An interesting feature of the EEROS framework is the safety layer embedded in HAL that allows to handle the safety behavior of the robot in emergency situations.

A different approach is used by *PODO* in Ref. 11, where the hardware abstraction is done at the level of the robot using a shared memory mechanism. Nevertheless no mention about interfacing different robot hardware has been found in their work.

The *OpenRTM-AIST* RT framework in Ref. 12 has the goal to build an independent middleware in order to improve the reusability of software on the independent platform. It uses a different approach, each hardware component is seen as an RT component with its internal state machine that handles the component's life cycle. The communication between components follows the publisher/subscriber model and it is realized using the data port.

Similar consideration to the *OROCOS* framework can be done for the not real-time (NRT) *YARP* middleware shown in Ref. 20, where the user has to implement a specific driver for each device family.

ROS, the popular component-based NRT framework described in Ref. 21, aims to provide a really high-level hardware abstraction by means of the *ROS-CONTROL*[b] packages. The robot hardware interface is designed in such a way to move the integration effort to the driver level and let the user easily read the state of the robot and send commands.

The RT *XBotCore* platform was initially designed to provide a bland-level robot hardware abstraction just for EtherCAT-based robot, making really difficult interfacing with any new robot hardware. Despite the EtherCAT abstraction, *XBotCore* was highly tied to the low level with no autonomous threading capability, in the sense that it relied on the custom low-level software to become part of the thread loop.

---

[a]http://www.orocos.org/stable/documentation/rtt/v2.x/doc-xml/orocos-components-manual.html#idp41997920.
[b]http://wiki.ros.org/ros_control.

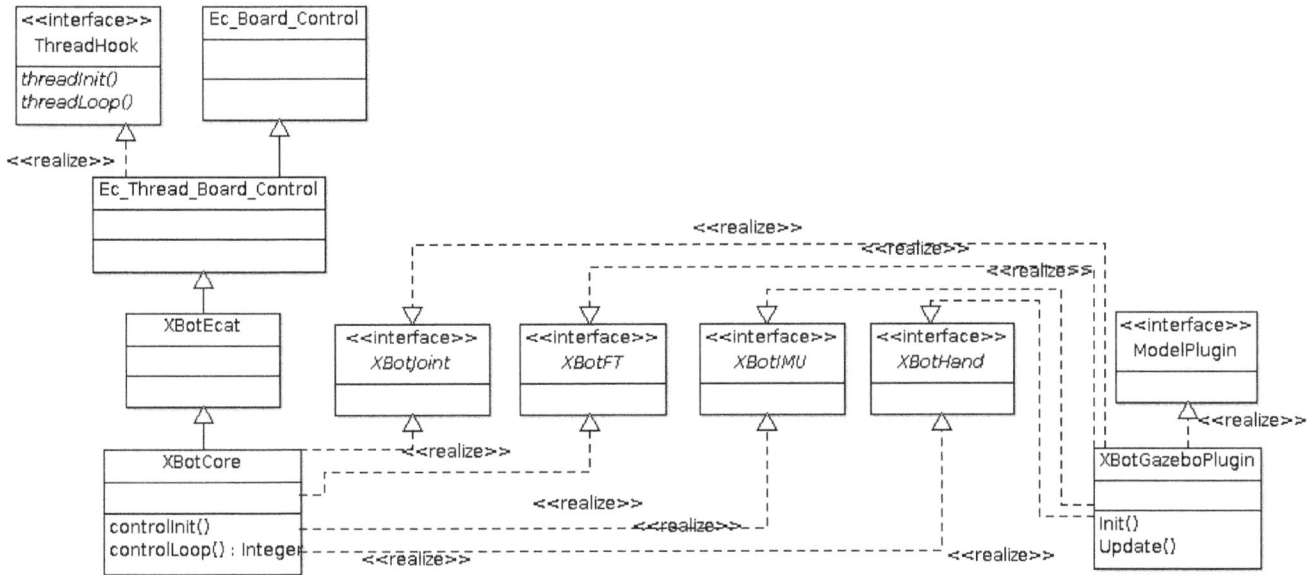

Fig. 2. Initial *XBot* software design as UML class diagram.

The proposed work tries to overcome the aforementioned issues making a really flexible system and improving software reusability. To achieve that, a strategy similar to ROS has been adopted allowing to interface easily with the code already written for ROS and as a result the advantage of having a choice between *RT* and *NRT* control system. However it is important to note that the RT capability depends on the low-level hardware and software layer.

3. R-HAL

In this section we will discuss both the initial and the current design choices of the *XBotCore* platform trying to underline pros and cons by comparing the two software designs.

3.1. *The former XBotCore design*

As mentioned in the previous section, the initial design of the *XBotCore* was not ready to easily plug a new robot hardware. Figure 2 shows the former software architecture where four main elements can be analyzed:

- XBotEcat, responsible to interface the low-level software control;
- XBotCore, designed to provide the implementation of the interfaces;
- Interfaces representing Joint, IMU, force–torque (FT) sensor, Hand;
- GazeboXBotPlugin, aims to emulate the XBotCore component acting as a Gazebo[c] simulator model plugin.

The above-described architecture allows to reuse part of the code but it is not completely optimal for a full reuse leading

to a code replication inside the *GazeboXBotPlugin* and in any new concrete implementation. Moreover the UML class diagram shows the lack of any hardware abstraction layer, since only the *XBotEcat* implementation is provided as an endpoint to the low-level control.

The described solution makes the *XBotCore* class an embedded component inside the low-level control software relying on the Template Pattern of the base class in order to be part of the control loop. The resulting system architecture has almost zero complexity but the flexibility issue increases the effort the user has to carry out to support any new hardware.

3.2. *The R-HAL approach*

The purpose of the new software design is to provide a middleware independent from the low-level hardware/software with autonomous threading capability.

The UML class diagram in Fig. 3 shows the software architecture with the R-HAL approach. In order to keep the UML diagram cleaner we decided not to show the list of methods inside each interface. For example the *XBotJoint* interface contains all the abstract methods to get and set the information related to the joints.

The core component is the *HAL* interface responsible to provide an endpoint to the low-level hardware/software layer in such a way to be flexible towards any new robots.

The HAL achieves its job by providing three abstract methods that each concrete implementation has to implement.

In detail, adopting a master–slave approach, the R-HAL life cycle is described by the following methods:

- *init*( ), useful in the initialization phase (open the connection, initialization of the data structures);

[c]http://gazebosim.org/.

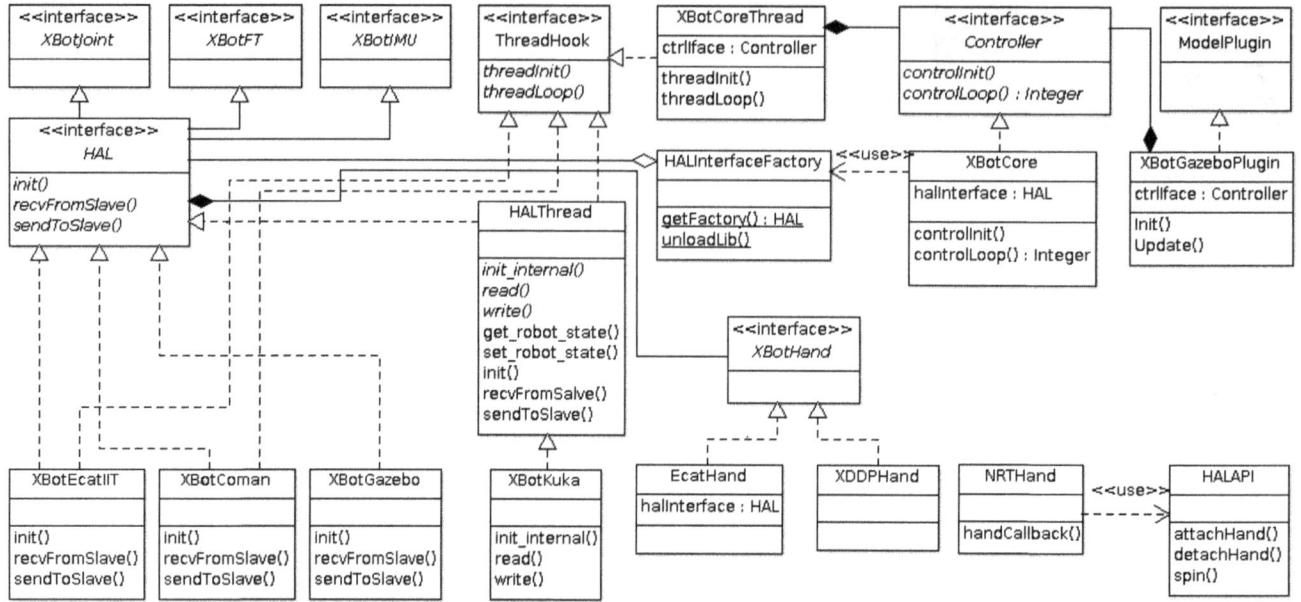

Fig. 3. R-HAL *XBotCore* software design as UML class diagram.

- *recvFromSlave( )*, designed to read all the data coming from the different types of slaves (joint, IMU and force–torque sensor) and fill the internal data structures;
- *sendToSlave( )*, deals with the communication to send the reference signals to the joint slaves.

The optional hand actuator is seen as a particular case of a joint where higher-level operation like grasping can be defined.

All the R-HAL implementations must provide the behavior for the aforementioned methods in addition to the interfaces that deal with joint, inertial measurement unit sensors and force–torque sensors. Optionally it is possible to add the type of the R-HAL implementation allowing the user to distinguish between simulation and real robot by just checking it at runtime.

R-HAL methods *init*, *recvFromSlave* and *sendToSlave* do not take any parameters because they rely on the internal shared data structures. Whoever wants to create a new R-HAL implementation has to deal with the creation of the suitable data structures. Moreover it is up to the user to handle the synchronization between threads if needed.

The current design leaves the user completely free regarding the integration of any new robot. Our design choice has been to have a separate thread handling the low-level robot control loop. The synchronization between the *XBotCore* thread and the *R-HAL* thread is needed to access safely the shared data structures.

Dealing with concurrent programming is very tedious and error-prone because advanced expertise is needed in order to implement the right synchronization pattern. We need to address that the built-in *HALThread* class is provided. The goal is to mask all the details about the thread synchronization giving to the user just two synchronized methods to get

and set the robot state (position, velocity, torque, stiffness and damping). Once more, the behavior for the initialization, reading and writing must be provided. It is important to note that all those methods get called inside the R-HAL thread.

By exploiting the model dynamic API mentioned in Ref. 1 it is possible to automatically shape the layout of the data structures by reading the correspondent *URDF* file related to the robot we want integrate inside the framework. This has been achieved through the use of the map data type.

We provide the implementations for handling EtherCAT (WALK-MAN, CENTAURO), Ethernet (COMAN) and KUKA LWR 4-based robots. Moreover the Gazebo R-HAL implementation allows to test the behavior in simulation.

The described design does not limit the user to add new behavior because it is just required to implement other set of interfaces. In particular the need to handle legacy code leads the EtherCAT and Ethernet implementations to not using the *HALThread* class. Although the same process could be used to integrate a new robot it would be easier to rely on the helper class provided. That is what has been done for the KUKA robot in such a way to exploit the model dynamic API.

All the R-HAL implementations are built as shared library loaded at runtime according to what was specified in a configuration file. In particular, the factory design pattern has been adopted to load/unload several implementations. The *XBotCore* class will load the specific R-HAL implementation getting the factory from the *HALInterfaceFactory* class where the symbols are resolved immediately to avoid slowdowns during the RT execution.

The *ThreadHook* class gives the threading capability such that it is possible to realize the control loop. The *XBotCore* class acts as a specific controller implementation and it interacts with the *HAL* by calling the suitable methods as shown in

```
void XBot::XBotCore::control_init(void)
{
    halInterface->init();
    init_internal();

    return;
}
int XBot::XBotCore::control_loop(void)
{
    int state = halInterface->recv_from_slave();
    if(state == 0)
      loop_internal();
    halInterface->send_to_slave();
}
```

Fig. 4. *XBotCore*–R-HAL components interaction.

```
if(_xbot_joint->get_link_pos(joint_id, aux)) {
    _link_pos[joint_id] = aux;
}

if(_xbot_joint->get_link_vel(joint_id, aux)) {
    _link_vel[joint_id] = aux;
}

if(_xbot_joint->get_torque(joint_id, aux)) {
    _torque[joint_id] = aux;
}
```

Fig. 5. *XBotCore*–Interfaces interaction.

Fig. 4. The controller can easily access to the joint, FT and IMU by just invoking the appropriate methods provided by the interfaces as partially shown in Fig. 5.

The task of *XBotCoreThread* class is to initialize and run the control loop of the specific controller implementation. The described abstraction layer allows to easily switch between simulation and real robot as shown in Fig. 1. The *XBotGazeboPlugin* class emulates the same behavior of the *XBotCoreThread* class relying on the *ModelPlugin* class to be part of the Gazebo internal loop.

The advantages of the presented architecture are evident and despite the high flexibility provided it does not increase considerably the complexity of the software design.

3.3. *Mixing RT–NRT controlled hardware*

The software architecture presented in the previous subsection is well designed in the scenario in which we are supposed to have either a fully RT or fully NRT environment. However how to mix NRT and RT controlled hardware is a requirement to be considered. One possible example of the aforementioned condition could be interfacing an RT hardware robotic arm with an NRT robotic end-effector (hand). A solution that exploits the Cross-Domain Datagram Protocol pipes (XDDP pipes) from *Xenomai*[d] (required for the *XBot*

[d]https://xenomai.org/.

RT architecture) has been adopted. The design idea is to have a mapping between the joint ID and *IXBotHand* interface to load at runtime the specific RT-based or NRT-based end-effector (hand) control implementation. The XDDP pipes are used to have publisher and subscriber towards the NRT side. The helper class *HALAPI* will be used in a generic NRT process to realize the communication and as a result the possibility to integrate end-effectors with different hardware and software capabilities.

4. Experiments

To validate and evaluate the performance of the XBotCore software platform, we performed a set of experiments on different robotic platforms as described in the following subsection. Moreover a comparison between the former software design and the R-HAL-introduced design is provided.

4.1. *Experimental setup*

The proposed software architecture has been validated on the following robotic platforms (Fig. 6):

- WALK-MAN robot, a full-size humanoid with 33 degree-of-freedoms (DOFs), four custom F/T sensors and one IMU. The WALK-MAN head is equipped with a CMU Multisense-SL sensor that includes a stereo camera, a 2D rotating laser scanner and an IMU.
- CENTAURO upper body prototype has 15 DOFs in Ref. 15. Each arm has seven DOFs. For the trunk there is an additional one DOF that permits the yaw motion of the entire upper body and extends the manipulation workspace of the robot. Two custom F/T sensors are placed at the wrists.
- COMAN robot, a 95-cm tall humanoid robot with 29 DOFs. Custom torque sensors are integrated into every joint to enable active stiffness control and six-DOF sensors are included at the ankle to measure ground reaction forces.

Fig. 6. Robotic platforms used during the experiments: on top left corner KUKA LWR 4+, on bottom left corner COMAN, on the center WALK-MAN robot and on the right CENTAURO upper body.

- KUKA LWR 4+, an industrial seven-axis jointed-arm robot. Each joint is equipped with a position sensor on the input side and position and torque sensors on the output side. The robot can thus be operated with position, velocity and torque control. Data transfer between the robot controller and an external computer is carried out using the Fast Research Interface (FRI).

All the robotic platforms were tested both on the Gazebo environment and on the real hardware by running an *XBot* plugin that performs a circular trajectory. Thanks to the dynamic API the experiments were executed in all the robots without code modification but just loading the configuration file for the specific robot platform. In particular, the OpenSoT control framework from Refs. 13 and 14 was used to solve whole-body control and inverse kinematics.

The software has been tested on Linux OS 3.18.20 with-Xenomai-ipipe-2.6.5 running on COM Express i7 CPU @ 2.30-GHz quad-core for WALKMAN and CENTAURO while the COM Express i7 CPU @ 1.7-GHz dual-core was used for COMAN and KUKA arms. RTnet, an open source hard real-time network protocol stack for Xenomai and RTAI, was used assuring an implementation of the UDP/IP, TCP/IP (basic features), ICMP and ARP in a deterministic way.

It is important to mention that very few changes have been done on the FRI API provided by KUKA in order to have the real-time capabilities. In particular the POSIX Xenomai skin has been used to easily wrap the UDP communication as real-time invocation. This feature is available for any kind of UDP robot leading to a successful reuse of the code. RTnet was used to load the deterministic version of UDP kernel module.

4.2. *Results*

In Fig. 7 we present the results of the R-HAL overhead analysis in terms of control period. We report the data of the

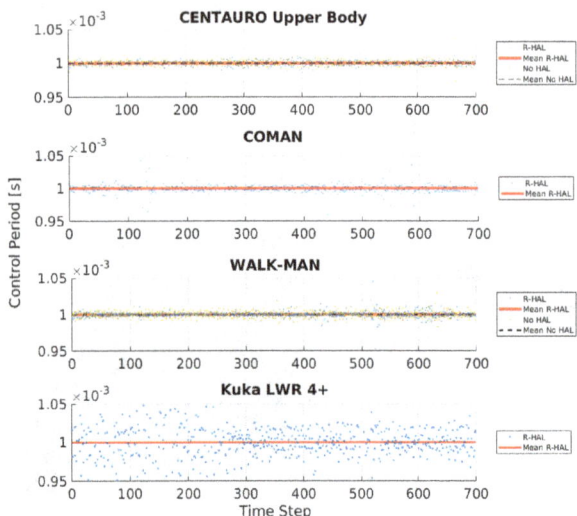

Fig. 7. Control periods comparison between the XBot architectures with R-HAL and without HAL.

Table 1. Single-core mean CPU usage (%).

	No HAL	HAL
CENTAURO upper body	9.3 ± 1.4	9.4 ± 1.9
WALK-MAN	18.2 ± 1.7	18.5 ± 1.5
COMAN	n.d.	12.2 ± 2.3
KUKA	n.d.	3.9 ± 1.0

experiments on the four robotic platforms described in Sec. 4.1. It is important to note that we requested a RT control frequency of 1 kHz (i.e., 1 ms control period) for all the robots. It is evident that no overhead has been introduced with R-HAL compared to the former XBot software design (no HAL) regarding the EtherCAT-controlled robots (CEN-TAURO, WALKMAN).

To complete our analysis, in Table 1, we show a comparison of the mean CPU usage times for the XBot platforms with and without R-HAL. The outcome is that R-HAL does not introduce any processing overhead.

5. Conclusions

The extensive uprise in robotic hardware of different forms is imposing a significant challenge to the robot software developers when they have to deal with robotics systems with hardware incompatibility making code development, porting and reuse highly inefficient. To address this barrier, hardware abstraction layers are necessary to mask the robot hardware variances and the associated hardware-dependent instructions to the robot from the application/control software developers, enabling efficient and unified code development and reuse among different robot hardware.

Towards this direction, we presented the concept of R-HAL, a new robot hardware abstraction layer as a way to improve the reuse of the code among different hardware robotic platforms. In particular, we leveraged on the power of the *XBot* software framework to develop and validate R-HAL on a range of different robotic platforms. A comparison between the initial software design and the new one has been performed showing the advantages and disadvantages of both architectures. The autonomous threading capability together with the R-HAL interface allows to easily adopt the *XBot* framework as a middleware for several different robots by exploiting the power of the software polymorphism. Moreover an example of integrating RT and NRT domains to support hybrid RT–NRT hardware has been shown leading to a very high flexibility, towards any type of low-level interface. The results demonstrate that the change in complexity does not affect the CPU load maintaining the desired execution bandwidth (control period, RT–NRT) in all robotic platforms.

Future work will focus on the design of a safety layer embedded in R-HAL by applying hardware-specific capability constraints to the generated command signals and defining the robot-specific workspace for safe human–robot collaboration. Furthermore, R-HAL will be validated on other research and industrial robotic platforms to rigorously evaluate its benefits. The XBot software platform is released a free and open source.[e]

Acknowledgments

The research leading to these results has received funding from the European Union Seventh Framework Programme [FP7-ICT-2013-10] under the Grant Agreement No. 611832 WALKMAN and European Unions Horizon 2020 Research and Innovation Programme under the Grant Agreement Nos. 644839 (CENTAURO) and 644727 (CogIMon). The authors would like to thank Luka Peternel for the support provided during the experiments.

References

[1] L. Muratore, A. Laurenzi, E. Hoffman, A. Rocchi, D. Caldwell and N. Tsagarakis, XBotCore: A real-time cross-robot software platform, *Proc. IEEE Int. Conf. Robotic Computing* (2017).

[2] T. J. Murray, B. N. Pham and P. Pirjanian, Hardware abstraction layer for a robot, US Patent 6889118 (2005).

[3] S. Jörg, J. Tully and A. Albu-Schffer, The hardware abstraction layer: Supporting control design by tackling the complexity of humanoid robot hardware, *Proc. IEEE Int. Conf. Robotics and Automation (ICRA)* (2014).

[4] G. Metta, G. Sandini, D. Vernon, L. Natale and F. Nori, The iCub humanoid robot: An open platform for research in embodied cognition, *ACM Proc. 8th Workshop Performance Metrics for Intelligent Systems* (2008).

[5] A. Parmiggiani, M. Maggiali, L. Natale, F. Nori, A. Schmitz, N. Tsagarakis, J. Victor, F. Becchi, G. Sandini and G. Metta, The design of the iCub humanoid robot, *Int. J. Humanoid Robot.* **9**, 1250027 (2012).

[6] N. G. Tsagarakis, S. Morfey, G. Cerda, L. Zhibin and D. Caldwell, Compliant humanoid COMAN: Optimal joint stiffness tuning for modal frequency control, *Proc. IEEE Int. Conf. Robotics and Automation (ICRA)* (2013).

[7] C. Semini, N. G. Tsagarakis, E. Guglielmino, M. Focchi, F. Cannella and D. G. Caldwell, Design of HyQ: A hydraulically and electrically actuated quadruped robot, *Proc. Inst. Mech. Eng. I, J. Syst. Control Eng.* **225**, 831 (2011).

[8] OMG, Object Management Group (2017), http://www.omg.org/.

[9] OMG- JASA, Hardware Abstraction Layer for Robotic Technology (HAL4RT) (2016), http://www.omg.org/spec/HAL4RT/1.0/Beta1/PDF/.

[10] EEROS, EEROS Wiki (2017), http://wiki.eeros.org/.

[11] J. Lim, I. Shim, O. Sim, H. Joe, I. Kim, J. Lee and J.-H. Oh, Robotic software system for the disaster circumstances: System of Team KAIST in the DARPA Robotics Challenge Finals, *Proc. Int. Conf. Humanoid Robotics* (2015).

[12] A. Noriaki, S. Takashi, K. Kosei, K. Tetsuo and Y. Woo-Keun, RT-Middleware: Distributed component middleware for RT (Robot Technology), *Proc. IEEE/RSJ Int. Conf. Intelligent Robots and Systems* (2005).

[13] A. Rocchi, E. M. Hoffman, D. G. Caldwell and N. G. Tsagarakis, OpenSoT: A whole-body control library for the compliant humanoid robot COMAN, *Proc. IEEE Int. Conf. Robotics and Automation (ICRA)* (2015).

[14] E. M. Hoffman, A. Rocchi, A. Laurenzi and N. G. Tsagarakis, Robot control for dummies: Insights and examples using Open-SoT, *Proc. 17th IEEE-RAS Int. Conf. Humanoid Robotics (Humanoids)* (2017).

[15] L. Baccelliere *et al.*, Development of a high performance bi-manual platform for realistic heavy manipulation tasks, *Proc. 2017 IEER/RSJ Int. Conf. Itelligent Robots and Systems* (2017).

[16] EtherCAT Technology Group, EtherCAT technology introduction (2012), https://www.ethercat.org/en/technology.html.

[17] N. G. Tsagarakis *et al.*, WALK-MAN: A high performance humanoid platform for realistic environments, *J. Field Robot.* **34**, 1225 (2016).

[18] S. Potra and G. Sebestyen, EtherCAT protocol implementation issues on an embedded Linux platform, *Proc. IEEE Int. Conf. Automation, Quality and Testing, Robotics* (2006).

[19] H. Bruyninckx, OROCOS: Design and implementation of a robot control software framework, *Proc. IEEE/RAS-EMBS Int. Conf. Biomedical Robotics and Biomechatronics* (2002).

[20] G. Metta, P. Fitzpatrick and L. Natale, YARP: Yet another robot platform, *Int. J. Adv. Robot. Syst.* **3**, 8 (2006).

[21] M. Quigley, K. Conley, B. P. Gerkey, J. Faust, T. Foote, J. Leibs, R. Wheeler, A. Y. Ng and L. Natale, ROS: An open-source Robot Operating System, *Proc. ICRA Workshop Open Source Software* (2009).

[e] https://github.com/ADVRHumanoids/XBotCore.

Robotic intelligence and computational creativity

Agnese Augello, Ignazio Infantino, Umberto Maniscalco,
Giovanni Pilato*, Riccardo Rizzo and Filippo Vella

ICAR-CNR, Italian National Research Council
Via Ugo La Malfa 153, 90146 Palermo, Italy
*giovanni.pilato@cnr. it

The paper illustrates a cognitive architecture for computational creativity based on the Psi model and the mechanisms inspired by dual-process theories of reasoning and rationality. In particular, three applications of computational creativity will be summed up: (a) a robot capable of executing creative paintings through a multilayer mechanism that implements an associative memory and is capable to properly mix elements belonging to different domains; (b) a robot aimed at producing a collage formed by a mix of photo-montage and digital collage: the artwork is created after a visual and verbal interaction with a human user; and (c) a humanoid robot capable of improvising a dancing choreography in real-time according to the listened music. First results in computational creativity show a set of potentialities to be explored that can shed light on human and artificial creativities and artificial intelligent systems.

Keywords: Robotic intelligence; computational creativity; cognitive architectures.

1. Introduction

Creativity is an inherent human aptitude that is essential for the development and invention of new ideas and concepts; it is generally defined as the ability to establish novel and high-quality associative solutions. The capability associating tasks, concepts, ideas, knowledge and experiences in a pertinent manner is usually considered among the most important factors when considering creativity.[1]

However, according to Boden, creativity is more than just novelty-producing thought: it is related to novel exploration and creation of mental representations. It requires a great amount of knowledge, conceptual thinking, the ability to make associations, problem-solving and reflective self-criticism.[2] It is a mysterious, high-level cognitive function aimed at exploring and finding novel and valuable solutions for a given problem.

Mednick[3] defines creativity as "the forming of associative elements into new combinations, which either meet specified requirements or are in some way useful."

On the other hand, Boden claims that there are three forms of creativity: a *combinatorial* one, making unfamiliar combinations, that can be generated either deliberately or unconsciously; an *exploratory* one, where new ideas are generated by means of an exploration of concepts in a conceptual space; and a *transformational one*: in this last case the idea is generated by transforming existing conceptual spaces.

Conceptual spaces for Boden are "structured styles of thought which are not originated by individuals, but they are included in the person's culture or they are sometimes taken from other cultures.

Among the first and foremost theories regarding creative processes, we can mention that one of Wallas[4] who identifies four stages of creativity: *preparation*, in which the problem is examined and investigated in all directions; *incubation*, i.e., voluntarily avoiding to think about the problem; *illumination*. i.e., when the novel idea spontaneously arises in mind; and *verification*, i.e., the conscious effort to check the actual validity of the new idea.

Rhodes[5] suggested to focus the attention on four "P"s namely person, process, place and product: (1) *person*: the person who creates, (2) *process*: the cognitive procedures involved in the creative process, (3) *place*: the environment, organizational culture, resources and best practices in which creativity occurs and (4) the *product* resulting from creative activity.

Moreover, among the first theories of creative process, Guilford[6] coined a distinction between convergent and divergent thinking, i.e., processes that induce a single solution and thinking activities that generate multiple possibilities.

Boden tried to formulate the creativity process into a computational framework.[2] Nowadays, the scientific field dealing with creativity comprises also a new branch named computational creativity[7,8] aimed at modeling and designing artificial system emulating to some degree the creativity process typical of human beings. Computational creativity can be also useful for understanding human creativity or it can be considered as a tool to support human beings in creative processes.

Creativity requires the application of basic cognitive processes to existing knowledge structures,[9] but understanding them is a great challenge, and this is the principal problem

*Corresponding author.

that makes very difficult the definition of a computational creativity system.[10]

At least three distinct groupings of creative domains can be identified in characterizing human creativity[11]: *expressive creativity* typically involving visual arts, writing, etc.; *performance creativity* that includes dance, drama, music, etc.; and that referred as to *scientific creativity* embracing the arising of inventions and generally the science field.[12]

Computational creativity is applied to many domains like visual arts,[13,14] music,[7,15–20] language,[21,22] poetry,[23,24] humor,[25,26] narrative,[27,28] mathematics[29] and cooking.[30,31,12]

Some authors have tried to generalize these approaches to an abstract theory of computational creativity.[8,32,33] Ventura in Ref. 34 also makes considerations about computational creativity in the context of the theory of computability. In Ref. 35 a conceptual framework for a philosophical discussion on the status of machine art and machine creativity has been introduced. In Ref. 36 artistic creativity is modeled as an iterated turn-based process, alternating between a conceptual representation and a material representation of the work-to-be. In Ref. 37 a computational framework for conceptual blending,[38] which in cognitive science is claimed as a fundamental ingredient for creative thinking, has been presented.

A cognitive model proposed in Ref. 39 considers the transformations that occur as the creative process is triggered and proceeds. After an initial phase which triggers an information retrieval process from memory, there is an analytical phase characterizing a focused form of thought which analyzes relationships of cause and effect.[40]

One of the main issues in the field of computational creativity is regarding the evaluation of the creative act. The evaluation of the creative process first is an internal assessment made by the artwork creator, and subsequently, it can involve an external judgment, given by an audience or a teacher.

Dealing with this issue, computational creativity researchers try to outline what are the elements to consider in the evaluation. As an example, Jordanous[41] indicates some key components which can be more easily managed in an artificial context, like emotional involvement; generation of results; domain competence; originality; social interaction and communication; variety, divergence and experimentation; value and evaluation.

Some computational creativity systems have a sort of self-evaluation mechanism that can enable the comparison between their artworks and some expected result, of course, external evaluation is more easy to obtain by analyzing feedbacks from users concerning some creative feature, such as those mentioned before. The evaluation results can be then used by the computational system to modify and improve the creative process. Cognitive models of the human mind can find common ground and experimental validation exploiting robotics, especially considering that recent robotic systems are available at a reduced cost and with increased performances. Researchers can integrate into robot's cognitive

architectures also the aspects relating to creativity, experimenting on different artistic domains such as dance, music, drama and painting. Among the possible cognitive architectures that can be exploited to this aim, the Psi one,[42,43] is particularly suited for the task, since it involves the concepts of emotion and motivation in cognitive processes, which are two important factors in creative processes.

An artificial creative agent should exhibit cognitive capabilities such as personality, habits, attitudes, etc., which are typical of a human being. The production of an artwork is the outcome of a creative act that depends on the experience and expertise of the artist, is also strongly influenced by motivation, emotions, learning and communicative capabilities. An agent, interacting with the environment, stores in its memory the most salient aspects of its perceptions. The stored concepts can be recalled when the agent experiences something similar, also if the new experience involves different sensing quantities. Products are the tangible results of ideas, and they are characterized by the media of expression, by the emotional impact on the subject who experiences the artwork and by their intrinsic value (e.g., aesthetic, utility and so on).

In this paper, we discuss a cognitive architecture for computational creativity inspired by the Psi model. In particular, three applications of computational creativity will be summed up: (a) a robot capable of executing creative paintings through a multilayer mechanism that implements an associative memory and is capable to properly mix elements belonging to different domains and whose creative process is inspired by dual-process theories of reasoning and rationality; (b) a robot aimed at producing a collage formed by a mix of photo-montage and digital collage: where the robot takes inspiration from a verbal interaction with a human user and the collection of Web items that she/he has published on his social network profile. The inspiration phase is strongly influenced by the semantic associations between the humanoid perceptions and the elements of the creative language of the artist: and (c) a humanoid robot capable of improvising a dancing choreography in real-time according to the listened music.

2. Cognitive Architectures and Computational Creativity

Robotic software architecture is a complex system that manages various aspects of an embodied artificial system. In the case of a creative robot, the design of such software structure has to take into account different levels of abstraction to manage its knowledge representation.

As previously stated, human creativity is both a mental process and the realization of a valuable outcome. Inspiration, learning, technique, culture, emotions and many other factors influence this human capability. To have the possibility to reproduce such mechanisms, the use of a cognitive architecture is desirable and could assure satisfying results.

A simplified approach to the artificial creativity could be to implement an algorithm able to produce random outcomes to explore the space of the possible results, and hoping to find a good one under a given *aesthetic* criteria. But without a *creative process* based on various human cognitive capabilities, the artificial creative system is subject to criticism, and it is not credible. In particular, the key aspects to highlight in such process are related to evaluation mechanisms. An internal evaluation process could be derived both from owned *artistic sensibility* and acquired *experience*. An external evaluation should cause the evolution of the behavior of the creative agent by considering the judgment of the humans (for example, a teacher, a critic or the final audience).

Dynamic and adaptive knowledge also needs to be incorporated, using internal representations that can take into account variable contexts, complex actions, goals that may change over time and capabilities that can extend or enrich themselves through observation and learning. On a broader perspective, the mental capabilities[44] of artificial computational agents can be introduced directly into a cognitive architecture or emerge from the interaction of its components. The approaches presented in the literature are different and range from cognitive testing of theoretical models of the human mind to the robotic architecture based on perceptual-motor components and purely reactive behaviors.[45,46] Currently, cognitive architectures have had little impact on real-world applications and limited influence in robotics. The aim and long-term goal is the detailed definition of the artificial general intelligence (AGI).[47] i.e., the construction of artificial systems that have a skill level equal to humans in general scenarios.

To make the cognitive architecture capable of generating behaviors similar to humans — and especially in the case of creativity — it could be essential to consider also the role of emotions. Reasoning and planning may be influenced by emotional processes and their representations as happening in humans. Ideally, this could be thought of as a representation of emotional states that, in addition to influencing behavior, also helps to manage the detection and recognition of human emotions. This fact is in agreement with the hypothesis that human creativity only makes sense when placed in the context of rich social interactions.[48–50]

Moreover, an artificial creative system has to be credible, and it is necessary to have standardized procedures to evaluate its creative behavior. In our analysis, we focus on two complementary approaches: SPECS proposed by Jordanous,[41] and FACE-IDEA proposed by Colton *et al.*[8] SPECS identifies 14 valuable parameters such as uncertainty, domain competence, intention and emotional involvement, originality, progression and development, social interaction, thinking and evaluation. FACE-IDEA aims to define some measurement methods, and an iterative cycle involving audience appreciation.

Taking into account such models, the cognitive architecture has the role to organize the processing flows that

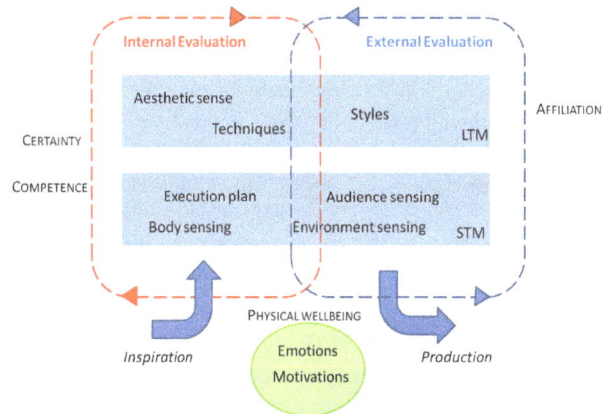

Fig. 1. The relevant entities for defining a creative cognitive architecture considering the evaluation mechanisms.

determine two relevant mechanisms: the internal process responsible for creative thinking and the external influences that determine its development aiming to produce a valuable artifact. The two mechanisms are cyclic and contribute to the development of the artificial creative agent. Figure 1 shows the cognitive creative framework, highlighting some relevant entities in the evaluation cycles. The internal evaluation aims to define an aesthetic sense of the agent, its repertories of techniques and the correct execution of the action required to produce the artifact. Moreover, in the case of a robotic agent (i.e., an embodied physical agent), also the body state could influence the internal evaluation of the creative process.

The parameters of Psi model named certainty and competence are strictly related to the internal mechanism of self-evaluation. The external evaluation aims to improve the styles and the behavior of the agent to perform successfully creative execution in a given environment (or domain). A learning phase is indispensable to determine the evolution of the agent, and a teacher evaluation allows the agent to improve its experience. In fact, the learning phase determines the storing of some entities in the persistent memory [long-term memory (LTM)]. The production combines experience with entities of the working memory (WM) [short-term memory (STM)] to consider the environment, the inner physical state of the robot and the audience reaction. Both internal and external evaluations contribute to determining emotions and motivations and cause variability of the results of production.

In the following sections, we describe three different computational creativity processes that we have defined and implemented according to the above-illustrated approach.

3. The Creative Painter

The creative painting process is based on the dual-process theory of mind.[51–53] This theory suggests that our cognition is governed by two types of interacting cognitive systems: rapid, automatic and associative processes of reasoning, referred to as S1 processes; and conscious, controlled and sequential processes referred to as S2. This approach has been considered plausibly applicable also in the area of creativity research,[54] where the manner in which S1 and S2 cooperate influences the differences in creative capabilities. A model of creativity based on the dual-process theory is the "four-quadrant model."[55] According to this model both S1 (implicit processing) and S2 (explicit processing) may conduct convergent or divergent strategies, to generate new ideas and to converge to the more appropriate one.

We exploited this model to implement the creative process of a computational artist.[56] The artwork creation is aimed at the completion of an image where one or more parts are missing, and it is carried out through phases of exploration, evaluation and, if necessary, reflection and replanning.

Figure 2 shows the four-quadrant model and the possible interactions (highlighted by bold arrows) involving S1 and S2 systems (operating in both convergent and divergent processes). Each process is labeled according to the associated kind of system: for example, an S1 system operating in a divergent process is executing an "Exploratory" task.

According to this model, both S1 (implicit processing), and S2 (explicit processing) may conduct convergent or divergent strategies to generate new ideas (a new solution to a problem), and to converge to the more appropriate solution as illustrated in Fig. 2. S1 can generate new ideas (EXP quadrant) by means of associative, combinatorial or transformational processes that can be therefore evaluated by S2 through an analytic strategy (AN quadrant), but it can also perform a tacit strategy to converge to the more appropriate solution, using knowledge formed in the past to quickly access default "previous best" behavior (TAC quadrant). Moreover, S2 can perform a divergent, reflective strategy to intentionally generate very distant points in solution space (REF quadrant).

The interactions among the four quadrants can contribute to the generation of creative solutions, but that can have inhibition effects, as discussed in another paper.[55] In Fig. 2 we have highlighted the interactions involved in the creative process proposed in our system.

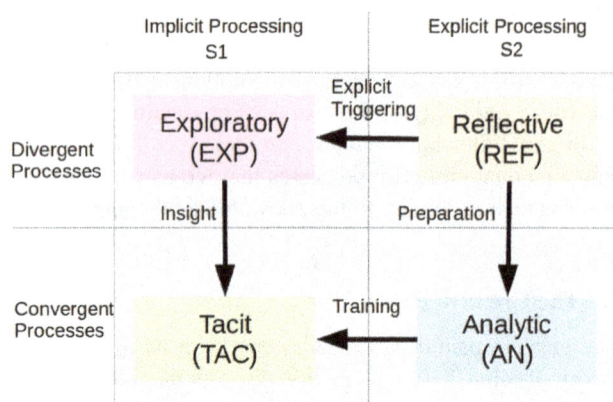

Fig. 2. Four quadrants: possible interactions involved in the creative process are highlighted in the figure.

We also analyzed how these S1 and S2 processes exploit the long-term memory and the working memory of the agent to make novel substitutions of the perceived image, by combining elements from different domains.

The creative process of the agent is based on the interaction of the LTM and WM.

During the training phase, the system learns and properly organizes its knowledge about artistic notions and the main execution techniques into the long-term memory. The term LTM refers to a memory space that stores information for a long time-span (even a lifetime). This memory is recognized in all cognitive theoretical views.[57] In our work, the LTM is an associative memory that stores whole images, image details or features useful for a creative process involving artwork or image creation. This knowledge is aimed at representing image features clustered by subject. Information in this memory is organized in domains, where each domain stores information on objects of the same kind.

The working memory is the memory area used by the running processes. In this memory, the pieces of information are stored for a short or medium time-span. In some cognitive theories, this memory is not separated from the so-called short-term memory.[57] In our model, WM is used to temporarily store the associations among separated domains. These associations come from the similarity of two pieces of information in separated domains. The WM is built by using a large SOM that clusters all the image details and features in all the domains. This will enable the substitution and exploration mechanism that is in the background of the creative process under consideration.

Let us consider that the agent has been trained with images belonging to two domains: the domains of faces and the flowers, and it has to complete an image representing a face. The agent can be more focused on a classical way to operate, deciding to complete the image with the most appropriate part, belonging to the face domain. But if, according to what emerges from its cognitive status (for example the set of urges), it is more inclined to find novel solutions, it can decide to relax a sort of focus of attention,[58,59] to explore other domains, such as the flowers domain, to retrieve different ideas that can fit with the rest of the image (this scenario is inspired by the well-known Arcimboldo paintings). The substitution can be obtained by using WM and LTM, as shown in Fig. 3. The WM creates a representation of the missing piece of the image, by clustering it and retrieving the suitable cluster center. This representation is used to retrieve, from the flowers domain of the LTM, a set of image pieces, candidate for the replacement.

What is obtained by this replacement can therefore be evaluated by S2 through an analytic strategy (AN quadrant), by also controlling how much detailed should be the check of the result obtained by the S1 process from the S2 system. AN monitors the mental execution of the artwork using the chosen association; for example, it can use a face recognition algorithm to check the presence of a recognizable face in the

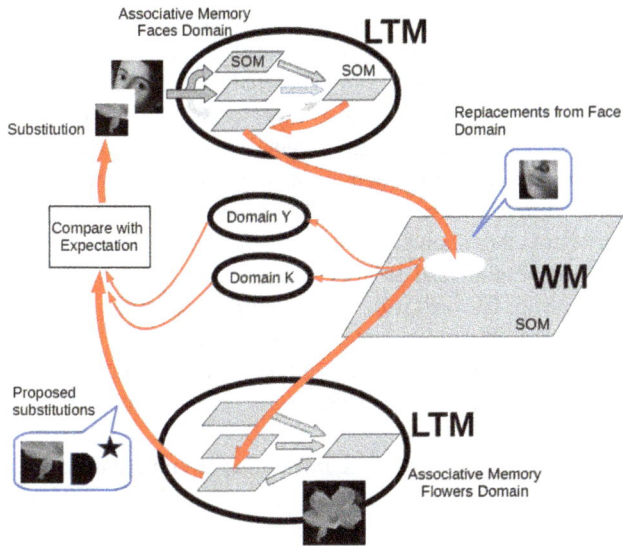

Fig. 3. The completion procedure outside the domain.

result of the substitution. This can be part of a preparation phase that develops reflective thinking. If the check returns a positive result, the system produces the painting. Otherwise, it can choose to perform a reflective strategy, a change of association domain to satisfy the internal model of the agent with respect to an internal model of the desired output. Moreover, S2 can perform a reflective strategy to

intentionally decide what points to consider in solution space (REF quadrant).

It can also choose to perform a tacit strategy to converge to the more appropriate solution, using knowledge formed in the past to quickly access default "previous best" behavior. In this case, the exploration is restricted to the more appropriate domain (TAC quadrant). The tacit strategy derives from a training process performed by analytics process. The possible interactions among the four quadrants can therefore contribute to the generation of different creative solutions.

4. Collages Creation Inspired by Human–Robot Interactions

Another application of the proposed architecture concerns the creation of artworks formed by a mix of photo-montage and digital collages. A robot, in particular, an NAO Aldebaran, takes inspiration from postural and verbal interactions with a user and his social web items to produce the collage. During the conversation, implemented using a chatbot module, the robot can find semantic associations between what the user says or posted in his social profile and the pictorial elements that are stored in its long-term memory. In particular, this "pictorial" knowledge is composed of a set of basilar sketches and artistic styles, which constitutes its basic repertoire and other pictorial elements provided by the user during the

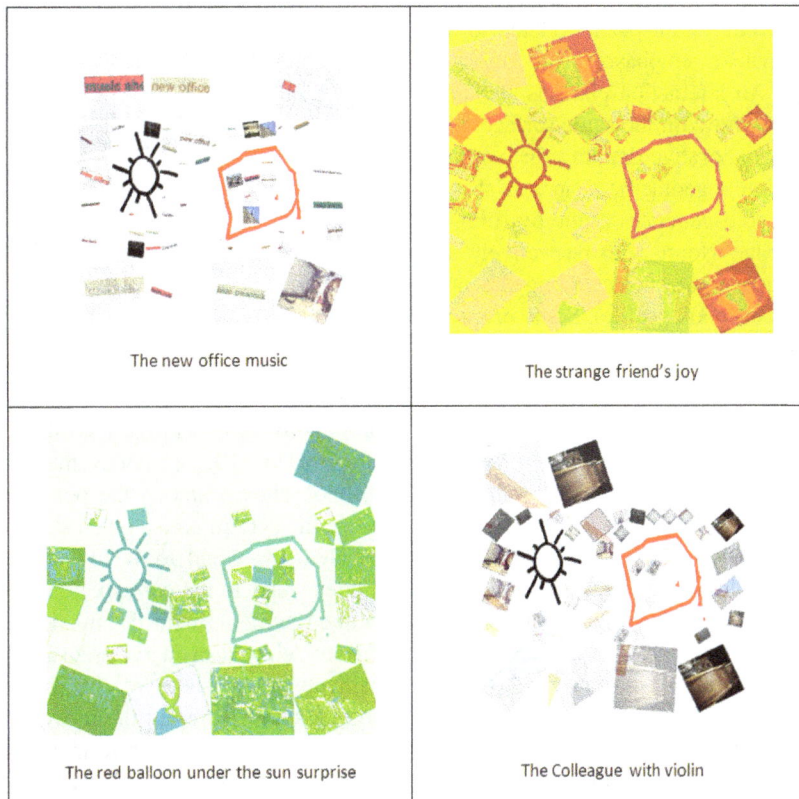

Fig. 4. Some of the created collages.

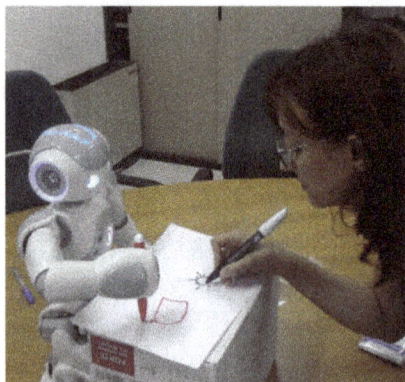

Fig. 5. Collaborative drawing of simples sketches.

interaction (for example, giving some images published onto its social network profile). The choice of the elements of the artistic repertoire to use in the creation of the collage depends on a semantic similarity between the natural language descriptions of such elements and the utterances produced during the conversation. To evaluate this similarity a semantic space, built through the well-known latent semantic analysis (LSA) approach,[60,61] is built, then, both the descriptions of the pictorial items and the utterances are mapped in the same space and compared, as described in another paper.[62] According to the similarity, the robot chooses those terms that best fit user profile. The recognition of human emotions by applying a sentiment analysis module[62] on the conversation influences the creative process of the robot since the chosen items are painted with a specific color palette related to the perceived emotion. This palette emphasizes the message represented by the collage. An interaction phase is also performed allowing the robot to draw on the same surface with the user and producing a collaborative drawing. In fact, the NAO is also able to draw simple geometric shapes as shown in Fig. 5. Finally, the humanoid generates a title by filling a simple title structure Adjective–Noun, with respectively the most frequent adjective's and noun characterizing the user conversation (or the user's Facebook profile). Some example of created collages is shown in Fig. 4.

5. The Humanoid Dancer

The dance is one of the capabilities in artistic domain that can be tested with a new generation robots.

Dance is a harmonic composition of movements driven by the music stimuli and by the interactions with other subjects. Dancing movements follow the rhythm of the music and are synchronous with the song progression. Therefore, both the timing and rhythm of the execution of the movements must be taken into account while trying to imitate human behavior.

The implementation of dancing capabilities in robots is not purely pursued for entertainment purposes. It provides new clues to deepen and improve various research thematics because it requires a robust learning phase that involves both

a real-time analysis of music and the choice of harmonious and suitable movements, and moreover social behavior and interaction capabilities.[63–65] The challenge lies in going beyond a preprogrammed dance executed by robots. A creative process should model the mental processes involved in human creativity to generate movements and taking into account different music genres. The robots' perceptions should influence the choice processes, and output of a learning process should lead to conceiving a personal artistic style, that could be reconsidered or refined after the audience evaluation. We have explored the possibility for a robot to improvise a choreography, building on actions that are either stored in a memory or derived from a continuous elaboration of previous experiences.

In our proposal, we took inspiration from human dance to create a dataset of movements that the robot can employ in his dance. The dataset, including also information about music features related to the sequences of movements, is used to train a variational encoder. This network allows obtaining a variation of the learned movements according to the listened music.

The resulting movements are new but are coherent with the learned ones and are well synchronized with the listened music. In the following, we illustrate two different solutions that exploit the above illustrated general framework and implement a creative robotic dance.

5.1. Creative dance through genetic algorithms

An implementation of robotic dance can be accomplished through a hidden Markov model that represents the dancing capability as a model with multiple hidden states that represent different actions to be performed. The hidden Markov model is essentially composed of two stochastic processes: the transition among states that drives the transition from a state to another and the emission of symbols that is a probabilistic function of the current state. For the transition, the so-called transition matrix, TM, measures the probability that the status of the system will evolve from the state i to the state j. The emission process can be represented by a set of emission probabilities from one of the N_s states, defining the probability of observing a symbol at given time. The emission matrix, EM, provides the probability that an output symbol, chosen among the possible M_s symbols, is emitted while the system is in a given state. An EM row has a sum of its elements equal to 1. In our system, we associated the hidden states to the corresponding basic movements, i.e., the emitted symbols with the music descriptions. According to this model, we can characterize a sequence of movements as a path along the states of the model so that the transitions among states represent the movements sequence.

The TM has a key role in the selection and the composition of the movements. The chosen movements are acquired from movements of human dancers, See Ref. 66, as a set of elementary movements.

In our system, the set of fixed elementary movements and TM are decided by a choreographer or an expert dancer. In future, autonomous mechanism can extend the set of movements in order to allow the robot to have more space to express its own creativity. An aesthetic measure could be used to drive the TM modifications: maximizing transition probabilities of harmonic movements and lowering probabilities of disliked movements.

The TM is independent of the music type while the link between music and movements is represented in the emission matrix. In our formulation, the emissions of the symbols are referred to as the music observations; since the emission is conditioned by the current state not all the music fit with the movements and vice versa. This representation inverts the relationship between cause and effect that one could conceive. Notwithstanding, we found this inversion profitable for the generation of the movements. The representation of the audio track, according to the given model of music, could be used as a sequence of emission of the Markov model. It is a sequence of observations, and it is used to estimate the best sequence of the corresponding states. Since they are not observable, probabilistic inference is done by the Viterbi algorithm.[67] The output of Viterbi algorithm corresponds to a series of states that are bound to a set of movements. Their composition creates the dance that the robot will perform when it listens to the input music. The desired result is to find out the status of the system, knowing only the observation of symbols. The music features, extracted to highlight the most relevant part for the dance, represent the coding of how the robot perceives music dividing the song into fragments and associating each part to nine different loudness categories. The output of the Viterbi algorithm corresponds to a sequence of movements, chosen by the robot, to act according to what it had heard.

To teach the robot to dance, we have used an approach inspired by the basic simple genetic algorithm (SGA)[68] and the interactive evolutionary computation (IEC).[69] The population of the genetic algorithm is made up of emission matrices, EMs, which constitute the associations of M movements (rows) and K music classes (columns). Each emission matrix is a real-valued chromosome, and it is composed of genes, which are the matrix rows. Each gene is constrained by the sum of all elements that must be equal to 1 since each element represents the probability of movement execution according to a perceived music class.

The population is composed of N individuals (chromosomes). At each evolution step a small subset of the initial chromosomes is selected to create the next generation. The crossover operation cuts the chromosome only at positions corresponding to the end of the rows preserving the coherence of the probabilistic interpretation. A selection process is carried on for selecting the movements that are more similar to an ideal dance proposed by the choreograph. The population is ordered according to an internal fitness function, and the best individuals are selected as parents for the next

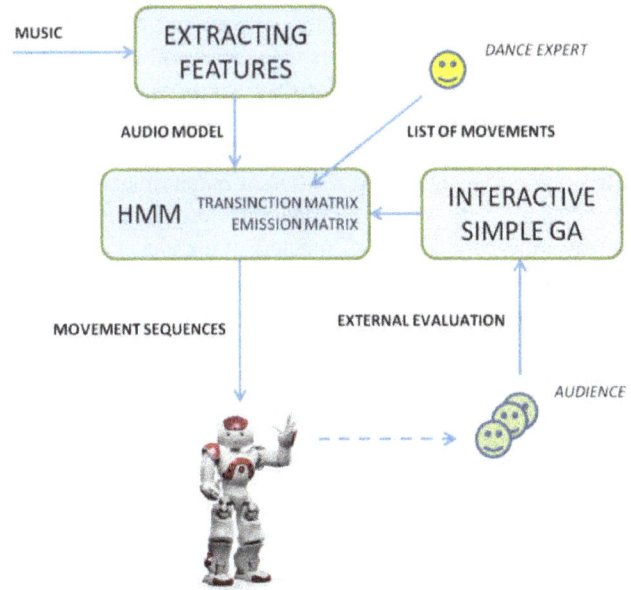

Fig. 6. Dance learning and performing with hidden Markov models.

generation. The most valuable movements are proposed to the master, and selected individuals are saved in a list of "privileged population" in case of positive judgment.[70] These individuals will survive for the next generations. The dances are shown by a 3D viewer, allowing the teacher to evaluate performances from the simulator without using the real robot.

When the privileged population reaches the desired cardinality, the robot stores its complete *repertoire* in the long-term memory, and it is ready to execute performances to submit for general audience evaluations. Through its cognitive architecture, the external evaluations affect the robot's behavior that will resort the priority of its movements to repeat the most appreciated steps. If it is necessary, the robot could try to make further improvements continuing evolution, eventually changing the internal fitness functions or introducing mutation operators.[71]

5.2. *Creative dance through variational encoder*

A second possibility to transform the acquired movement into a synthetic dance is to adopt a neural network that is trained with interesting examples and then use the acquired information to improvise a new dance.[72]

An example of application can be carried on with an autoencoder network. The autoencoders are neural networks that are trained to represent in the hidden layer all the data that have been previously exposed to. At the same time, for their particular topology, they can reconstruct the input starting from the hidden representation. The data is stored in a "lossy" way since the reconstructed x is not perfectly identical to the original one and this difference is determined by the chosen distance or "loss" function.

For this kind of networks, it is assumed that the dealt information can be faithfully approximated with a Gaussian distribution so that the encoder network can map the input samples into the parameters of the latent space. They uniquely identify the Gaussian distribution from which randomly sampled points z are then extracted. In the reconstruction phase, when the distinctive parameters of the Gaussian curve are altered a different output mapping of the decoder network is obtained and therefore modified movements are generated. The parameters of the model are trained taking into account the reconstruction loss that forces the decoded samples to match the initial inputs. Furthermore, the net is trained aimed at minimizing the Kullback–Leibler divergence between the learned latent distribution and the prior distribution. This elaboration avoids an overfitted learning of the original dataset.

Since the latent space of the network is Gaussian, the values of loudness and variance have to be transformed through the inverse cumulative distribution function of the Gaussian to produce coherent values with the latent space. The transformed value is then injected into the network's latent space to generate the movements related to the perceived music.

Through this system, a single movement is generated for any music beat. Given the position of the beat and the interval between two consecutive beats, the robot can execute one movement for each detected interval. The network, using as an input the intensity and variance of the music interval, outputs a configuration of joints that represents one single movement. The movements that the network predicts are dependent on the processed music features.

For the above-described computations, a set of music information must be taken into account. For the system we created a music perception module that has been used to extract relevant features from the listened music. Usually, audio processing requires complex algorithms and many computational resources do not allow the robot to have a fast reaction to its auditive stimuli. For this reason, we have chosen to detect only musical beats and intensities of sound using fast algorithms. We employ ESSENTIA library,[73] to obtain the relevant

information from the perceived audio signal. In detail, the features extracted are: beats locations, beats interval, beats per minute (BPM) and confidence of detected beats.

The generation of the movements starts with the extraction of some features from the music input, in particular, the system extracts the loudness and the variance of the rhythmic sound that will be used as input of the latent space of the variational autoencoder to generate the movements according to the perceived music.

The dance performance is driven by a music perception module that extracts relevant features from the listened music. Usually, audio processing requires complex algorithms and many computational resources do not allow the robot to have a fast reaction to its auditive stimuli. For this reason, we have chosen to detect only musical beats and intensities of sound using fast algorithms.

6. Conclusions

In this paper, we have discussed the role of cognitive architectures for computational creativity. We have focused our attention on both the cognitive capabilities and the internal and external evaluation mechanisms that from the bases of a computational creativity architecture. These mechanisms determine emotions and motivations that lead to a process aimed at obtaining variability of the products resulting from the "creative" activity, e.g., a painting, a collage or a dance.

We have instantiated this general framework based on evaluation of three robotic systems that play the role of a creative painter, a creative collage-maker and a humanoid dancer, respectively.

Our experience demonstrates that considering the physicality of a system, like a robot, as well as evaluation, motivations and emotions play an essential role in the realization of a true computational creativity framework.

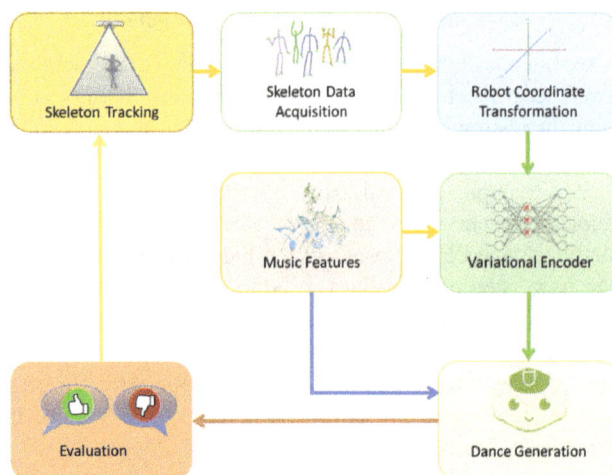

Fig. 7. Dance creation through a variational encoder.

References

[1] O. Gross, H. Toivonen, J. M. Toivanen and A. Valitutti, Lexical creativity from word associations, *Proc. 2012 Seventh Int. Conf. Knowledge, Information and Creativity Support Systems (KICSS)* (2012), pp. 35–42.

[2] M. A. Boden, *Artificial Intelligence: Handbook of Perception and Cognition* (Academic Press, 1996).

[3] S. A. Mednick, The associative basis of the creative process, *Psychol. Rev.* **69**, 220 (1962).

[4] G. Wallas, *The Art of Thought* (J. Cape, 1926).

[5] M. Rhodes, An analysis of creativity, *The Phi Delta Kappan* **42**, 305 (1961).

[6] J. P. Guilford, Creative abilities in the arts, *Psychol. Rev.* **64**, 110 (1957).

[7] G. A. Wiggins, Searching for computational creativity, *New Gener. Comput.* **24**, 209 (2006).

[8] S. Colton, J. W. Charnley and A. Pease, Computational creativity theory: The FACE and IDEA descriptive models, *Proc. Second Int. Conf. Computational Creativity* (2011), pp. 90–95.

[9]T. B. Ward, Creative cognition as a window on creativity, *Methods* **42**, 28 (2007).

[10]W. Duch and M. Pilichowski, Experiments with computational creativity, *Neural Inf. Process., Lett. Rev.* **11**, 123 (2007).

[11]S. H. Carson, J. B. Peterson and D. M. Higgins, Reliability, validity, and factor structure of the creative achievement questionnaire, *Creat. Res. J.* **17**, 37 (2005).

[12]F. Pinel, L. R. Varshney and D. Bhattacharjya, A culinary computational creativity system, *Computational Creativity Research: Towards Creative Machines*, Chapted 16 (Springer, 2015), pp. 327–346.

[13]S. Colton, The painting fool: Stories from building an automated painter, *Computers and Creativity*, Part I, Chapter 1 (Springer, 2012), pp. 3–38.

[14]D. Norton, D. Heath and D. Ventura, Finding creativity in an artificial artist, *J. Creat. Behav.* **47**, 106 (2013).

[15]M. A. Boden, The turing test and artistic creativity, *Kybernetes* **39**, 409 (2010).

[16]A. Cardoso, T. Veale and G. A. Wiggins, Converging on the divergent: The history (and future) of the international joint workshops in computational creativity, *AI Mag.* **30**, 15 (2009).

[17]D. Cope and M. J. Mayer, *Experiments in Musical Intelligence*, Vol. 12 (A-R Editions, Madison, 1996).

[18]P. McCorduck, *Aaron's Code* (W. H. Freeman & Co., New York, 1991).

[19]R. B. Dannenberg, A vision of creative computation in music performance, *Proc. Second Int. Conf. Computational Creativity* (2011), pp. 84–89.

[20]A. Eigenfeldt and P. Pasquier, Negotiated content: Generative soundscape composition by autonomous musical agents in coming together: Freesound, *Proc. Second Int. Conf. Computational Creativity* (2011), pp. 27–32.

[21]L. Gatti, M. Guerini, C. B. Callaway, O. Stock and C. Strapparava, Creatively subverting messages in posters, *Proc. Third Int. Conf. Computational Creativity* (2012), pp. 175–179.

[22]T. Veale and Y. Hao, Comprehending and generating apt metaphors: A web-driven, case-based approach to figurative language, *Proc. 22nd Nationsl Conf. Artifical Intelligence* (2007), 1471–1476.

[23]P. Gervás, Exploring quantitative evaluations of the creativity of automatic poets, *Proc. Workshop Creative Systems, Approaches to Creativity in Artificial Intelligence and Cognitive Science and 15th European Conf. Artificial Intelligence* (2002).

[24]J. Toivanen *et al.*, Corpus-based generation of content and form in poetry, *Proc. Third Int. Conf. Computational Creativity* (2012).

[25]R. Manurung, G. Ritchie, H. Pain, A. Waller, D. O'Mara and R. Black, The construction of a pun generator for language skills development, *Appl. Artif. Intell.* **22**, 841 (2008).

[26]O. Stock and C. Strapparava, The act of creating humorous acronyms, *Appl. Artif. Intell.* **19**, 137 (2005).

[27]R. P. y Pérez and M. Sharples, Three computer-based models of storytelling: Brutus, minstrel and mexica, *Knowl.-Based Syst.* **17**, 15 (2004).

[28]M. O. Riedl and R. M. Young, Story planning as exploratory creativity: Techniques for expanding the narrative search space, *New Gener. Comput.* **24**, 303 (2006).

[29]S. Colton, Computational discovery in pure mathematics, *Computational Discovery of Scientific Knowledge*, Chapter 9 (Springer, 2007), pp. 175–201.

[30]R. G. Morris, S. H. Burton, P. Bodily and D. Ventura, Soup over bean of pure joy: Culinary ruminations of an artificial chef, *Proc. Third Int. Conf. Computational Creativity* (2012), pp. 119–125.

[31]L. R. Varshney, F. Pinel, K. R. Varshney, A. Schörgendorfer and Y.-M. Chee, Cognition as a part of computational creativity, *Proc. 2013 12th IEEE Int. Conf. Cognitive Informatics & Cognitive Computing (ICCI* CC)* (2013), pp. 36–43.

[32]G. Ritchie, Some empirical criteria for attributing creativity to a computer program, *Minds Mach.* **17**, 67 (2007).

[33]G. A. Wiggins, A preliminary framework for description, analysis and comparison of creative systems, *Knowl.-Based Syst.* **19**, 449 (2006).

[34]D. Ventura, Can a computer be lucky? and other ridiculous questions posed by computational creativity, *Proc. Int. Conf. Artificial General Intelligence* (2014), pp. 208–217.

[35]M. Coeckelbergh, Can machines create art? *Philos. Technol.* **30**, 285 (2017).

[36]P. Dahlstedt, Turn-based evolution as a proposed implementation of artistic creative process, *Proc. 2012 IEEE Congr. Evolutionary Computation (CEC)* (2012), pp. 1–7.

[37]M. Eppe, E. Maclean, R. Confalonieri, O. Kutz, M. Schorlemmer, E. Plaza and K.-U. Kühnberger, A computational framework for conceptual blending, *Artif. Intell.* **256**, 105 (2018).

[38]G. Fauconnier and M. Turner, *The Way We Think: Conceptual Blending and the Mind's Hidden Complexities* (Basic Books, 2008).

[39]L. Gabora and D. Aerts, Contextualizing concepts using a mathematical generalization of the quantum formalism, *J. Exp. Theor. Artif. Intell.* **14**, 327 (2002).

[40]L. Gabora, Revenge of the neurd: Characterizing creative thought in terms of the structure and dynamics of memory, *Creat. Res. J.* **22**, 1 (2010).

[41]A. Jordanous, A standardised procedure for evaluating creative systems: Computational creativity evaluation based on what it is to be creative, *Cogn. Comput.* **4**, 246 (2012).

[42]C. Bartl and D. Dörner, Psi: A theory of the integration of cognition, emotion and motivation, *Proc. 2nd European Conf. Cognitive Modelling* (1998), pp. 66–73.

[43]I. Infantino, G. Pilato, R. Rizzo and F. Vella, I feel blue: Robots and humans sharing color representation for emotional cognitive interaction, *Biologically Inspired Cognitive Architectures 2012*, Chapter 30 (Springer, Berlin, 2013), pp. 161–166.

[44]D. Vernon, G. Metta and G. Sandini, A survey of artificial cognitive systems: Implications for the autonomous development of mental capabilities in computational agents, *IEEE Trans. Evolut. Comput.* **11**, 151 (2007).

[45]B. Goertzel and J. Pitt, M. Ikle, C. Pennachin and L. Rui, Glocal memory: A critical design principle for artificial brains and minds, *Neurocomputing* **74**, 84 (2010).

[46]P. Langley, J. E. Laird and S. Rogers, *Cognitive Architectures: Research Issues and Challenges*, Technical Report (eds.) (2002).

[47]B. Goertzel and C. Pennachin, *Artificial General Intellgence*, Congnitive Technologies Series (Springer, Berlin, 2007).

[48]M. Csikszentmihalyi, Society, culture, and person: A systems view of creativity, *The Systems Model of Creativity*, Chapter 4 (Springer, 2014), pp. 47–61.

[49]K. Sawyer, *Group Genius: The Creative Power of Collaboration* (Basic Books, 2007).

[50]R. K. Sawyer, The emergence of creativity, *Philos. Psychol.* **12**, 447 (1999).

[51]K. E. Stanovich and R. F. West, Individual differences in reasoning: Implications for the rationality debate? *Behav. Brain Sci.* **23**, 645 (2000).

[52] J. S. B. Evans and K. E. Frankish, *In Two Minds: Dual Processes and Beyond* (Oxford University Press, 2009).

[53] D. Kahneman, *Thinking, Fast and Slow* (Farrar, Straus and Giroux, New York, 2011).

[54] P. T. Sowden, A. Pringle and L. Gabora, The shifting sands of creative thinking: Connections to dual-process theory, *Think. Reason.* **21**, 40 (2015).

[55] R. Tubb and S. Dixon, A four strategy model of creative parameter space interaction, *Proc. Fifth Int. Conf. Computational Creativity* (2014), pp. 16–22.

[56] A. Augello, I. Infantino, A. Lieto, G. Pilato, R. Rizzo and F. Vella, Artwork creation by a cognitive architecture integrating computational creativity and dual process approaches, *Biol. Inspired Cogn. Archit.* **15**, 74 (2016).

[57] N. Cowan, What are the differences between long-term, short-term, and working memory? *Prog. Brain Res.* **169**, 323 (2008).

[58] L. Gabora, Cognitive mechanisms underlying the creative process, *Proc. 4th Conf. Creativity & Cognition* (2002), pp. 126–133.

[59] A. Augello, I. Infantino, G. Pilato, R. Rizzo and F. Vella, Combining representational domains for computational creativity, *Proc. Fifth Int. Conf. computational Creativity* (2014), pp. 272–275.

[60] T. K. Landauer, P. Foltz and D. Laham, An introduction to latent semantic analysis, *Discourse Process.* **25**, 259 (1998).

[61] G. Pilato and G. Vassallo, TSVD as a statistical estimator in the latent semantic analysis paradigm, *IEEE Trans. Emerg. Top. Comput.* **3**, 185 (2015).

[62] A. Augello, I. Infantino, A. Manfré, G. Pilato and F. Vella, Analyzing and discussing primary creative traits of a robotic artist, *Biol. Inspired Cogn. Archit.* **17**, 22 (2016).

[63] J.-J. Aucouturier *et al.*, Cheek to chip: Dancing robots and AI's future, *IEEE Intell. Syst.* **23** (2008).

[64] A. Augello, I. Infantino, A. Manfrè, G. Pilato, F. Vella and A. Chella, Creation and cognition for humanoid live dancing, *Robot. Auton. Syst.* **86**, 128 (2016).

[65] K. Shinozaki, A. Iwatani and R. Nakatsu, Concept and construction of a robot dance system, *Proc. Int. Workshop and Conf. Photonics and Nanotechnology* (2007), pp. 29–34.

[66] J. Koenemann and M. Bennewitz, Whole-body imitation of human motions with a Nao humanoid, *Proc. 2012 7th ACM/IEEE Int. Conf. Human-Robot Interaction (HRI)* (2012), p. 425.

[67] L. R. Rabiner, A tutorial on hidden Markov models and selected applications in speech recognition, *Proc. IEEE* **77**, 257 (1989).

[68] M. D. Vose, *The Simple Genetic Algorithm: Foundations and Theory*, Vol. 12 (MIT Press, 1999).

[69] H. Takagi, Interactive evolutionary computation: Fusion of the capabilities of EC optimization and human evaluation, *Proc. IEEE* **89**, 1275 (2001).

[70] A. Manfrè, A. Augello, G. Pilato, F. Vella and I. Infantino, Exploiting interactive genetic algorithms for creative humanoid dancing, *Biol. Inspired Cogn. Archit.* **17**, 12 (2016).

[71] A. Augello, I. Infantino, G. Pilato, R. Rizzo and F. Vella, Creativity evaluation in a cognitive architecture, *Biol. Inspired Cog. Archit.* **11**, 29 (2015).

[72] A. Augello, E. Cipolla, I. Infantino, A. Manfre, G. Pilato and F. Vella, Creative robot dance with variational encoder, arXiv:1707.01489 [CS.AI].

[73] D. Bogdanov, N. Wack, E. Gómez, S. Gulati, P. Herrera, O. Mayor, G. Roma, J. Salamon, J. R. Zapata, X. Serra *et al.*, ESSENTIA: An audio analysis library for music information retrieval, *Proc. Int. Society for Music Information Retrieval Conf.* (2013), pp. 493–498.

Subject Index

Author Index